Springer Tracts in Civil Engineering

Springer Tracts in Civil Engineering (STCE) publishes the latest developments in Civil Engineering - quickly, informally and in top quality. The series scope includes monographs, professional books, graduate textbooks and edited volumes, as well as outstanding PhD theses. Its goal is to cover all the main branches of civil engineering, both theoretical and applied, including:

- Construction and Structural Mechanics
- Building Materials
- Concrete, Steel and Timber Structures
- Geotechnical Engineering
- Earthquake Engineering
- Coastal Engineering; Ocean and Offshore Engineering
- Hydraulics, Hydrology and Water Resources Engineering
- Environmental Engineering and Sustainability
- Structural Health and Monitoring
- Surveying and Geographical Information Systems
- Heating, Ventilation and Air Conditioning (HVAC)
- Transportation and Traffic
- Risk Analysis
- Safety and Security

Indexed by Scopus

To submit a proposal or request further information, please contact:
Pierpaolo Riva at Pierpaolo.Riva@springer.com (Europe and Americas) Wayne Hu at wayne.hu@springer.com (China)

T. G. Sitharam · Sreevalsa Kolathayar ·
Ravi S. Jakka · Vasant Matsagar
Editors

Theory and Practice in Earthquake Engineering and Technology

Springer

Editors
T. G. Sitharam
Department of Civil Engineering
Indian Institute of Technology (IIT) Guwahati
Guwahati, Assam, India

Sreevalsa Kolathayar
Department of Civil Engineering
National Institute of Technology (NIT) Karnataka
Mangalore, Karnataka, India

Ravi S. Jakka
Department of Earthquake Engineering
Indian Institute of Technology (IIT) Roorkee
Roorkee, Uttarakhand, India

Vasant Matsagar
Department of Civil Engineering
Indian Institute of Technology (IIT) Delhi
Hauz Khas, Delhi, India

ISSN 2366-259X ISSN 2366-2603 (electronic)
Springer Tracts in Civil Engineering
ISBN 978-981-19-2323-4 ISBN 978-981-19-2324-1 (eBook)
https://doi.org/10.1007/978-981-19-2324-1

This Springer imprint is published by the registered company Springer Nature Singapore Pte Ltd.
The registered company address is: 152 Beach Road, #21-01/04 Gateway East, Singapore 189721, Singapore

Preface

The editors are pleased to present this text to the readers on theory and practice in earthquake engineering and technology. Extensive research work is conducted on earthquake engineering and allied areas across the world, quite naturally because of its relevance to the life safety of people. In the present text, the authors have provided a broad spectrum of research works carried out under several disciplines and sub-disciplines of earthquake engineering and technology, from theory to practice, starting from the rudiments thereof. Matsagar (2022) has initiated the discussion on earthquake engineering and technology by setting the tone of the deliberations within this text. Under the overarching umbrella of earthquake engineering and technology, systematic categorization of various disciplines and sub-disciplines has been made, while introducing them, some ongoing research works have also been presented in this opening chapter. The subsequent chapters cover several aspects of geology, seismo-tectonics, regional seismology, geotechnical earthquake engineering, soil-structure interaction, structural engineering, and dynamic response control of structures.

Application of site response studies in seismic hazard microzonation and ground characterization has been presented by Shukla (2022) in the next chapter. In engineering seismology, site response study helps in modeling the effects of the near surface layers of soil on earthquake ground motions, which is an important requirement in seismic hazard microzonation. Ground response studies carried out world over have been reviewed herein, and the discussion is then made in the Indian context.

Ravi-Kiran and Jakka (2022) have presented seismic design of shallow foundations giving the conceptual principles and design methodologies. They have as well deliberated on the current Indian practices, in the framework of codes. Foundation design for structures that are resilient under seismic activities poses several challenges to a geotechnical earthquake engineer owing to the complexities involved in the dynamic soil-structure interaction (SSI) problems. A geotechnical engineer is required to design foundation addressing all such complexities and yet provide design adequately safe under the anticipated forces. Such design procedure for shallow foundations, in particular, has been dealt with in this chapter.

During an earthquake event, adjacent structures may collide with each other, which is called seismic pounding, generating high impact forces in the structural members. Such seismic-induced pounding of structures and its mitigation through some connecting links have been proposed by Dey-Ghosh and Aviral-Kumar (2022). The authors have recommended to relook the current provisions of maintaining adequate gap between adjacent buildings and bridge girder in order to preclude possibility of seismic pounding, else connecting links in the form of energy absorbing dampers have been recommended.

Soil-Structure Interaction (SSI) plays a crucial role in evaluating seismic response of structures, which cannot be overlooked, especially when the underlying soil is flexible. The influence of the nonlinear SSI on yielding of pile embedded in stratified soil has been discussed in detail by Bhattacharjee and Borthakur (2022). Estimating yield moment of a pile embedded in stratified soil by conducting static pushover analysis has been shown in this chapter by modeling the pile-soil system in the open source simulation tool, OpenSees.

When soil loses its shear strength under the earthquake-induced dynamic excitation due to increased pore water pressure soil liquefaction is considered to have occurred. Satyam and Priyadarsini (2022) have fittingly discussed development of liquefaction susceptibility maps for Vishakhapatnam City in India that would be useful in the infrastructure development projects in future. The hazard maps developed based on the liquefaction severity and potential indexes are projected to help solving engineering problems in the considered study area.

Several earthquake response modification and control techniques and devices have been invented of late. One among such dynamic response control methods is seismic base isolation of structures. Base isolation systems are also proven effective for seismic response control of masonry dome (Kakade et al., 2022). The authors have adequately shown the effectiveness of the base isolation system employed in dynamic response control of masonry dome when subjected to earthquake base excitation. Especially, development of tensile stresses in the masonry dome has shown to be reduced advantageously.

With new developments in knowledge and knowhow, rapid changes are being made in the codes and standards. Such changes mandate reassessment of structures on new scale, parameters, or norms, and subsequently adopting measures to make them code compliant. Sarmah et al. (2022) have presented a method for rapid retrofitting of reinforced concrete (RC) columns using iron-based shape memory alloy, Fe-SMA to achieve enhanced seismic performance. This new technique helps in seismic retrofitting of RC columns by winding of thermally prestressed and actively controlled, Fe-SMA strips around the columns.

Considerable research contributions have been made at the Indian Institute of Technology (IIT) Roorkee in developing earthquake early warning system. Relevance of earthquake early warning system in India has been deliberated by Ashok-Kumar et al. (2022), especially emphasizing the need for addressing the blind zone region within 100 km radius from the epicenter. Nonetheless, research efforts are further required to be taken in making the earthquake early warning system more reliable and advancing/increasing the time for making the prediction, even in the near source regions. Artificial Intelligence (AI), Machine Learning (ML), Deep Learning (DL), and Neural Network (NN) techniques are finding their enormous applications in the earthquake early warning system to this effect, after such a pilot project has been completed at IIT Roorkee.

Local authorities require guidance on city planning, earthquake risk mitigation, and response actions taken after occurrence of an earthquake. After an occurrence of unfortunate earthquake estimating the loss incurred is also required, which is a daunting task. How to carry out such estimation of losses has been presented by Meslem et al. (2022), who have taken an example on earthquake loss information system developed for the North-Eastern City of Guwahati, Capital of Assam in India. The procedure laid down in this chapter can suitably be applied for a city before/after occurrences of earthquake.

Occurrence of earthquakes is unpredictable in terms of time and intensity in different regions. However, for important mega-projects, such as river-valley projects, estimation of probable intensity of earthquakes at the proposed site or location is quite essential. To this effect, a Probabilistic Seismic Hazard Analysis (PSHA) for hydropower project sites in the Himalayan Region has been carried out and presented by Srivastav and Satyam (2022) in their chapter. With an aim to quantify the rate of exceeding certain specified earthquake ground motion level at a specific project site, the PSHA technique has been employed for three chosen hydropower project sites located in Uttarakhand, Himachal Pradesh, and Jammu and Kashmir in India, which will be helpful in generating a site-specific seismic hazard map.

In heavy industries and structures, the structure-equipment-piping interactions are important considerations under the earthquake excitation (Reddy, 2022). Failures are likely to occur not only in the parent (host) structure but also either in equipment or in the piping systems, which sometimes are categorized as lifeline structures. The forces induced in the structure, equipment, and piping depend upon how they are interacting with each other. It is notable that such dynamic interactions vary largely based on whether the secondary structures are acceleration-sensitive or displacement-sensitive. Thereby, seismic designs of the primary and secondary structures are greatly influenced by the structure-equipment-piping interactions, as discussed in this chapter.

Advanced seismic design approaches are being adopted in modern design codes and standards. In this context, the Performance-Based Seismic Design (PBSD) of Reinforced Concrete (RC) structures discussed by Gwalani and Singh (2022) becomes highly relevant. In the near future, the Bureau of Indian Standards (BIS) is anticipated to publish standard guidelines on the PBSD of code-compliant reinforced concrete buildings. Hence, this chapter, describing the concept of the PBSD and how to apply it in earthquake-resistant design of buildings, appropriately illustrating the Nonlinear Static Analysis (NSA) and Nonlinear Time History Analysis (NLTHA) methods, is very timely and useful to the structural designers.

Borah et al. (2022) have carried out a comparative analysis of the Standard Spectral Ratio (SSR) and Horizontal to Vertical Spectral Ratio (HVSR) methods for site response analysis. They have concluded that the SSR method provides a more accurate and conservative estimate of site amplification response as compared to the HVSR, which therefore can suitably be adopted for the seismically active North-Eastern Regions in India.

It is indeed pleasing to see that a wide variety of topics have been dealt with in these chapters of the book. We feel that these contributed chapters in this book have elaboratively highlighted tenets of theory and practice in earthquake engineering and technology aptly. Therefore, we believe that the latest developments in earthquake engineering and allied disciplines presented through these 14 chapters will prove to be highly informative to the readers and pave ways for further research.

Guwahati, India T. G. Sitharam
Mangalore, India Sreevalsa Kolathayar
Roorkee, India Ravi S. Jakka
Hauz Khas, India Vasant Matsagar

Contents

Editors and Contributors

About the Editors

Prof. Dr. T. G. Sitharam is the Director of Indian Institute of Technology Guwahati, Assam since July 2019. He is a member of the Science and Engineering Research Board (SERB), Established through an Act of Parliament: SERB Act 2008, Department of Science & Technology, Government of India. He is a Senior Professor in the Department of Civil Engineering, Indian Institute of Science (IISc), Bangalore, and served IISc for more than 27 years. He was Chairman of the Board of Governors at IIT Guwahati during 2019–2020 for more than a year. He was the former Chairman, Research Council, CSIR- CBRI (Central Building Research Institute, Roorkee). He is holding the position of Director (additional charge) of Central Institute of Technology, Kokrajhar, Assam (A Deemed to be University), since May 2021. Over the last 35 years, he has carried out research and development in the area of geotechnical and infrastructure engineering, seismic microzonation and soil dynamics, Geotechnical earthquake engineering and has developed innovative technologies in the area of earth sciences, leading to about 500 technical papers, 20 books with Google scholar H-index of 47 and I-10 index 137 with more than 7175 citations. He has guided 40 Ph.D. students and 35 Master's Students. He was listed in the world's top 2% of scientists for the most-cited research scientists in various disciplines by Stanford University in 2020. His name appeared again in the top 2% of scientists IN Elsevier by Stanford University in 2021. His broad area of research falls into Geotechnical Infrastructure engineering, earth sciences, hydrology, seismology, and natural hazards. He has carried out Seismic microzonation of many urban centers in India and he is an authority on seismic microzonation and site effects. Presently, he is the President of the Indian Society of Earthquake Technology and he was the chairman of the 7th International conference on recent advances in geotechnical earthquake engineering and soil dynamics held in July 2021. He is the chief editor of the International Journal of Geotechnical Earthquake Engineering, (IJGEE), PA, USA since 2010. He is the Editor-in-chief, Springer Transactions in Civil and Environmental Engg series, Book Series, Singapore. He is the Fellow of ASCE, Fellow of ICE (UK), Diplomat of Geotechnical Engineering (D.Ge.) from Academy of Geo-Professionals, ASCE; Fellow of IGS, ISET, ISES, and many other Societies. He is also a certified professional engineer and a chartered engineer from the Institution of Engineers.

Prof. Dr. Sreevalsa Kolathayar pursued M.Tech from IIT Kanpur, Ph.D. from Indian Institute of Science (IISc), and served as International Research Staff at UPC BarcelonaTech Spain. He is presently Assistant Professor in the Department of Civil Engineering, National Institute of Technology Karnataka, Surathkal, India. Dr. Sreevalsa has authored five books, edited 16 book volumes, and published 100 research papers. He is Associate Editor of two International Journals and acted as a reviewer for many. His research interests include Seismic Hazard Assessment, Local Site effects, Disaster Risk Reduction, Earth reinforcement, and Water geotechnics. He is currently the Secretary Indian chapter of the International Association for Coastal Reservoir Research (IACRR), and Executive Committee Member of the Indian Society of Earthquake Technology (ISET). In 2017, The New Indian Express honored Dr. Sreevalsa with South India's Most Inspiring Young Teachers Award. He is the recipient of the ISET DK Paul Research Award from the Indian Society of Earthquake Technology (ISET) in 2018. He is on the roster of two technical committees of ASCE Geo-Institute. Dr. Sreevalsa received the "IEI Young Engineers Award" by The Institution of Engineers (India), in recognition of his contributions in the field of Civil Engineering. He served as the Organizing Secretary to the 7th International Conference on Recent Advances in Geotechnical Earthquake Engineering.

Prof. Dr. Ravi S. Jakka is working as an Associate Professor in the Department of Earthquake Engineering, Indian Institute of Technology, Roorkee, and serving as Editor to the International Journal of Geotechnical Earthquake Engineering & ISET Journal of Earthquake Technology. He graduated in Civil Engineering from Andhra University Engineering College in the year 2001. He has obtained masters and doctorate degrees from IIT Delhi in the years 2003 and 2007 respectively. His areas of interest are Dynamic Site Characterization, Soil Liquefaction, Seismic Slope Stability of Dams, Landslides, Foundations & Seismic Hazard Assessment. He has published over 100 articles in reputed international journals and conferences. He has supervised over 35 Masters Dissertations and 7 Ph.D. thesis, while he is currently guiding 10 Ph.D. Thesis. He has received prestigious DAAD and National Doctoral fellowships. He has obtained University Gold Medal from Andhra University. He also received the 'Young Geotechnical Engineer Best Paper Award' from the Indian Geotechnical Society. He was instrumental in the development of the Earthquake Early Warning System for northern India, a prestigious national project. He was also the Organizing Secretary to 7th International Conference on Recent Advances in Geotechnical Earthquake Engineering.

Prof. Dr. Vasant Matsagar is currently serving as Dogra Chair Professor in the Department of Civil Engineering at Indian Institute of Technology (IIT) Delhi. He obtained his doctorate degree from Indian Institute of Technology (IIT) Bombay in

2005 in the area of seismic base isolation of structures. He performed post-doctoral research at the Lawrence Technological University (LTU) USA, in the area of Carbon Fiber Reinforced Polymers (CFRP) in bridge structures. His current research interests include structural dynamics and vibration control; multi-hazard protection of structures from earthquake, blast, fire, and wind; finite element methods; Fiber-Reinforced Polymers (FRP) in prestressed concrete structures; and bridge engineering. He has published over a hundred international journal papers, hundreds of international conference manuscripts, a book, six edited proceedings, and has filed for two patents. His current Google Scholar citations are 3,794, having an h-index of 29. He has guided students at both Undergraduate (UG) and Postgraduate (PG) levels in their bachelor's and master's projects and doctoral research. Besides student guidance, he is actively engaged in sponsored research and consultancy projects at national and international levels. He is also involved in teaching courses in structural engineering, e.g., structural analysis, finite element methods, numerical methods, structural stability, structural dynamics, design of steel and concrete structures to name a few.

Contributors

Bhattacharjee Arup Department of Civil Engineering, Jorhat Engineering College, Jorhat, Assam, India

Borah Sasanka Department of Civil Engineering, Assam Engineering College, Guwahati, Assam, India

Borthakur Bidisha Department of Civil Engineering, Jorhat Engineering College, Jorhat, Assam, India

Chamoli B. P. Indian Institute of Technology Roorkee, Roorkee, India

Deb S. K. Civil Engineering Department, IIT Guwahati, Assam, India

Dutta A. Civil Engineering Department, IIT Guwahati, Assam, India

Ghosh Aparna (Dey) Department of Civil Engineering, Indian Institute of Engineering Science and Technology (IIEST) Shibpur, Howrah, India

Goswami Diganta Department of Civil Engineering, Assam Engineering College, Guwahati, Assam, India

Gwalani Payal Department of Earthquake Engineering, IIT Roorkee, Roorkee, India

Jakka Ravi S. Department of Earthquake Engineering, Indian Institute of Technology Roorkee, Roorkee, Uttarakhand, India

Kakade Pushkar G. College of Engineering Pune, Pune, India

Kumar Ashok Indian Institute of Technology Roorkee, Roorkee, India

Kumar Aviral Department of Civil Engineering, Indian Institute of Engineering Science and Technology (IIEST) Shibpur, Howrah, India

Kumar Pankaj Indian Institute of Technology Roorkee, Roorkee, India

Kumar Shukla Atindra Meteorological Department, Ministry of Earth Sciences, Government of India, New Delhi, India

Lang Dominik Stiftelsen NORSAR, Kjeller, Norway

Lindholm Conrad Stiftelsen NORSAR, Kjeller, Norway

Madhekar Suhasini N. College of Engineering Pune, Pune, India

Matsagar Vasant A. Professor and Dogra Chair, Department of Civil Engineering, Indian Institute of Technology (IIT) Delhi, Hauz Khas, New Delhi, India

Mazumdar Kiron Geological Survey of India, Kolkata, India

Meslem Abdelghani Stiftelsen NORSAR, Kjeller, Norway;
Norwegian University of Life Sciences, Ås, Norway

Munot Hema K. College of Engineering Pune, Pune, India

Nandyala Ravi Kiran Centre of Excellence in Disaster Mitigation and Management, Indian Institute of Technology Roorkee, Roorkee, Uttarakhand, India

Pandey Bhavesh Indian Institute of Technology Roorkee, Roorkee, India

Pathak Jayanta Department of Civil Engineering, Assam Engineering College, Guwahati, Assam, India

Priyadarsini Swathi Department of Civil Engineering, Lords Institute of Engineering and technology, Hyderabad, India

Rathore Govind Indian Institute of Technology Roorkee, Roorkee, India

Reddy G. R. VJTI, Mumbai, India

Sarmah M. Civil Engineering Department, IIT Guwahati, Assam, India

Satyam Neelima Discipline of Civil Engineering, Indian Institute of Technology Indore, Simrol, Indore, India

Singh Yogendra Department of Earthquake Engineering, Indian Institute of Technology Roorkee, Roorkee, India

Srivastav Ambika Geotechnical Engineering Laboratory, Earthquake Engineering Research Centre, International Institute of Information Technology Hyderabad, Hyderabad, India

Chapter 1
Earthquake Engineering and Technology

Vasant Matsagar

Abstract Earthquake engineering and technology has been dealt with by introducing different sub-disciplines thereof, contemporary practices, and the latest developments. Upon giving background to open-up various topics, the expertise required and professionals involved in various disciplines have been systematically presented by giving information exchanges between the domains and sub-disciplines. Genesis of earthquakes from seismology viewpoint is given by explaining plate tectonics and its numerical modelling along with applications. Causes of earthquake in Himalayan subduction zone are described, upon describing seismic waves and their propagation from a medium. Gradually changing over the discussion from geology and seismology to geotechnical and structural earthquake engineering, seismic (dynamic) soil-structure interaction (SSI) has been introduced. The discussion on infrastructure resilience pave way for earthquake disaster management. Therefore, ongoing efforts being made in developing seismic design philosophy based on performance-based seismic analysis has then been elaborated. To this effect, static pushover analysis procedure has been explained. Subsequently, some of the advanced dynamic response modification devices have been presented in greater details, including different types of damping devices, tuned mass dampers, seismic base isolation systems, new seismic protection means, and their practical applications have then been shown. In case of the semi-active control devices, this chapter also has presented structural control algorithms and latest techniques thereof. Finally, the need for future research in earthquake engineering has been highlighted for benefit to students and researchers.

V. Matsagar (✉)
Professor and Dogra Chair, Department of Civil Engineering, Indian Institute of Technology (IIT) Delhi, Hauz Khas, New Delhi 110 016, India
e-mail: matsagar@civil.iitd.ac.in

T. G. Sitharam et al. (eds.), *Theory and Practice in Earthquake Engineering and Technology*, Springer Tracts in Civil Engineering,
https://doi.org/10.1007/978-981-19-2324-1_1

1.1 Background and Opening Remarks

The frequency of minor to major earthquakes happening around the world is quite high as shown in Table 1.1. In so far as India is concerned, typically the northern part of India is severely affected because of the number of earthquakes happening every year, thus it mandates the civil infrastructure to be resistant to the earthquake-induced forces. From Table 1.1, it is clear that earthquakes with a magnitude of about five to six on the Richter scale happen almost daily somewhere or the other. The earthquakes occurring on the mainland have severe consequences on the manmade civil infrastructure (Fig. 1.1).

Therefore, civil engineers are naturally concerned with the safety of the infrastructure against earthquake-induced forces. As indicated in Fig. 1.2, a number of

Table 1.1 Frequency of occurrence of earthquakes [3]

Richter magnitude	About once every	Example
9.0+	20 Years	2011, Japan earthquake and tsunami
8.5–8.9	10 Years	2010, Chile earthquake and tsunami
8.0–8.4	1 Year	2008, China earthquake
7.5–7.9	3 Months	1906, San Francisco earthquake
7.0–7.4	1 Month	2010, Haiti earthquake
6.5–6.9	1 Week	1989, Loma Prieta, 1994, Northridge (LA)
6.0–6.4	3 Days	2014, Napa, 2011, New Zealand
5.5–5.9	1 Day	2007, Alum Rock (San Jose), 2011, Virginia

Fig. 1.1 Post-earthquake scenario [17]

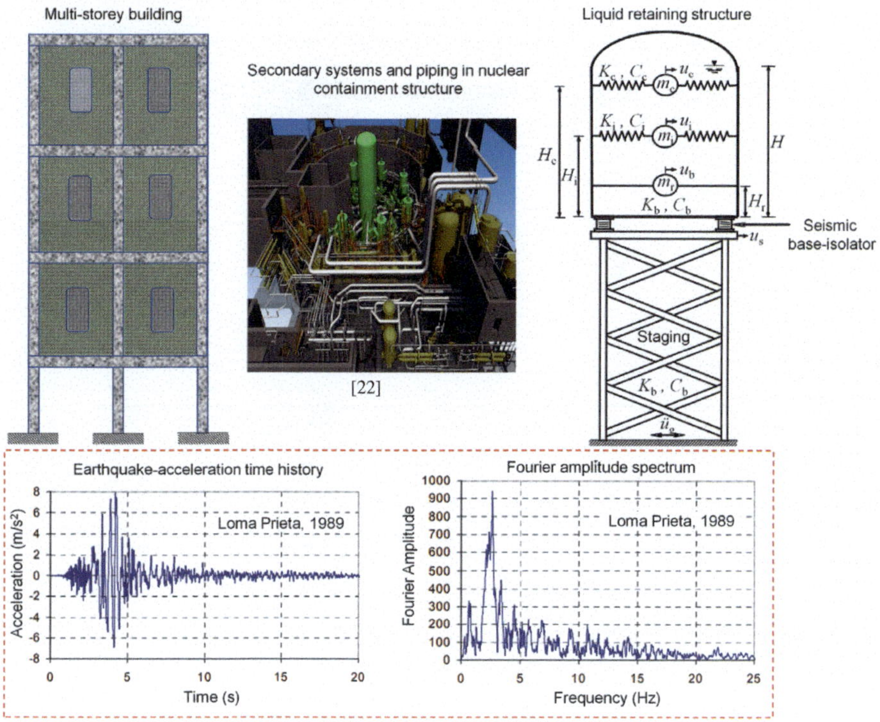

Fig. 1.2 Civil engineering infrastructure in day-to-day life

structures ranging from buildings to towers and various other types of structures are built for residential or commercial use. Further, nuclear power plants and wind turbine towers are vital for energy generation. The lifeline structures for water supply such as the liquid storage or containment tank and their corresponding pipeline and piping systems need to remain functional always. Hence, earthquake or seismic analysis and design of the civil infrastructure is an essential requirement. The analysis of structures subjected to earthquake ground motions is carried out either in the time domain, i.e., on a time scale, or in the frequency domain, i.e., on the frequency scale, where the loading function or the forcing function is the input given to the structures to determine their dynamic response.

Another important lifeline structure, as shown in Fig. 1.3, is the bridge, which is an essential component of the road and rail network for surface transportation. After any calamity, engineers must ensure that the bridges remain functional so that rehabilitation and rescue operations can be carried out effectively. Any loss of transportation infrastructure cannot be afforded during and after the earthquake. Any amount of downtime (out of service duration) for rehabilitation of these lifeline structures is highly undesirable because these lifeline structures play a vital role in the recovery of normalcy after the havoc caused by earthquakes. Thus, the earthquake

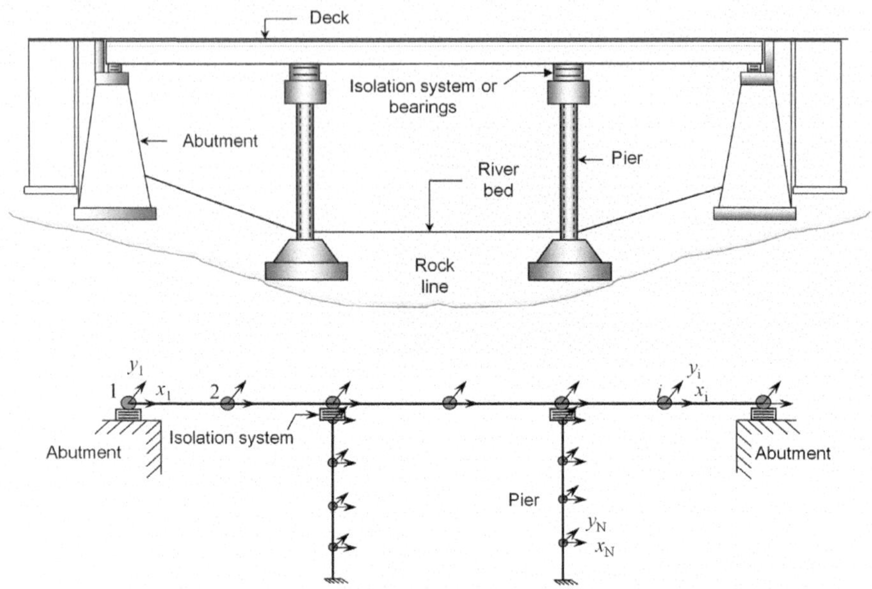

Fig. 1.3 Typical bridge structure [9]

resilience of bridges should be maximized under the anticipated earthquake-induced actions.

1.2 Expertise and Professionals Involved

As illustrated in Fig. 1.4, let us recall one of the most well-known stories heard in childhood about how six people perceived an elephant when blindfolded. For example, the blindfolded person who feels the ears of the elephant assumes that it is a fan. Another person observes the legs of the elephant and concludes that it is probably the trunk of a tree. Another examines the tail and thinks that it is most likely a rope, and so on.

On a similar note, earthquake engineering has various facets corresponding to different disciplines, as shown in Fig. 1.5; fortunately, it is well understood by engineers. It begins with the seismology or Earth sciences disciplines dealing with the origin of the earthquakes and their propagation. Then the geotechnical earthquake engineering typically deals with soil dynamics and the foundation design against earthquakes and considers the soil-structure interaction (SSI). However, in structural engineering, there are four major fields: earthquake-resistant (analysis and) design of structures, structural (dynamic or vibration) response control, performance-based earthquake engineering, and structural dynamics (nonlinear vibrations), all of which will be discussed in the later sections of the book. Both geotechnical and structural

Fig. 1.4 An elephant and six blindfolded people

Fig. 1.5 Different disciplines in earthquake engineering and technology

engineers rely on the principles of structural dynamics (engineering vibrations) while solving the problems related to earthquake engineering.

It is important here to understand the fine difference between earthquake engineering and earthquake technology! In earthquake engineering, sound theoretical basis is adopted for all the studies conducted, whereas in earthquake technology, real-life applications of those theories are pursued. Thus, earthquake engineering has a rather theoretical orientation as against the applied orientation in earthquake technology. Earthquake engineering works on the development of new analysis and design methods, which are then implemented to solve specific technical problems in

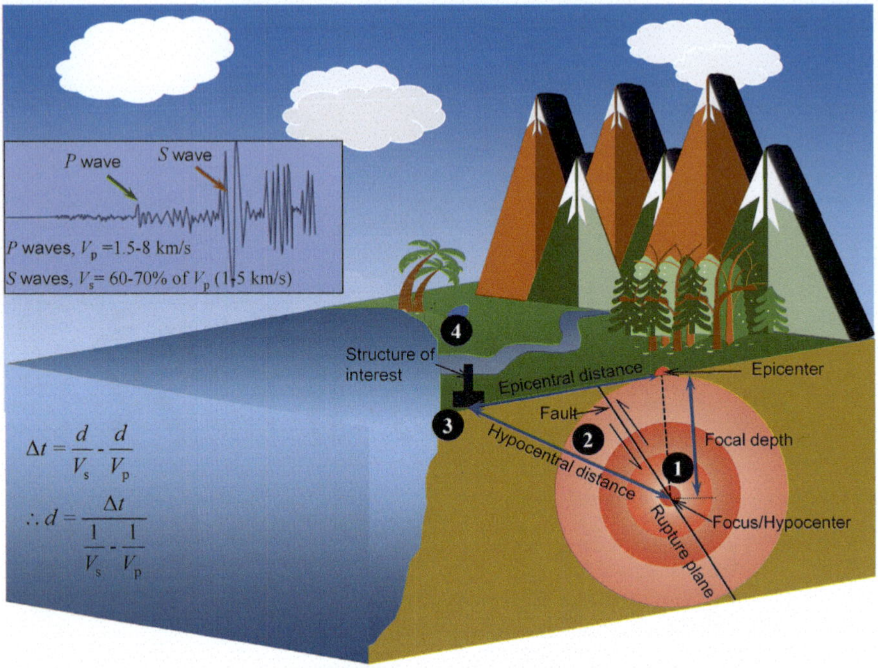

Fig. 1.6 Basic terminologies used in earthquake engineering and technology

the technology field. Earthquake engineering focuses on research, development, and design; on the other hand, earthquake technology involves testing, construction, and field applications. Thus, technology is the practical application of engineering knowledge. In the present book, both aspects of earthquake engineering and technology have been construed with clarity.

Figure 1.6 shows the origin of earthquakes, where ① depicts the location of the fault where the fracture process initiates and rupture takes place, known as the focus or hypocenter. The seismic waves propagate in all directions from the focus. Eventually, the interest of structural engineers remains on how the structures of interest are affected due to the seismic activity. In Fig. 1.6, the focus of the earthquake is numbered as ① and the structure which is of interest is numbered as ④. Between those two points (① and ④), the competence of professionals in at least four different interdisciplinary fields is required, as indicated in Tables 1.2 and 1.3. Professional expertise in geology, seismology, geotechnical engineering, and structural engineering is required to deal with the various aspects of earthquake engineering and technology. Typically, the geotechnical and structural engineers work hand-in-hand in carrying out soil-structure interaction (SSI) studies. However, they require inputs from geologists and seismologists on the magnitudes and the site-specific characteristics (intensities) of the earthquake, how the seismic waves propagate to the considered site, and what kind of soil amplification occurs in the site. Once such information is delivered

Table 1.2 Professionals involved in various activities related to earthquake engineering and technology

Professional	Activities
Geologists	Tectonic movements
	Geological studies
Seismologists	Seismotectonics
	Microzonation
Geotechnical engineers	Soil dynamics/rock mechanics
	Geotechnical earthquake engineering
Structural engineers	Earthquake engineering and technology
	Civil built infrastructure

Table 1.3 Expertise in various disciplines of earthquake engineering and technology

Activity	Scheme	Authors	Information
Macrozonation	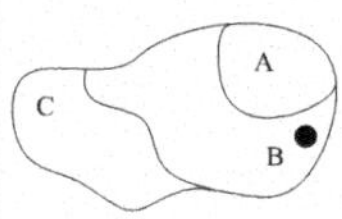	• Geologists	• Type of possible earthquakes
		• Seismologists	• Magnitude
			• Intensities
Microzonation	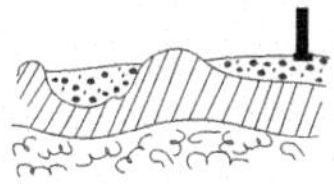	• Geologists	• Source position
		• Seismologists	• Magnitude
			• Intensities
			• Attenuation
			• Duration
Site condition	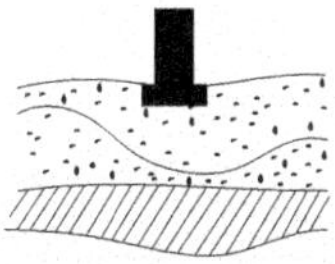	• Geologists	• Soil stratification
		• Geotechnical engineers	• Framing in soil type
			• Duration
			• Time history records
			• Design spectrum
Structure characteristics	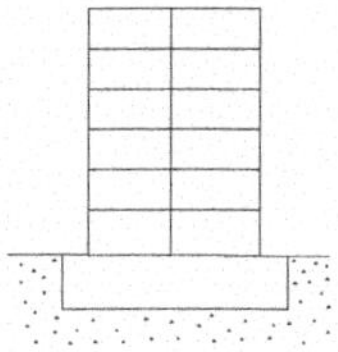	• Geotechnical engineers	• Level of protection
		• Structural engineers	• General configuration
		• Mechanical engineers	• Materials
		• Architects	• Foundation type
		• Builders	• Structure type
		• Owner	

to the geotechnical and structural engineers, vibration responses of soil, foundation, and the structural system are evaluated, which facilitate the earthquake-resistant design of the structures. Further, structural engineers focus on how to effectively control or limit the structural response so that no catastrophic or devastating failure of the structure takes place. Finally, the aim is to ensure that the dynamic response of the structure meets all the specified requirements, such as functionality, safety, and serviceability requirements.

1.3 Seismology: Why Do Earthquakes Occur?

1.3.1 Genesis of Earthquakes

The genesis of earthquakes primarily involves the expertise of geologists to comprehend the phenomena associated. While seeking answers for why and how an earthquake happens, typically a comparison can be made with what happens in a hotpot during the preparation of tea! When tea is prepared inside the hotpot, convective currents are formed in the water, wherein the current flows from the bottom of the pot toward the top because of the difference in temperatures (temperature gradient) within the water. The heat is directly applied at the bottom of the hotpot, therefore the water in the lower portion gets heated and rises upwards; then the colder water toward the top gets displaced downwards. Following this, the temperature of the underlying water rises again, and the cycles continue to form the convective current. Almost similar to what happens inside the teapot, convective currents are formed within the Earth due to the different temperature zones of the Earth underneath the lithosphere. The temperature of the innermost liquid zone of the Earth is extremely high, while the solid outer soil surface which is in contact with the atmosphere is relatively colder, and semi-molten lava is present in between these two zones (Fig. 1.7). Due to the

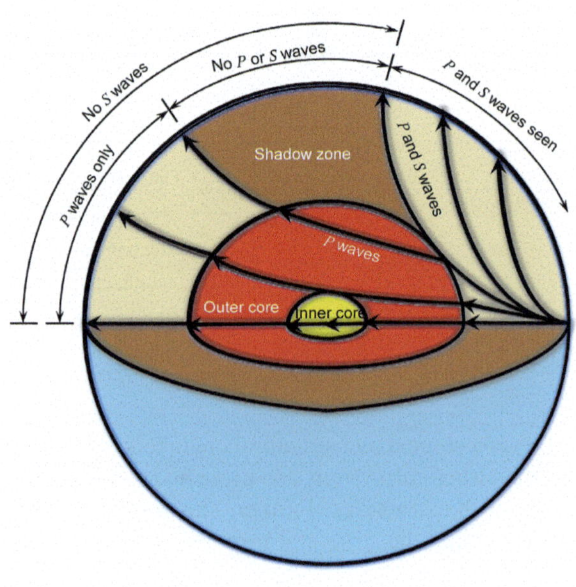

Fig. 1.7 Propagation of seismic waves

variations in the temperature, the convective currents are developed in the liquid and semi-molten zones, which in turn give rise to the various phenomena leading to the movement of the Earth's surface (in the form of movement of tectonic plates, as will be discussed later). Once these movements due to seismic activity take place at a point on the Earth's surface, different types of waves propagate on the Earth. Though in the next section the occurrence of an earthquake has been elaborated further, Fig. 1.7 shows how some of these primary and secondary waves are transmitted from the focus toward the different parts of the Earth's surface. The propagation of these waves depends on how far away the considered site is from the focus, and through which media it propagates. Interestingly, the shadow zone shown in Fig. 1.7 is formed due to the absence of any shear strength offered by liquid and semi-molten lava. Further discussion on the types of waves and the particle movements will be made subsequently to enhance the understanding of the origin of earthquakes.

Now, due to the cooling down of the Earth's crust, shrinkage takes place, thus the tectonic plates are formed, which are shown in Fig. 1.8. The tectonic plates keep on moving due to the formation of the convective currents within the molten or semi-molten lava underneath. Owing to the movement of the tectonic plates, particularly the movement of one plate across another along the plate boundaries, there are a number of earthquakes taking place round the year. If the attention is shifted to the Indian plate, as shown in Fig. 1.9, the Indian plate is continuously moving toward the Eurasian plate. This is the reason why a lot of earthquakes occur in the Himalayan region, and hence additional attention needs to be given across this region in the Indian peninsula.

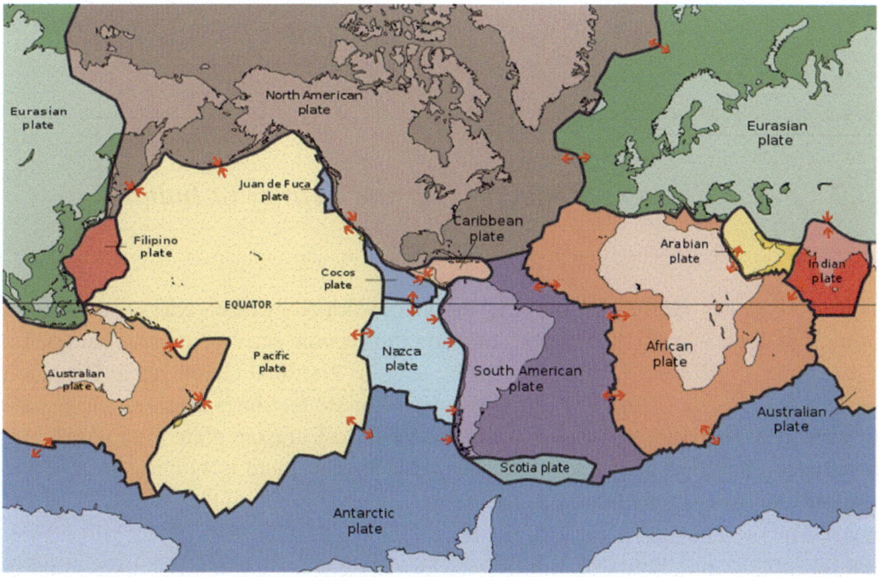

Fig. 1.8 Major tectonic plates on the Earth's surface [11]

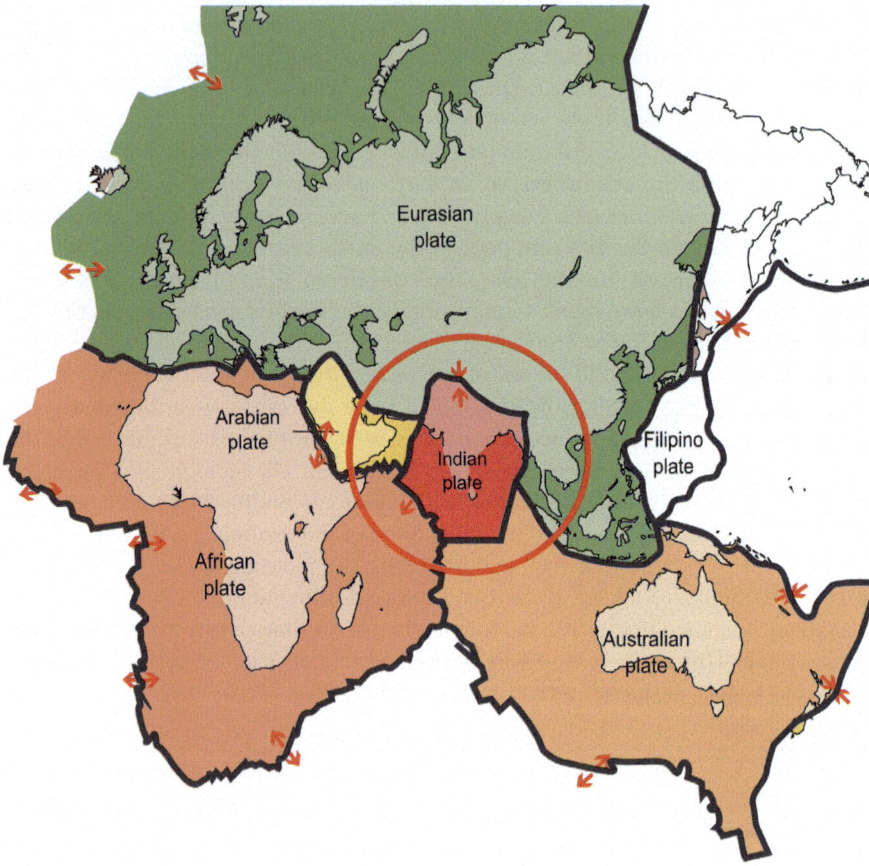

Fig. 1.9 Relative movement of the Indian plate [6]

1.3.2 Finite Element Model for Seismic Activity in Indian Plate

Figure 1.10 shows a finite element (FE) model developed by researchers at the Indian Institute of Technology (IIT) Madras to study the seismic activity in the Indian plate. Researchers at different institutes in India, such as National Geophysical Research Institute (NGRI), are similarly investigating the plate tectonics and seismic activities caused in the Himalayan region due to the Indo-Eurasian plate interaction. All these researchers are trying to model how the Indian plate is moving toward the Eurasian plate and thus obtain the effects of this movement on the seismicity of the region. The FE analysis is carried out wherein the entire plate is modeled with certain specific support and boundary conditions. The FE simulations are able to accurately predict the movement of the Indian tectonic plate, which is found to be approximately 12 mm per year. Thus, the FE simulations could successfully give an idea about how

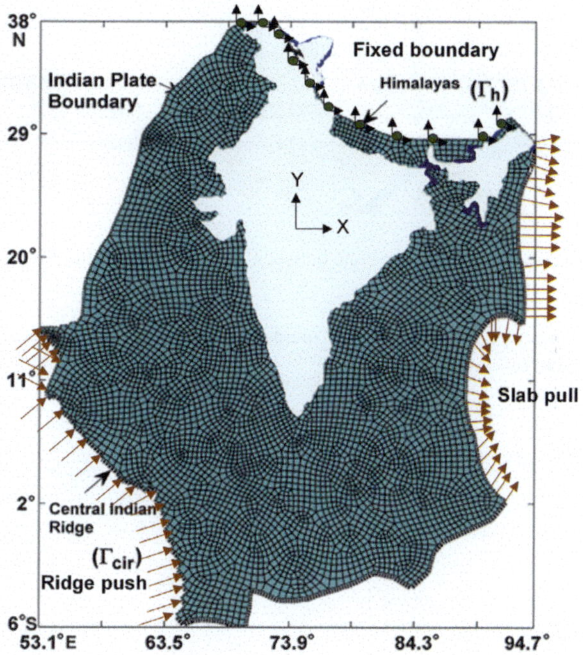

Fig. 1.10 Finite element (FE) model for seismic activity in Indian plate (modified from [8])

the tectonic plates are moving, and hence the nature of the earthquakes is anticipated. Subsequently, the same knowledge base is utilized to model the earthquake phenomenon, i.e., the ground motion is simulated. These days such simulations are being carried out by using advanced data analytics tools like machine learning or deep learning approaches. Either specific source modeling or blind source modeling approaches are followed for predicting the ground motions at a particular location.

1.3.3 Cause of Earthquake in Himalayan Subduction Zone

The primary cause of the earthquakes in the Himalayan and the Trans-Himalayan region of India is the formation of the subduction zone between the Indo-Australian plate and the Eurasian plate, as illustrated in Fig. 1.11. The Indo-Australian plate is continuously subsiding beneath the Eurasian plate, thus raising the heights of the Himalayan mountains in this process and causing earthquakes due to the abrupt motion at the boundaries of the two plates.

A number of other geotectonic features can also contribute to earthquake events. For example, there are some regions where an oceanic spreading ridge exists, or even there are some active volcanic areas, where earthquakes occur as a result of some of these geotectonic features, as shown in Fig. 1.12. At the plate boundaries, there

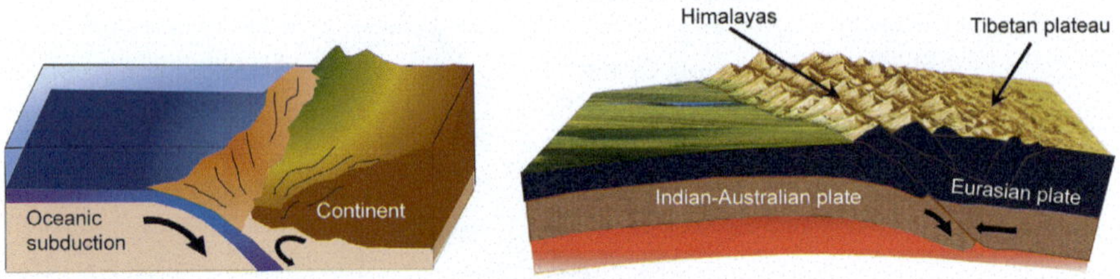

Fig. 1.11 Subduction zone between Indo-Australian and Eurasian plates

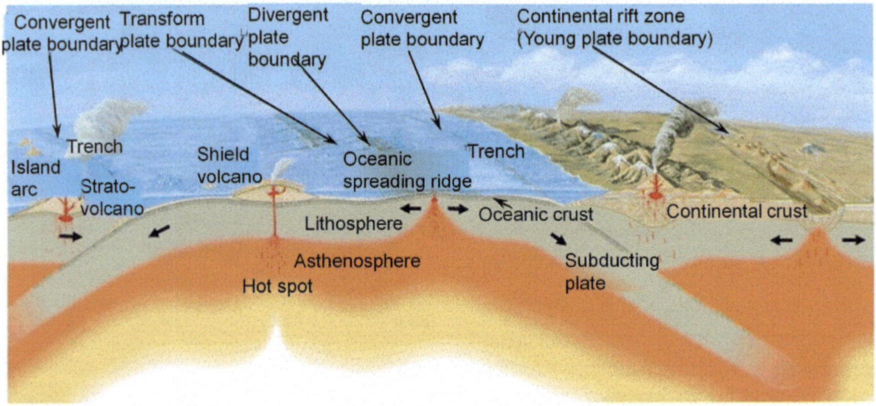

Fig. 1.12 Seismotectonic features [24]

are different kinds of movements taking place relative to each other, which are categorized as convergent, divergent, and transform boundaries. The movement of the plate boundaries is convergent when the plates are approaching toward each other. The movement of the plate boundaries is divergent when the plates are moving away from each other. When seen from the top, if the plate boundaries are moving in the transverse direction relative to each other, they are known as transform boundaries. All these seismotectonic features give rise to earthquakes having different characteristics based on how the accumulated energy is released in the process. Typically, seismologists study such features and determine the characteristics of earthquake ground motion and also determine the features of the wave propagation. Further, macro- and microzonation of the region facilitate the collection of ground motion data and its assimilation for strong motion studies.

As mentioned earlier, earthquakes originate due to the rupture at active fault locations when the tectonic plates move relative to each other. The tectonic movements build up strains in the rock masses. Typically, during these continuous tectonic movements, when the strain developed in the rocks exceeds the strain capacity (ultimate strength) at any fault location, the rocks break in a brittle manner, causing a sudden rupture and release of the accumulated energy. In this process, a large amount of energy is released. This is explained in geology using the elastic rebound theory. The exact location at which such a rupture on the fault location initiates is called focus (hypocenter) and its projection on the ground surface is called the epicenter. The distance below the ground surface between the epicenter and the focus is called the focal depth.

In order to protect the built environment, i.e., the civil infrastructure from the wrath of earthquakes, discerning the underlying physics involved in the genesis of earthquakes is crucial. The quantification of the energy released at the origin of the earthquake (i.e., at the focus) is made in terms of the magnitude of the earthquake, which is typically measured using the Richter scale. From Fig. 1.6, it is known that the distance from the epicenter to the location of the structure of interest is called the epicentral distance, and the distance from the hypocenter to the structure of interest is known as hypocentral distance. The intensity of ground shaking experienced at the structure of interest depends mainly on the epicentral distance. Hereby, the readers should note that the magnitude and intensity of the earthquake are two distinct features, and are required to be clearly distinguished and appropriately used in structural analysis and design. The magnitude of the earthquake is obtained from the energy released at the source of the earthquake, irrespective of the location of the structure of interest. On the contrary, the intensity of the earthquake depends on the location of the considered structure relative to the epicenter of the earthquake; intensity is indirectly related to the amount of energy released at the source. During the process of transmission of the seismic waves (body waves and surface waves) from the focus to the structure of interest, if the propagation is monitored then it can be used for giving an early warning. This is possible because the different seismic waves have varied speeds of propagation in the ground medium. An early warning to the residents is thereby given by calculating the difference in the time of arrival of the different types of seismic waves (Fig. 1.6). In the Indian context, sensors and instruments are installed by researchers from the Indian Institute of Technology (IIT) Roorkee, the sensors are placed ranging from the Himalayan region to the most populated northern cities such as the National Capital Territory (NCT) Region of Delhi. To understand the concept of prediction of major earthquakes, the different types of seismic waves are needed to be studied, as illustrated in Figs. 1.6 and 1.13.

There are broadly two types of seismic waves: body waves and surface waves. Further, the body waves are of two types depending on how the soil particles move. If the particles move in the form of compression and rarefaction, i.e., particles are moving in the direction of wave propagation, then such types of body waves are called pressure or primary (P) waves. However, in the shear or secondary (S) waves the particles move up and down relative to each other, i.e., the particles move in the transverse direction with respect to the direction of transmission of the seismic waves. In both these waves, the movement of particles is traceable in the cross-section of the medium through which they propagate.

Generally, the surface waves are most destructive to the structures; however, their propagation velocity is relatively much lower than that of the body waves, particularly that of the primary (P) waves. The primary waves propagate faster as compared to the secondary waves and the other surface waves. The surface waves are of two types: the Love waves, where the particles move sideways (when viewed from the top, it

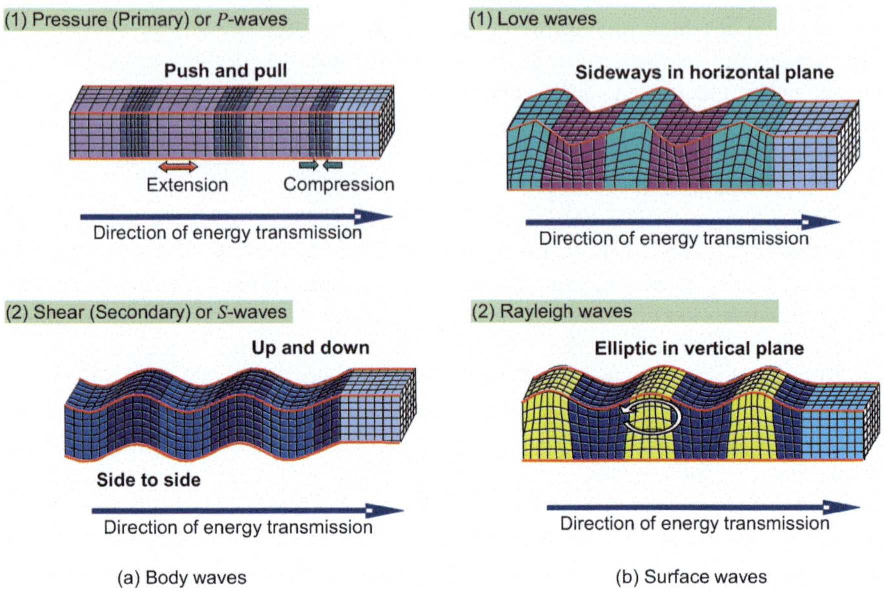

Fig. 1.13 Types of seismic waves [16]

looks like a serpentine motion), and the Rayleigh waves, where the particles move in an elliptical orbit (in the cross-sectional view). These characteristic information about the P and S waves are exploited for predicting or giving early warnings to the users or the citizens, even maybe on their cellphones or smartphones.

For example, in the earthquake time history shown in Fig. 1.6, the arrival of the primary and secondary waves at a location can be deciphered. The duration between the arrival of the P and S waves is calculated, and this information from at least three locations is used in locating the epicenter using the method of triangulation. Furthermore, the hypocentral distance can also be calculated by a similar assessment. Further, using this knowledge, a prior warning could be given to the residents of a heavily populated region, that a significant earthquake might be approaching, and eventually some invaluable lives could be saved.

Now, shifting our focus toward the structural design aspects. It is very important to consider what are the characteristics of the ground motions felt at a particular location (site) such as what should be peak ground acceleration (PGA) taken in the structural design. The ground acceleration experienced at a particular location decreases with the increase in the distance of the site from the epicenter (and even the focus); such a decrease is referred to as attenuation of the earthquake ground motion. Several ground attenuation relationships have been proposed by researchers for various regions. Certain ground motion attenuation relationships have been proposed for the Indian peninsular region as well. In Fig. 1.14, such attenuation relationships are shown which depict how the peak ground acceleration (PGA) in the horizontal direction

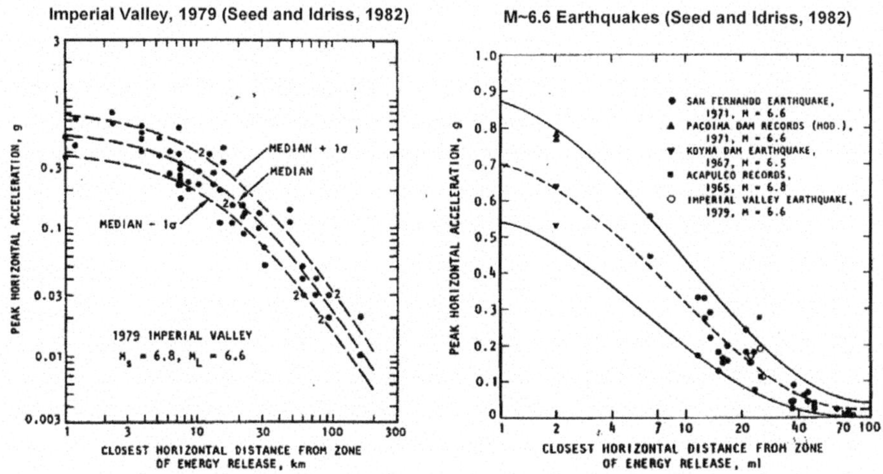

Fig. 1.14 Ground motion attenuation relationships [19], Seed and Idriss [23]

reduces away from the focus or the fault rupture location. Naturally, in the areas in closer vicinity of the epicenter, higher PGA is experienced as compared to that at farther distances.

For peninsular India, the attenuation relation for the PGA and the spectral acceleration (S_a) for the rock site is given by Raghukanth and Iyenger [19] as

$$\ln y = c_1 + c_2(M - 6) + c_3(M - 6)^2 - \ln R - c_4 R + \ln(\varepsilon)$$

where y is the ratio of spectral acceleration at bedrock level to acceleration due to gravity (S_a/g), M refers to the moment magnitude, ε and R denote the errors in the regression and hypocentral distance, respectively. The coefficients c_1, c_2, c_3, and c_4 are specified by Raghukanth and Iyenger [19].

1.4 Seismic Soil-Structure Interaction (SSI)

Now, switching over from the domain of geologists and seismologists to geotechnical engineers, the information given by the former is used by the latter to analyze site-specific earthquake ground motions. Based on the subsoil condition at a site of interest, the geotechnical engineers examine the effect of soil characteristics on the incoming earthquake waves. While obtaining the final input ground motion to the structure, the time period of the soil underneath and the stratifications or layer properties are important parameters of interest.

Usually, dynamic soil-structure interaction (SSI) needs to be considered during an earthquake, as well as the liquefaction potential of the soil should also be duly assessed. Different analytical or numerical approaches are followed for modeling the dynamic SSI, ranging from simplistic (approximate) to detailed (rigorous) modeling of the dynamic soil properties (Fig. 1.15a–d). In simplistic two-dimensional (2D) modeling, as depicted in Fig. 1.15a, spring-dashpot models can be employed for the deep foundations, by assigning suitable mass, spring coefficients, and damping coefficients, depending on the type of soil or rock underneath [25]. The modulus of subgrade reaction governs the dynamic parameters assigned in the spring-dashpot-type discrete modeling of the soil.

The various arrangements of the mass, springs, and dashpots in the discrete models (Fig. 1.15b) are rather simple to model in the commonly used commercial software packages such as SAP2000®, ETABS®, STAAD-Pro, etc. However, a detailed and accurate analysis for important structures, such as nuclear power plants, dams, or other strategically important structures, is conducted by either three-dimensional (3D) elastic half-space or continuum type of modeling approaches based on rigorous

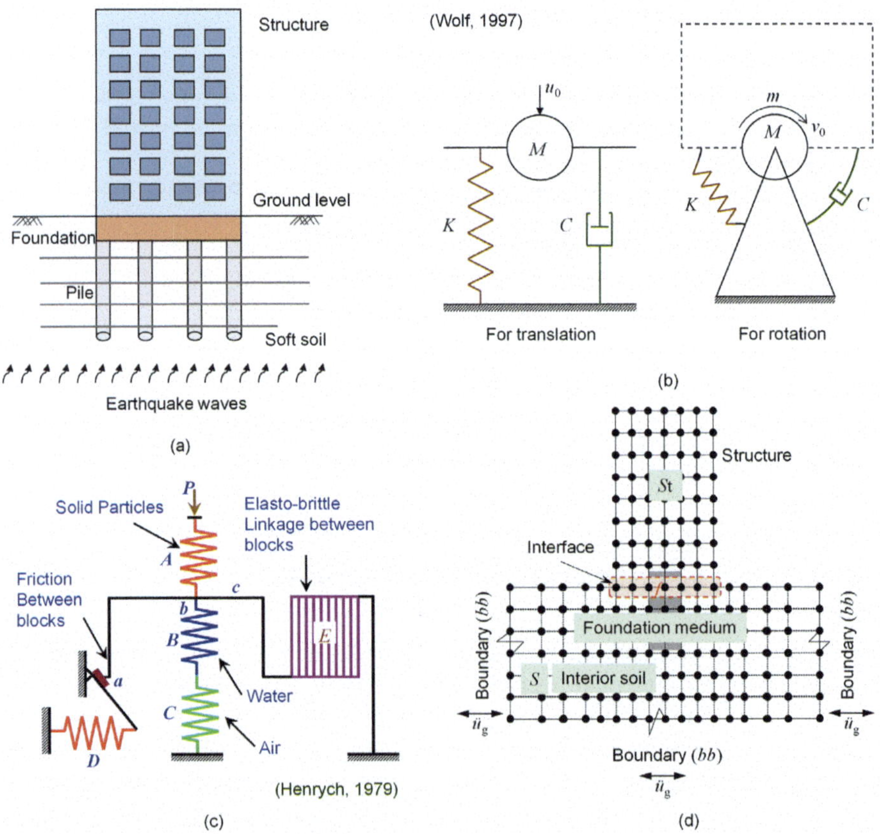

Fig. 1.15 **a** Seismic SSI in a multistory building with pile foundation, **b** spring-dashpot-mass models for translation and rotation, **c** rheological model for SSI, and **d** finite element (FE) model of soil-structure system for direct analysis methods

3D FE analysis. One of these models is shown in Fig. 1.15d, wherein the SSI problem is modeled using the FE approach. Among the simplified and detailed models, some other soil modeling approaches address the shortcomings of these two models, in terms of accuracy and computational time. These mathematical models proposed by geotechnical engineers consider various phenomena taking place within the soil mass, such as inter-particle friction, elastoplastic interactions, etc. Using this modeling approach, the behavior of the foundation in case of any earthquake event can also be studied. Moreover, individual solid particles of the soil are also modeled and the interaction between them is considered depending on whether it is a friction interaction or an elastoplastic interaction (Fig. 1.15c) [5]. Saturated soil exhibits a different response than unsaturated soil; therefore, some of these models take moisture content and air content into consideration appropriately. Nevertheless, spring-dashpot models are still more popularly used in day-to-day structural designs when SSI is to be accounted for.

1.5 Earthquake Disaster Management

Before discussing the role of structural engineers on how they deal with earthquake ground motions and the forces that are induced on the structures thereby, the concepts of earthquake disaster management should be known. While delving into risk-based earthquake disaster management, some terms like hazard, vulnerability (fragility), and risk are introduced. Hazard may be defined as the probability of occurrence of the earthquake at a particular location, whereas the return period of an earthquake is the duration in which the same intensity of the earthquake may occur again. Figure 1.16 shows seismic hazard curves, which are simply the plots of the annual probability of exceedance versus the earthquake intensity felt, defined in terms of PGA. Based on this relation, a distinction is established between low seismic hazard and high seismic hazard.

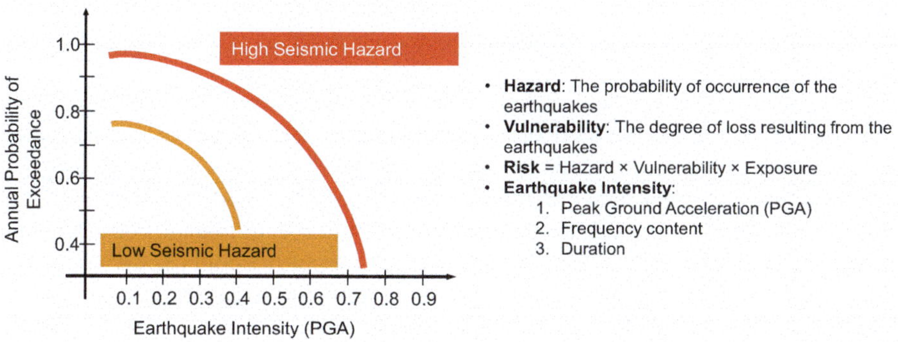

Fig. 1.16 Basic terminologies used in earthquake disaster management

The intensity of an earthquake is basically defined in terms of peak ground acceleration (PGA), frequency content, and the duration of the earthquake. While generating the seismic hazard curves, the PGA has been used as an intensity measure here; however, the other intensity parameters could also be used to obtain similar plots. Generally, for important structures, the earthquake intensity measure quantified in terms of frequency content is more suitable. The intensity of an earthquake based on duration accounts for the number of cycles of repetitive loading applied to a structure; therefore, for long-duration earthquakes, it is a necessary intensity measure. However, currently, the Indian codes mostly deal with the PGA as an intensity measure of the earthquake, and quite appropriately so.

Subsequently, the vulnerability of the structures for varying earthquake intensities is determined in terms of PGA, as well as the mode of failure that may occur is predicted. If the probability of failure of the structures is plotted against the intensity measure of the earthquakes (PGA, in this case), these are known as the fragility curves, indicating how fragile the concerned structures are to a given intensity of an earthquake. Based on the fragility curves, the seismic risk to which the civil infrastructure is exposed can be calculated. Such seismic risk calculations are conducted by initially developing the fragility curves of the considered structure (also called vulnerability curves sometimes), and subsequently calculating the hazard at the specific site and also exposure level. To sum up, the plots of the probability of failure of the structure versus the earthquake intensity (PGA or spectral acceleration) are the fragility curves. There is another school of thought in the research community which considers the vulnerability curves different from fragility curves; in their opinion, the former also accounts for the risk and cost implications.

Eventually, in achieving the ultimate objective of an earthquake resilient society, earthquake hazard mitigation strategies are adopted based on the seismic risk posed by the considered structure. In seismically developed countries such as Japan and the west coast of America, extensive studies have been conducted to quantify the seismic risk posed by the built environment and also on the measures to mitigate the earthquake hazard. One of the reasons for calculating fragility or vulnerability is to quantify the failure probability of a particular structure in a compact form. Though fragility and vulnerability are generally used interchangeably, to be specific the structural damage is considered in the former whereas monetary and human losses are considered in the latter.

For routine earthquake-resistant design of structures, Indian Standard (IS) 1893 (Part 1 to 5): 2016 [1] provides procedures and guidelines. Keeping the debate aside, whether this is a code or a standard, IS 1893 mainly recommends using a linear response spectrum, either for the seismic coefficient method or for the complete response spectrum analysis of structures. The response spectrum plot is obtained by conducting dynamic analysis on a linear single degree of freedom (SDOF) system. The peak displacement response of an SDOF system with a specific time period is determined when the SDOF system is subjected to a particular earthquake. By gradually increasing the time period of the SDOF system, ranging from relatively stiffer to more flexible structures, the peak structural displacement responses could be plotted against their corresponding time periods, this is known as the displacement spectrum curve for the specific excitation. Further, each ordinate of the displacement spectrum could be multiplied by their corresponding natural frequency to obtain the pseudo velocity spectrum. Similarly, the pseudo acceleration spectrum is derived by multiplying the square of the natural frequencies with the ordinates of the displacement spectrum curve. Avoiding the computational burden of evaluating the actual velocities and acceleration of all the considered structural systems, the pseudo velocity spectrum and the pseudo acceleration spectrum plots provide a good estimate of the energy dissipated by the system and the base shear attracted by the system, respectively. A valuable modification to the conventional response spectra curves is the tripartite plot, which combines all the three, acceleration (A), velocity (V), and displacement (D) response spectra into a single plot. One such example of a tripartite plot of the Imperial Valley, 1940 earthquake is shown in Fig. 1.17. Note that the term spectra is the plural form of spectrum. In the case of actual earthquake ground motions, the response spectrum has several peaks and valleys (i.e., it is not smooth). Hence, the response spectra are obtained for various site-specific earthquakes, and the plots are normalized and averaged to convert those into a single curve. Further, the graph is smoothened and scaled according to the different soil conditions, to obtain the design response spectrum as given in IS 1893 (Part 1) 2016 (refer to Fig. 1.18).

The Indian seismic code (Part 1 of IS 1893: 2016) has provided the design response spectra depending on the type of rock or soil available at the particular site. From Fig. 1.18, it can be observed that there are two types of design spectra available in the code, one is for the equivalent static approach and the other for the response spectrum approach for earthquake analysis of structures. The equivalent static approach is rather simple, which is an extension of the past practice followed in an earthquake engineering discipline. Structural engineers used to take about 10% of the total weight or seismic weight of the structure in the horizontal direction for designing

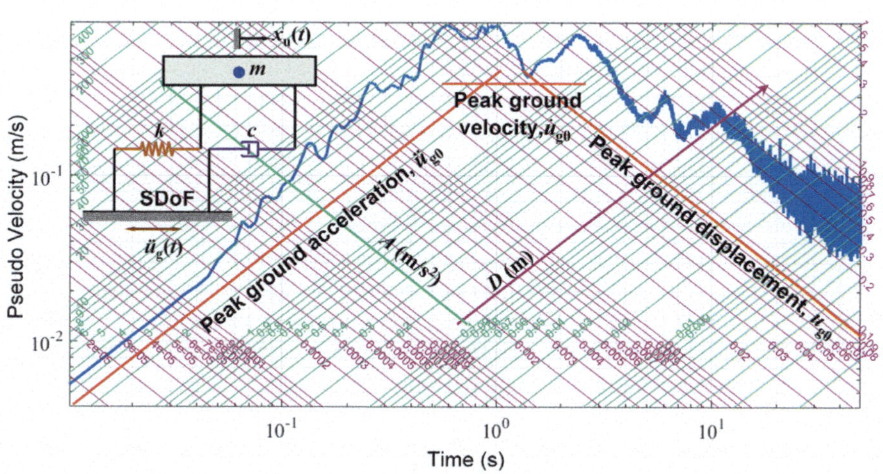

Fig. 1.17 A-V-D response spectra on a tripartite plot with 5% damping ratio (Imperial Valley, 1940)

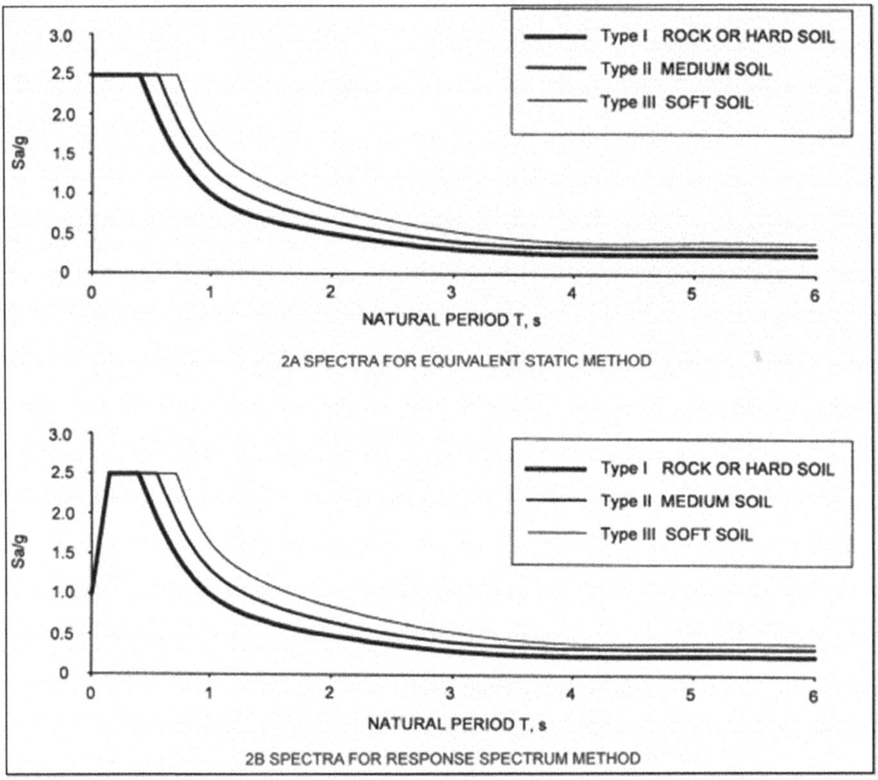

Fig. 1.18 Design response spectra as per IS 1893 (Part 1): 2016

lateral load resisting elements, such as columns. An improvement from such a rule of thumb approach, now the seismic base shear is calculated using the equivalent static method by considering only the first mode of vibration of the structure. The seismic coefficient method (SCM) is one such equivalent static approach for earthquake-resistant design of structures, as suggested in IS 1893: 2016 (Part 1). Nonetheless, some researchers even consider it to be a dynamic approach because the response spectra are obtained by conducting dynamic analysis.

For certain important structures such as the nuclear containment structures, simplified seismic analysis based on the generic spectrum or the design spectrum is not suitable. In that case, a site-specific design spectrum must be used, and a detailed structural model is prepared to account for all nonlinearities. From site-specific design spectra, spectrum-compatible earthquake time histories are generated to conduct nonlinear time history analysis of the important structures to evaluate their seismic response. For obtaining the site-specific design spectra, certain methods are available, for example, GovindaRaju et al. [4] have suggested one such method. Herein, important guidelines to calculate the site-specific design spectra are provided (Fig. 1.19a) by taking into account the shear wave velocity profile, the modulus reduction curve,

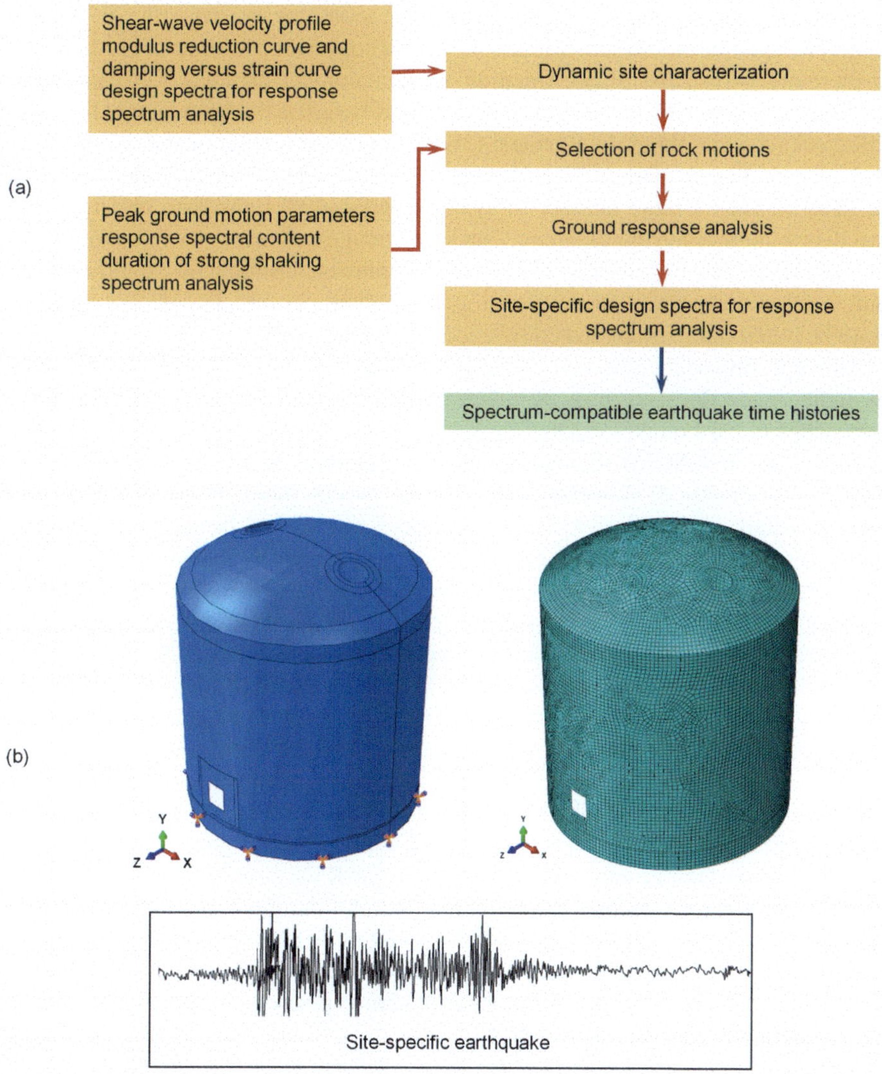

Fig. 1.19 Site-specific earthquake response analysis for important structures: **a** site-specific design spectrum and spectrum-compatible ground motions, inspired from Govinda Raju et al. [4], **b** inspired from [18] studies

and the damping ratio curve of the soil profile. Peak ground motion parameters, response spectra content, duration of strong shaking, and spectrum analysis of the considered ground motion are used to select the strong motions. Eventually, the site-specific design spectra for response spectrum analysis are derived. Further, this site-specific response spectrum could be utilized to carry out the spectrum-compatible earthquake nonlinear time history analysis. Synthetic or artificial earthquake ground motions can be generated easily by using some tools; for example, the WAVGEN code developed at the Indian Institute of Technology (IIT) Kanpur could be used. Furthermore, seismic fragility estimation of a containment shell has been carried out at the Indian Institute of Technology (IIT) Bombay based on the formation of through-wall cracks using a three-dimensional (3D) finite element (FE) model. Conducting stochastic studies on earthquake response of structures in the time domain demands the generation of a large number of synthetic ground motions with certain statistical distribution or attributes. Hence, such practical limitations prohibit using these state-of-the-art seismic analysis and design techniques in routine design offices. For routine seismic design of structures, simplified albeit conservative design methods are to be relied upon as advocated by codes and standards.

Although important structures require nonlinear time history analysis, currently the Indian code does not provide much details about it. In IS 1893: 2016, three approaches to earthquake-resistant design of structures are mentioned: (a) seismic coefficient method (SCM), (b) response spectrum method (RSM), and (c) time history (TH) analysis method. However, only the first two methods, SCM and RSM, are given in detail within IS 1893 (Part 1): 2016. The SCM is basically an approximate method which adopts a static approach to solve a dynamic problem. The SCM considers only the fundamental (first) mode of vibration of the structure to evaluate its seismic response. In the SCM, the time period of the structure is estimated based on some empirical formulas, which depend on the characteristics of the structure such as the type of infill used. Since the SCM is an approximate approach, naturally the method is made more conservative. The SCM is quite popularly used for the earthquake-resistant design of routine structures. A better, theoretically improved approach as compared to the SCM is the response spectrum analysis (RSA) or response spectrum method (RSM), which also accounts for the higher vibration modes. For classically damped (considering damping to be proportional to the mass and/or stiffness of the structure) linear structures, the RSM is quite suitable, especially when lumped mass modeling has been adopted. All the modes of vibration which contribute toward greater than 90% of the modal mass participation can be considered to be contributing to the total dynamic response of the structure. Nonetheless, even in the RSM, the superposition of the modal response is statistical and thereby approximation is adopted inherently.

The RSM approach is based on the linear response spectrum, therefore it is not suitable for the nonlinear structures. For non-classically damped structures, where external damping may have been added, the RSM is unsuitable, unless the desired nonlinear response spectrum is made available. In such cases, the nonlinear time history analysis (NLTHA) is the most suitable and advantageous method to evaluate the seismic response of the structure. The different kinds of advanced engineering materials used in structures introduce nonlinearity to the structure, and hence suitable nonlinear time history analysis must be carried out. In the case of the nonlinear time history analysis, diverse material models or behaviors can effectively be modeled while taking into account non-classical damping, i.e., including the externally added damping. Moreover, the geometric nonlinearity and large displacement effects can be suitably accounted for while conducting the NLTHA. When higher degrees of nonlinearity exist, such as in the case of the pounding of structures, the NLTHA is adopted suitably. In seismic pounding of adjacent (closely spaced) structures, the structures collide because of oscillation in opposite directions, i.e., out-of-phase, thus the structures hit each other at different time instances and at different locations. Thereby, the contact behavior of the two structures after pounding can be appropriately modeled with smaller analysis step sizes. The NLTHA can also account for possible spatial and temporal distributions of inelasticity within the structure.

Figure 1.20 depicts a simplified dynamical model of a nonlinear structure where the stiffness or damping of the structure could be nonlinear and this nonlinearity can be accounted for by conducting a nonlinear time history analysis. For important structures, a finite element-based (FE) modeling approach is adopted along with spectrum-compatible time history of the earthquake ground motion, and the problem is solved using either numerical integration methods or state-space analysis method. Here, as the equations are nonlinear, a sound numerical approach is required for conducting the time history analysis. Some of the commercially available software such as SAP2000®, ETABS®, etc., offer nonlinear response history analysis facilities. Future versions of the Indian standards are expected to give more details

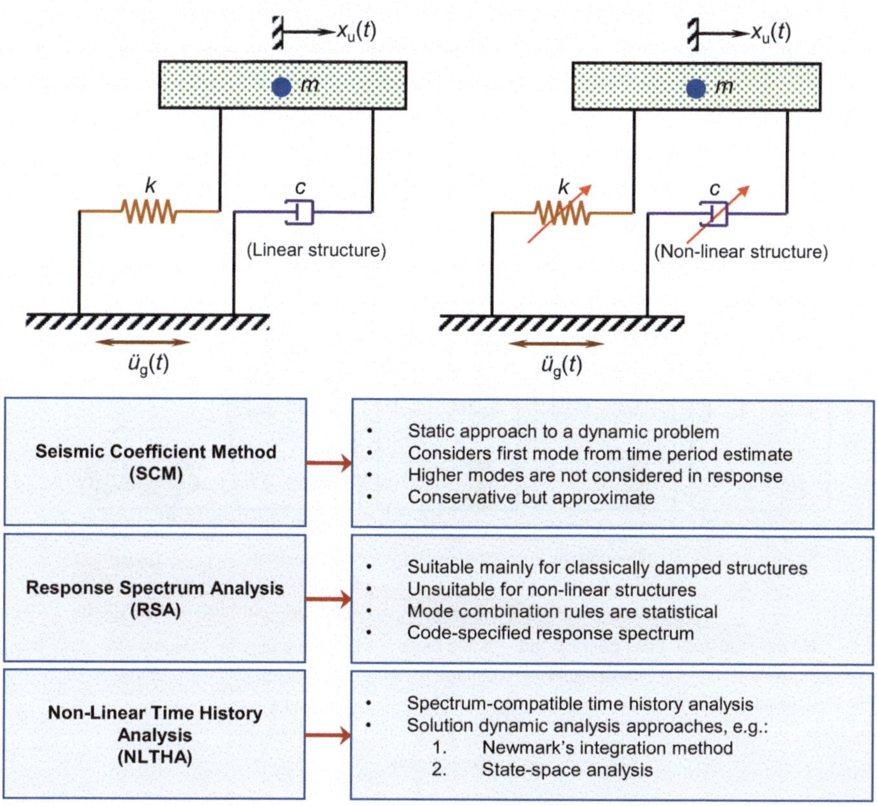

Fig. 1.20 Different analysis techniques for linear and nonlinear time history analysis

and recommendations on how to conduct the nonlinear response history analysis of structures.

1.6 Seismic Design Philosophy: Performance-Based Seismic Analysis

Performance-based earthquake engineering (PBEE) is an emerging discipline in earthquake analysis and design of structures. Performance-based seismic design (PBSD) of structures is a new philosophy which is being followed in some of the seismically advanced countries. The upcoming version of the IS code on seismic design of structures will also have provisions/guidelines concerning performance-based seismic design (PBSD). However, it is also believed that the current Indian Standard (IS) code accounts for some of the performance levels which are a part of PBSD indirectly. To understand the PBSD, it is important to know about the seismic performance levels, which in turn depend on the damage states of the structure. Figure 1.21 explains the different damage states of a structure pictorially. If the considered building remains operational with no damages after being hit by a major earthquake, and all the systems are fully functional even during the earthquake, then it is called an operational level performance. No (or negligible) damage is experienced by the structure in the seismic performance of operational (O) level, i.e., continued functionality. In the next case, the structure is subjected to a more severe earthquake event than that at the operational level, and some minor damage is seen in the building which is serviceable and does not jeopardize the immediate use of the structure. Then,

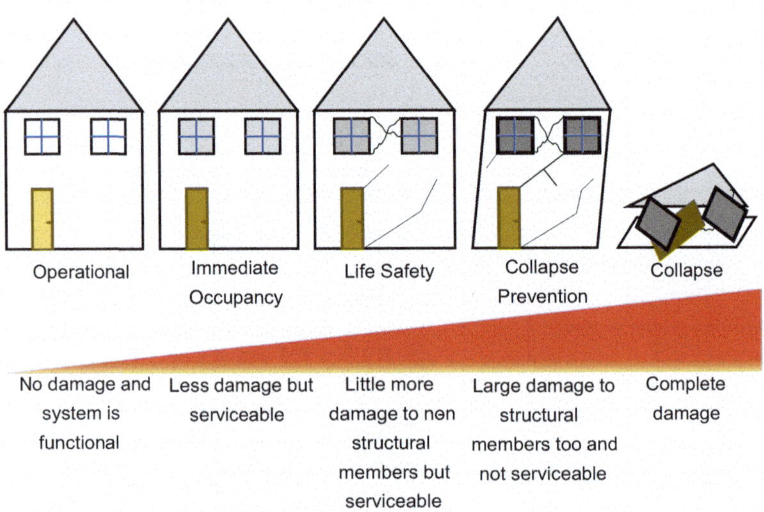

Fig. 1.21 Damage states and performance level thresholds

the structure can be immediately occupied after the earthquake event and it is known as the immediate occupancy (IO) category in the performance-based seismic design. With further increased intensity of an earthquake, more damage is experienced by the structure, i.e., some non-structural members might have failed, which calls for repair work. Nevertheless, life safety has been ensured; hence, this seismic performance level of the structure is called the life safety (LS) level. In the fourth damage state of the structure, severe or large damages are expected, even in the lateral load resisting elements, such as the columns and shear walls, due to the increased intensity of the earthquake. In this case, it may not be advisable to strengthen or rehabilitate these structural elements after the earthquake; however, since sudden collapse of the structure is prevented, it is known as the collapse prevention (CP) performance level. Eventually, for an earthquake of intensity significantly higher, complete damage and collapse of the entire structure might occur; however structural design needs to preclude it (beyond considered design level event).

In the PBSD, the major three categories of damage states are considered: light damage, moderate damage, and severe damage; and, correspondingly, immediate occupancy (IO), life safety (LS), and collapse prevention (CP) seismic performance levels of the structures are considered. Depending upon the type of structure being designed and its intended use under the design basis of the earthquake, the IO level, LS level, and CP level are decided as seismic performance. For example, a hospital building should at least satisfy the IO level of seismic performance, if not the O level. In order to ensure that a structure meets these seismic performance levels, depending on the anticipated earthquake ground motion it experiences, nonlinear analysis is conducted by subjecting the structure to the different levels, i.e., different intensities of an earthquake. The outcome of this analysis is seismic performance evaluation of the structure, which is possible by conducting a pushover analysis.

1.7 Static Pushover Analysis

There are two different ways of conducting static pushover analysis: (a) pushing the structure laterally by applying some forces and monitoring the displacement experienced, or (b) applying some displacement in a specific manner and monitoring the force generated, as shown in Fig. 1.22. The first approach is called the force-based approach and the second one is called the displacement-based approach. Conventionally, the force-based approaches have been followed, however nowadays displacement-based approaches are becoming more popular due to their advantages over the former. Displacement-based approaches are suitable for capturing the post-peak response, particularly for the highly inelastic behavior of the structure. In static pushover analysis, horizontal forces are applied (quasi-statically) in a predefined pattern on the structure, and load–deflection curve is obtained for the entire structure. For example, the distribution of the horizontal load applied to the structure, to conduct the pushover analysis, could be similar to the first mode shape. For the

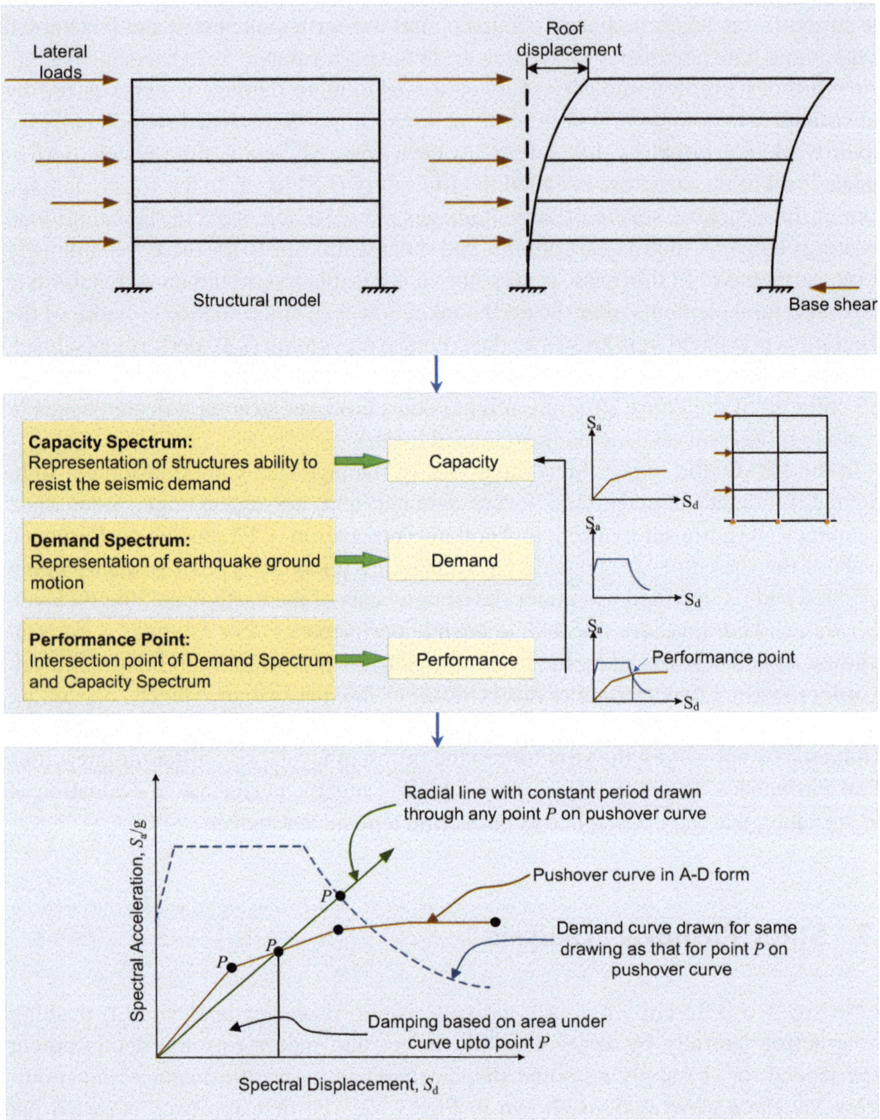

Fig. 1.22 Static pushover analysis [7]

applied horizontal load, the peak deflection is calculated and the load–deflection curve is plotted (Fig. 1.22).

In Fig. 1.22, a multistory building is applied with a certain pattern of horizontal forces and the corresponding displacement of the building is noted. Then, gradually the applied horizontal load is increased and with each increment the deflection of the building is noted. At a particular level of horizontal loading first hinge forms in the structure. With a further increment in the horizontal load, subsequent hinges keep forming until the collapse mechanism is formed, i.e., the structure becomes unstable. However, for large structures, only a few initial hinge formations are adequate to conclude the analysis. In this process of lateral pushing, the structure experiences high inelastic deformations, and a nonlinear load–deflection curve is obtained, either through a force-based approach or a displacement-based approach. This nonlinear load–deflection curve gives an idea about the seismic load-carrying capacity of the entire structure, hence it is called a capacity spectrum. It is a representation of how well a structure can resist the seismic forces which are induced in it. The load–deflection curve is converted into a spectral acceleration (S_a) and spectral displacement (S_d) curve, by dividing the applied horizontal force by the seismic or inertial mass, as the inertial force is a product of mass and acceleration.

The response spectrum of an earthquake is generated by plotting peak responses of SDOF systems with varying time periods and constant damping. The acceleration response spectrum plotted for an earthquake in terms of S_a and S_d (refer to Fig. 1.22) is an indication of the seismic demand on a structure. Hence, this curve is also called the demand spectrum, signifying the requirement imposed from the earthquake-induced loading on the structure. The two curves, capacity spectrum and demand spectrum, are then overlapped, and their intersection is called the performance point. Such overlapping of the two curves, i.e., capacity and demand spectrum curves to examine the seismic performance is shown in the spectral acceleration versus spectral displacement curve, shown in Fig. 1.22. Based on this performance point obtained, the seismic performance of a structure under consideration can be assessed, as to whether it meets the immediate occupancy (IO) criterion, life safety (LS) criterion, or the collapse prevention (CP) criterion (refer to Fig. 1.21).

Nowadays, performance-based seismic design (PBSD) is carried out popularly through pushover analysis, especially for designing some of the important structures. Nevertheless, probabilistic seismic vulnerability techniques or design approaches have been introduced and practiced based on an enhanced understanding of the behavior of structures under seismic actions, and also taking into consideration the probability of occurrence of structural failures due to the earthquake-induced loading. However, codes and standards are yet to completely adopt such latest probabilistic seismic assessment approaches. The seismic analysis of structures seen thus far in the codes are majorly deterministic (possibilistic); however, earthquake events are random in nature as well as the structural properties are inheretntly uncertain. For instance, the characteristics of an earthquake and its probability of occurrence are unpredictable in nature; every earthquake event is different from the other, and during each earthquake every structure experiences it differently. On the other hand, the material and geometric properties of the structural members are also uncertain. For example, the grades of concrete and steel considered in the design of the structures may not be precisely the same as the grades of concrete and steel achieved/used in the field; such variation in the properties needs to be considered through modeling the uncertainties stochastically. Seismic analysis and design approaches are being proposed for accounting some of these randomness in the ground motion and the

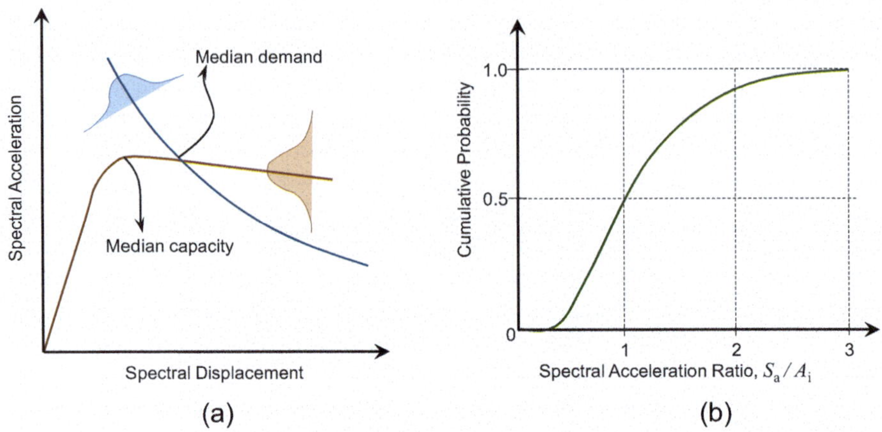

Fig. 1.23 **a** Capacity-demand acceleration-displacement spectra with randomness and/or uncertainty in structural behavior and ground motion response, **b** normalized fragility curve

uncertainty in the structural response. By considering such probabilities, the spectral acceleration and spectral displacement curves can be plotted with some probabilistic distribution, which means that there is a particular range in which demand and capacity vary, as seen from the two curves in Fig. 1.23a. Thus, seismic performance can be evaluated with a certain confidence level depending upon the structural and excitation parameters. The probability of failure of a structure, i.e., the chances of the structure reaching a particular limit state of failure, is calculated with the increasing peak ground acceleration (PGA) or spectral acceleration (S_a); the curve obtained as such is called fragility curve as shown in Fig. 1.23b. The probability of failure increases as the intensity of earthquake excitation increases, which is expressed via some intensity measure (IM) such as PGA or S_a.

Further, in Fig. 1.24, cumulative probability or probability of failure of the structure is obtained when the intensity measure (IM), i.e., either PGA or S_a is increased. Figure 1.24 also shows the relation of the fragility curves with the earthquake return period and confidence interval. Further, it is observed from the fragility curves (Fig. 1.24) that as the intensity of the earthquake increases, the probability of failure also increases. In the construction of a fragility curve the cumulative probability density function is expressed by

$$F(S_a) = \Phi\left[\frac{1}{\beta_c} \ln\left(\frac{S_a}{A_i}\right)\right]$$

where Φ = standard log-normal cumulative distribution function; S_a = spectral acceleration amplitude; A_i = median spectral acceleration necessary to cause i^{th} damage to occur; β_c = normalized composite log-normal standard deviation, incorporating aspects of uncertainty and randomness for both capacity and demand.

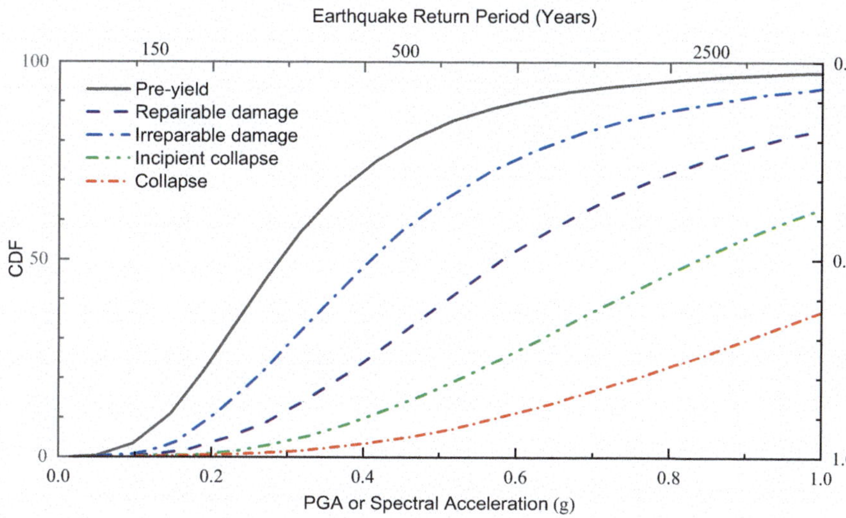

Fig. 1.24 Fragility curves for ductile design [12]

Naturally, the probability of some minor repairable damages to occur is higher as compared to the occurrence of irreparable damages or incipient collapse, especially at lower IM. In performance-based engineering, such fragility curves are extremely important to decide the damage states corresponding to the performance levels. Furthermore, some of the advanced techniques or latest innovations in structural earthquake engineering are employed herein to reduce the probability of failure even at high-intensity levels of earthquake and account for the uncertainties. Such seismic fragility analysis, conducted on the probabilistic scale by duly considering the randomness in excitations and uncertainties in parameters of structure, forms the backbone of the performance-based seismic design (PBSD) of structures.

One example of such an approach used in conducting seismic fragility analysis with uncertainty quantification is shown in Fig. 1.25. In the proposed framework, dynamic analyses of SDOF or MDOF system are conducted for either a unidirectional earthquake excitation or multi-directional earthquake excitations. In this probabilistic approach, in addition to the input structural parameters being uncertain, the input excitations are also probabilistic. Several random and probable samples of ground motion are generated to conduct the dynamic analysis of the SDOF or MDOF systems to determine their probability of failure in different modes, and the fragility curves are plotted by taking the uncertainties into consideration. The analysis steps shown in Fig. 1.25 in the form of a flowchart help in conducting this kind of seismic fragility analysis with uncertainty quantification. Upon calculating the capacity of the structure, the demand from the earthquake excitation is determined, and whether the structure has experienced failure in any mode is verified. If some failure is experienced, under the action of a particular intensity of earthquake, a point on the fragility curve is obtained corresponding to that IM. If no failure is

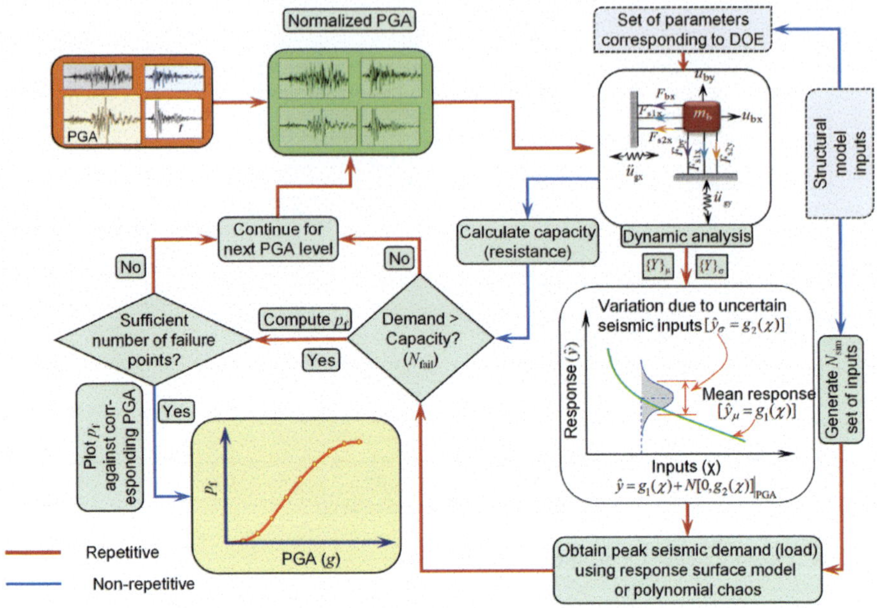

Fig. 1.25 Seismic fragility analysis with uncertainty quantification [21]

observed, the next increased intensity level of the earthquake is considered in the analysis, and this procedure is repeated. Subsequently, the probability of failure curve is plotted by joining the failure points and later vulnerability analysis can be carried out. To quantify the uncertainty, the dynamic parameters of the structure, dependant on material and geometric properties observed in the field, are varied. For example, possible variation (probability distribution) in the grade of concrete and that of the steel reinforcement grade is considered for calculating the seismic response of the structure.

1.8 Advanced Dynamic Response Modification Devices

There are other earthquake-resistant design approaches where the dynamic response of the structure is modified or controlled by installing mechanical devices or dampers. Such relatively sophisticated devices are primarily designed to modify the dynamic response of the structures, i.e., to reduce the amplitude of the vibration response which may be in terms of displacement or acceleration of the structure. A limited number of such devices have been employed in India; however in the rest of the world, especially in seismically developed countries, the use of dampers is quite prevalent. Japan being a seismic-prone country, such vibration response control devices are widely used in built infrastructure in Japan. Approximately, there are more than 2000 base-isolated buildings, more than 300 structures equipped with passive response dampers, and more than 40 active-controlled buildings in Japan. Comparatively, in India, there are about 10 base-isolated buildings, about 5 buildings installed with passive response control dampers, and the number of active control devices installed in structures in India is unknown.

The structural response control devices are typically categorized as (a) passive control devices, (b) active control devices, (c) semi-active control devices, and (d) hybrid control devices. Some of these devices are listed below; although it is not an exhaustive list, it is an indicative or representative list.

i. Tuned mass dampers (TMDs) and active tuned mass dampers (AMDs)
ii. Viscous, visco-elastic, and friction dampers
iii. Buckling-restrained bracings
iv. Base-isolation devices
v. Smart structures or systems or devices or dampers
vi. Semi-active and controllable dampers

Figure 1.26 depicts one simple example of a classical damper, known as the tuned mass damper (TMD). The dynamic properties of the main structure in Fig. 1.26a are governed by its mass (m_s), stiffness (k_s), and damping (c_s). If the main mass is dynamically excited, it will experience some vibration. The amplitude of this vibration will increase if the frequency of the excitation is near its natural frequency, i.e., in the resonance condition. Now, let's consider a small auxiliary mass (m_d) is installed on or attached to the main mass via a spring with stiffness, k_d, and a damper with a damping coefficient, c_d (Fig. 1.26b). Also, the stiffness of the auxiliary mass is adjusted in such a manner that the frequencies of the two masses match with each other, and then it is observed that the auxiliary mass vibrates quite vigorously under the same dynamic excitation, thereby actually reducing the dynamic response of the primary structure. Therefore, the displacement response of the main structure is controlled. This auxiliary mass is known as the tuned mass damper (TMD). The

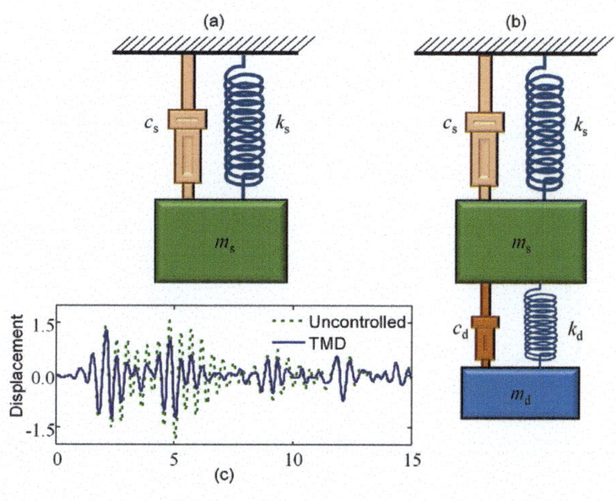

Fig. 1.26 **a** Uncontrolled structure, **b** tuned mass damper-controlled structure, **c** performance of the TMD

reduction in the dynamic response of the primary structure is achieved due to the tuning of the natural frequencies of the two connected masses, which forced them to oscillate in opposite phases to each other. The efficacy of a TMD in reducing the dynamic displacement response of the main structure is shown in (Fig. 1.26c) on time scale.

Essentially, these control devices provide a means to dissipate or absorb the energy imparted on the structure during an earthquake ground motion. When such energy management devices are not provided in the structure, the structural members like the lateral load resisting elements are required to dissipate or absorb this input energy via vibration of the members or via damage (formation of plastic hinges) in the members. In the conventional approach, the lateral load resisting members are designed so as to dissipate this input energy to the structures to make them earthquake-resistant. The nonlinearity (ductility) introduced in the structures improves its seismic performance considerably. Further, in capacity-based design, some structural members are identified to be designed with a higher level of nonlinearities, facilitating the formation of plastic hinges at specific predetermined locations. Thereby, the input energy is primarily absorbed by those members through large damages, thus protecting the other members from being damaged. Thus, in these tactfully engineered structures, the seismic performance of the structure is dictated by the designer such that all these nonlinearities are concentrated at certain locations or predefined members, whereas the rest of the structural members remain linear. Typically, the force–displacement (H-d) behavior of the ductile and non-ductile members is seen to be similar to what is shown in Fig. 1.27. However, if the entire input energy is required to be dissipated or absorbed by the structural members, such as columns and shear walls, they are liable to get damaged and they may even fail. Therefore, the mechanical devices become a viable alternative to either dissipate or absorb such input energy. In that case, the main structural members are not mandatorily required to be designed with such high nonlinearities (ductility) from an earthquake-resistant design viewpoint. Such design approach potentially avoids damage induced in the structural members, while dissipating or absorbing a significant amount of energy in the devices designed

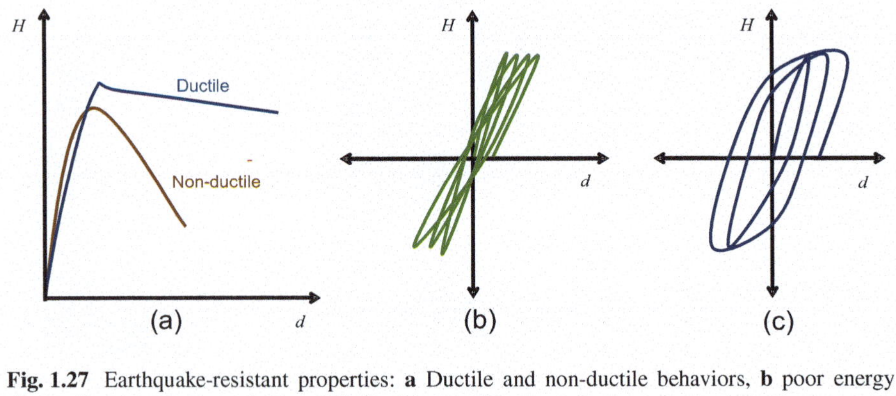

Fig. 1.27 Earthquake-resistant properties: **a** Ductile and non-ductile behaviors, **b** poor energy dissipation capacity, **c** good energy dissipation capacity

for the purpose. Thus, some specific devices are installed in the structures to cater to the seismic input energy. Also, after the end of an earthquake event, those external devices which get damaged due to the earthquake could conveniently be replaced with new ones. Repair or replacement of these control devices is much easier than repairing the structural members which have experienced seismic damage. These control devices could be replaced without even affecting the functionality of the structure.

In Fig. 1.27a, the area under the horizontal force (H) to displacement (d) plot in a non-ductile member is lesser than that in the ductile members, which indicates poor energy absorption. In an earthquake-resistant design structure, it is always better to achieve ductile behavior in the members so that adequate warning is received before the complete failure of the member. In the capacity design method, some of the earmarked members are designed such that their seismic capacity curve exhibits ductile behavior, while other structural members remain in their linear elastic zone, which automatically makes the area under the force–displacement curve larger, i.e., showing higher energy dissipation, as shown in Fig. 1.27c. Similarly, for the response control devices, a good energy dissipation or absorption capacity is intended as compared to the conventional materials and structures. A stable force–displacement curve having a larger area enclosed is an indication of higher damping, and is thus preferred in earthquake engineering. The supplemental damping devices which are used in seismic response control of structures are typically designed to exhibit bulkier hysteretic behavior when subjected to cyclic loading (refer to Fig. 1.28). Among several possible types of hysteresis loops, Fig. 1.28 depicts the elastic perfectly plastic (EPP) model which is seen in some types of response control devices, such as a friction pendulum system. Some other devices are characterized by elasto-plastic hardening (EPH) type of models, whereas some other devices are based on friction models. These are only a few of the different models of energy dissipation exhibited by the damping devices which are commonly used in practice.

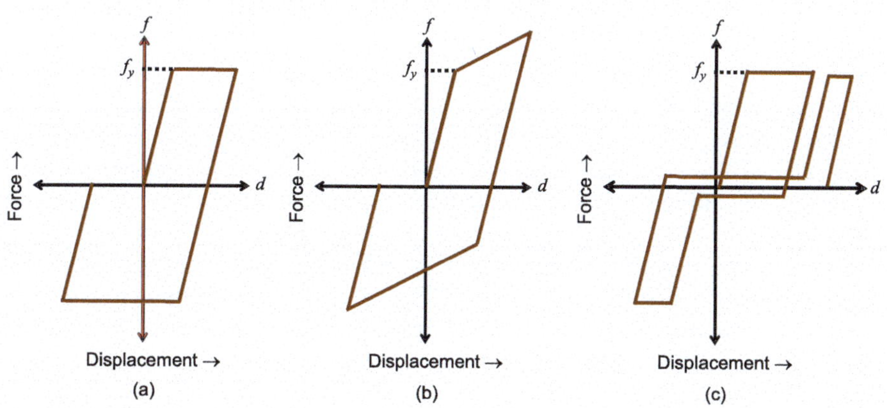

Fig. 1.28 Hysteresis models: **a** Elastic perfectly plastic (EPP) model, **b** elasto-plastic hardening (EPH) model, and **c** slip (Coulomb dry friction) model

Some of the most popular structural response control devices are shown in Fig. 1.29. Figure 1.29a,b depicts a fluid viscous damper (FVD) in which viscous fluid is contained in the cylinder and the piston moves when there is a relative displacement between the two ends of this damper. This can be compared to the shock absorbers used in automotive vehicles! In India, FVDs have been installed in Apollo Hospital Building in Delhi. Also, Airport Authority Building at Indira Gandhi International (IGI) Airport is equipped with the FVDs. However, the use of the FVDs is still relatively less popular in India as compared to the other seismically developed countries. In the visco-elastic damper, in addition to the damping through viscosity, an elastic force is also present in the restoring force behavior of the damper (Fig. 1.29c). The addition of damper(s) in a structure directly influences its dynamic properties such as damping and/or stiffness, i.e., the damped natural frequencies (also, time periods) would be different from the original undamped natural frequencies of the structure.

Some of the friction-type dampers (Fig. 1.29d, e) are also quite popular in the seismically advanced countries such as Canada and USA. Pall friction dampers are very commonly used in North America in the form of cross-bracings. In India, there are a few structures where some of these friction-type damping devices have been used; however, the number of such structure is very small. Sumitomo damper is another commonly adopted friction damper, which is similar to the fluid viscous damper in geometry though working on the principle of friction. Apart from that, at the Indian Institute of Technology (IIT) Delhi a few innovative types of dampers are developed at a laboratory scale. One such novel friction-type damping device is shown in Fig. 1.30. This friction-based damping device has an additional spring whose stiffness is designed to be nonlinear in nature. These dampers are either provided in the structural frame of the building or may even be used beneath the base of the building to reduce the large displacements in the base floor of base-isolated structures.

1.9 Seismic Base Isolation

At the Indian Institute of Technology (IIT) Delhi, a number of novel seismic base-isolation devices have been studied and innovative/novel base-isolation systems are also developed. For example, a new base-isolation system is constructed with the help of bidirectional elliptical rolling rods (ERRs), as shown in Fig. 1.31. This device has been introduced to overcome some of the shortcomings of the existing base-isolation systems. Figure 1.32 shows the design and arrangement of the ERRs and the developed mathematical model thereof (Fig. 1.33). Essentially, these rods detach the building structure from the vibrating ground and thereby protect it from the fury of earthquakes. Thus, the ERRs actually isolate the structure from the ground (Figs. 1.31 through 1.33).

In the seismic base isolation of structures (Fig. 1.34), the control or response modification is achieved by providing flexibility and damping to the structure. It is relatively more popular than the use of dampers in improving the seismic performance of

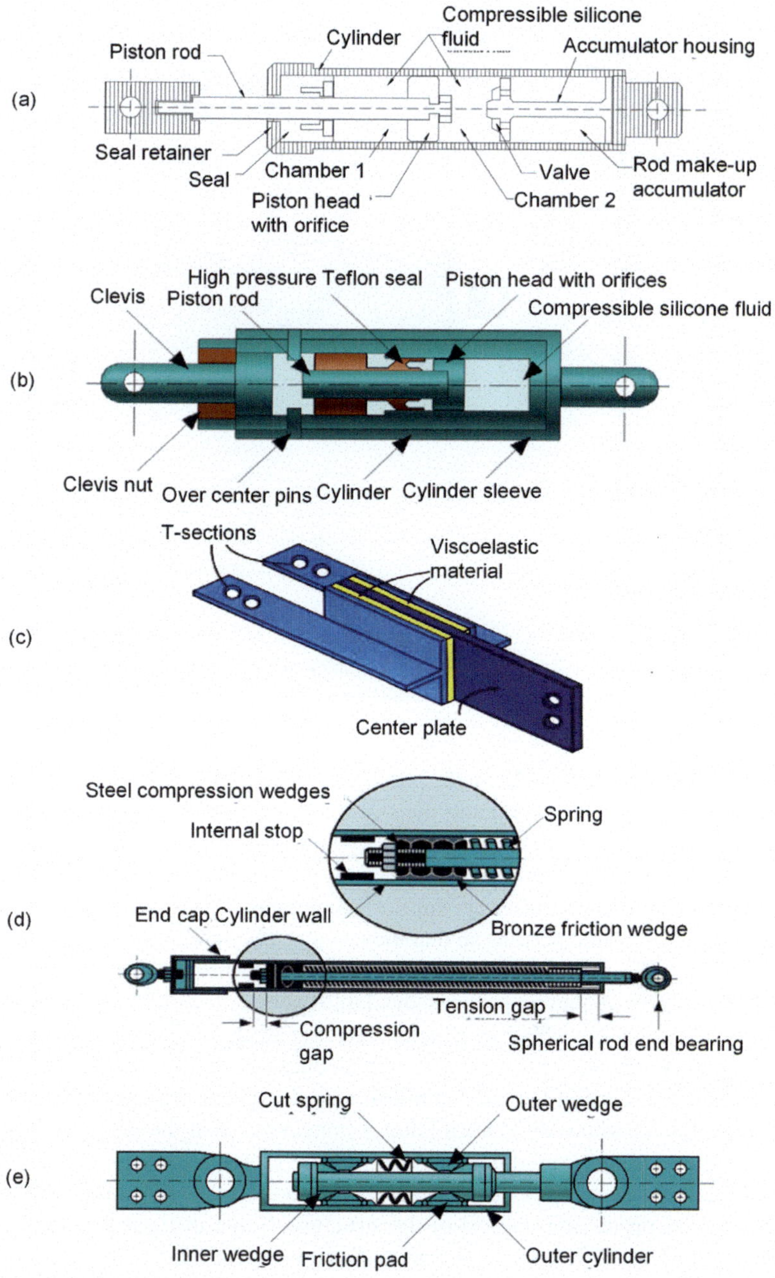

Fig. 1.29 Structural response control devices: **a** Fluid viscous damper [10], **b** pressurized fluid damper manufactured in Taylor Devices, Inc. [13], **c** visco-elastic damper [13], **d** energy-dissipating restraint manufactured in Fluor Daniel, Inc. [13], and **e** uniaxial friction dampers manufactured in Sumitomo Metal Industries, Ltd [13]

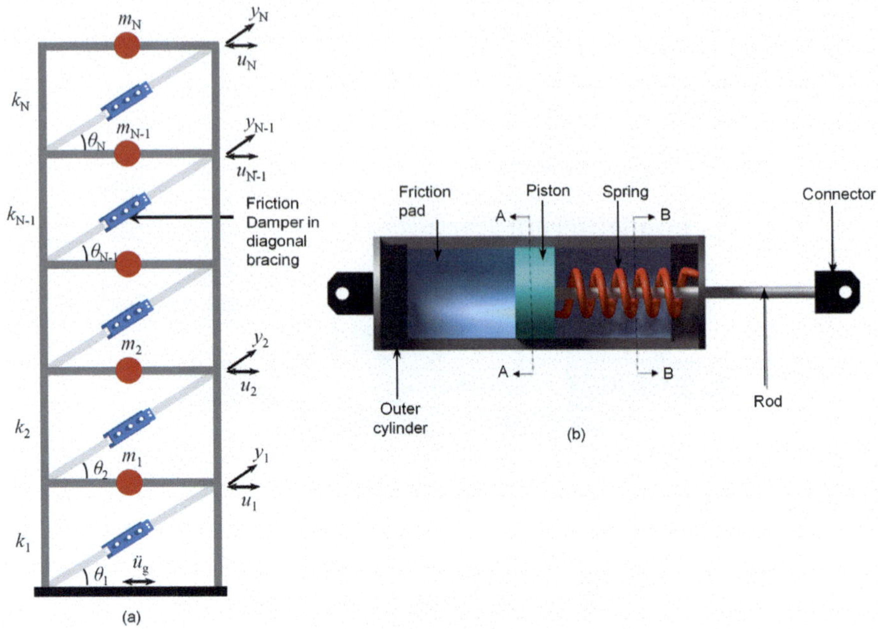

Fig. 1.30 **a** Mathematical models of *N*-story friction damper frame (FDF), and **b** interior view of the new friction-based damping system

a structure. Base isolation works on the principle of elongating (lengthening) the time period of the structure, i.e., reducing the structure's natural frequency. It is comparable with trees, in terms of dynamic property, which typically do not fall or fail during earthquakes because they are flexible, and sway largely during the earthquakes. In seismic base isolation, structures are similarly safeguarded by making them flexible, by lengthening their time periods so that they sway considerably during earthquakes but do not fail. The desirable flexibility is introduced through the use of various types of bearing or isolation systems available, some of which are quite widely used, such as laminated rubber bearings (LRBs) or lead core rubber bearings (also called New Zealand, NZ systems). The effectiveness of the base-isolation systems is due to changing the fundamental cantilever mode of vibration of the structure to the isolation mode, where the deformation is concentrated at the isolation level and the rest of the structure remains almost rigid. The factory-made isolation systems under the building experience most of the strain and therefore they take the wrath of the earthquakes, protecting the rest of the structure. Some of these bearings shown in Figs. 1.35, 1.36, 1.37, 1.38, 1.39 and 1.40 show various types of base-isolation systems which facilitate isolating the superstructure from the shaking ground and dissipating the seismic energy at the isolator level. These base-isolation systems help in shifting the period of the structures toward a more flexible side, thereby the spectral acceleration ordinates are reduced (refer to Fig. 1.34). During the earthquake event, the isolators may experience some damage; however, it is very easy to replace

Fig. 1.31 Mathematical model of flexible shear-type multistory building mounted on elliptical rolling rods (ERRs) [20]

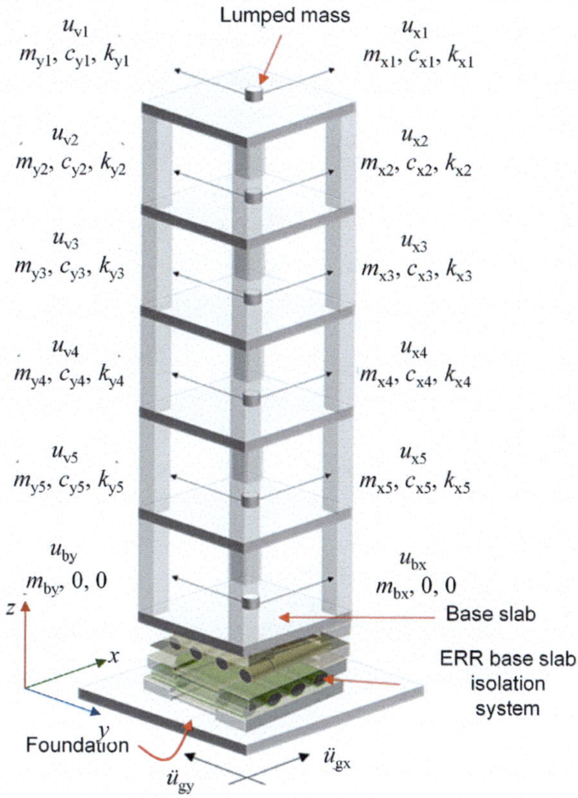

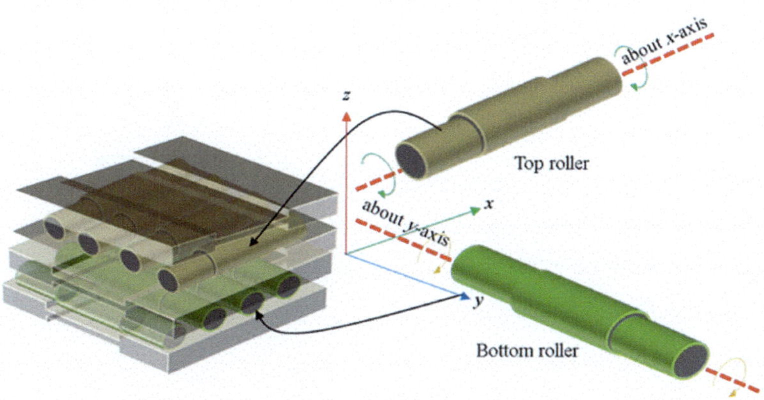

Fig. 1.32 Conceptual 3D views of the ERR with grooves [20]

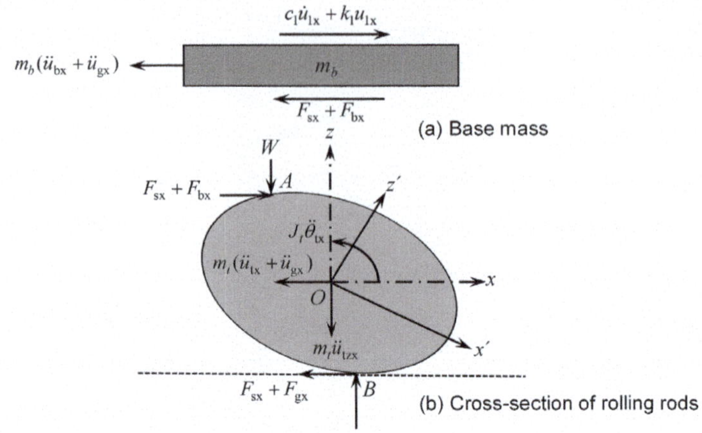

Fig. 1.33 Relationship between displacements of the base mass and ERR in x–z plane

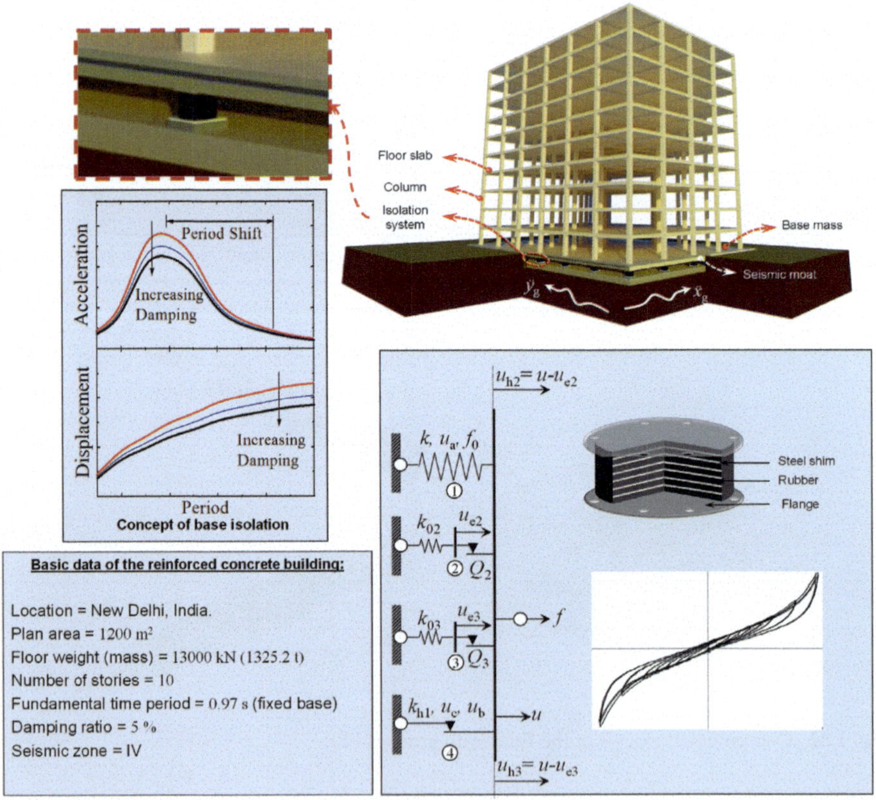

Fig. 1.34 Typical seismic base-isolated structure [28]

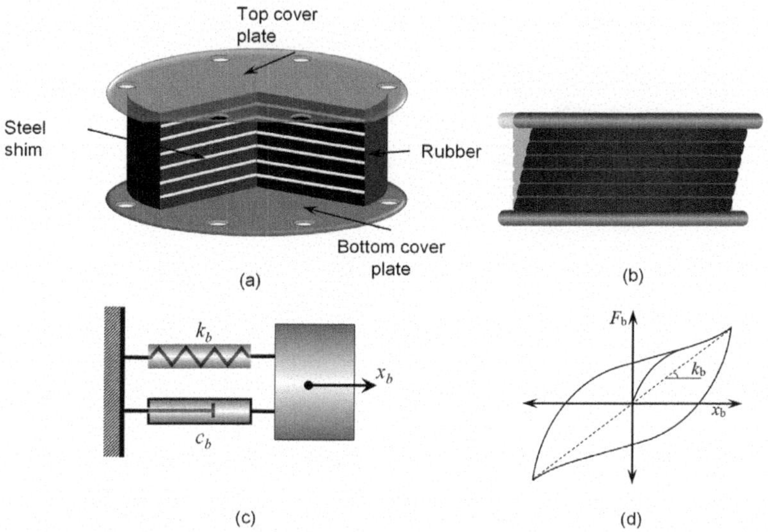

Fig. 1.35 **a** Laminated rubber bearing (LRB), **b** front view of LRB, **c** schematic diagram of LRB, **d** force–deformation behavior of LRB

them once the earthquake is over. Also, inspection and periodic maintenance of the isolation systems are convenient. Up to a certain level of earthquake excitation, for which the isolation system is designed, the superstructure is maintained to have a linear elastic behavior. It implies that no damage is experienced in the superstructure; and therefore the response reduction factor (R) could be kept inapplicable (i.e., equal to 1) to avoid any inelasticity and maintain continuous operational (O) level seismic performance. Thus, by employing the advanced base-isolation technology, the operational (O) level or immediate occupancy (IO) level of seismic performance is conveniently achievable, which will not cause any disruption in the functioning of the structure even after a major earthquake. Furthermore, seismic retrofitting by using the base-isolation technology is possible, which has a number of advantages, such as the superstructure intervention is almost nil, hence the original aesthetics of the structure is maintained intact; and also, the retrofitting operation can be carried out without affecting the functioning of the building [15].

A large number of base-isolation systems have been devised and implemented in real-life constructions. These isolation systems are broadly categorized as (a) elastomeric bearings, (b) sliding systems, and (c) rolling systems. Laminated rubber bearings (LRB), as shown in Fig. 1.35, are the most popular among all and are now widely available in India. Their mathematical modeling is carried out by using a simple spring and dashpot model and they provide a stable and bulky force–deformation hysteresis loop. A large amount of energy is dissipated by the laminated rubber bearing, which is quantified through the area enclosed within its force–deformation loop. Another popular bearing used across the world, including India, is a rubber bearing with a lead core inside, known as the lead core rubber bearing (New

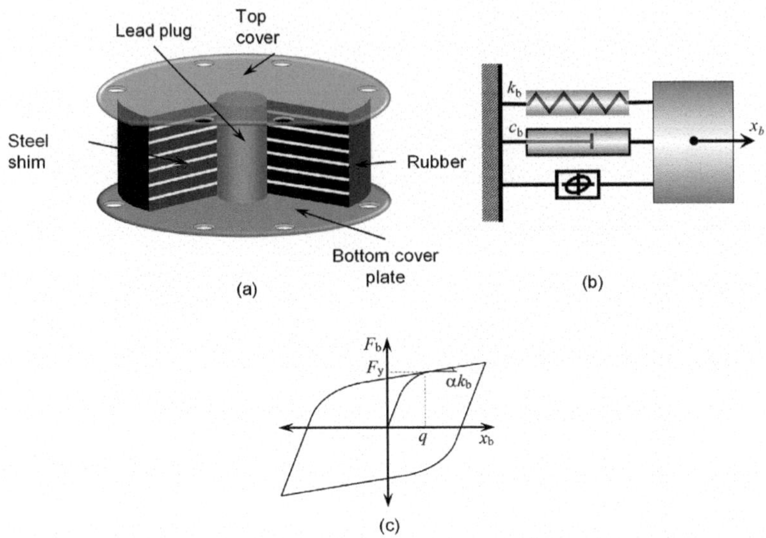

Fig. 1.36 **a** Lead core rubber bearing (New-Zealand, NZ system), **b** schematic diagram of NZ system, **c** force–deformation behavior of NZ system

Zealand, NZ system), as shown in Fig. 1.36. The central lead core provides the designed initial stiffness to the isolator so that it does not oscillate much during the ambient winds or small earthquake tremors. Additionally, the lead core dissipates energy, and even after rupture due to an earthquake event it recrystallizes afterward so that it is again available for sustaining the subsequent earthquakes, which means, the NZ system is not needed to be replaced after every earthquake.

The other isolation systems based on the sliding friction phenomenon are categorized as sliding isolation systems. Figure 1.37 depicts a pure-friction (PF) system, which is typically used in parallel or in addition to the other two isolation systems shown earlier (laminated rubber bearing and lead core rubber bearing) to reduce the cost of isolating the structure. Usually, the LRBs are costlier than pure-friction systems. The PF system is also called sliders, providing mostly velocity-independent low friction force, with no restoring capability. In bridges, such sliding isolation systems (without or with guides) are quite popular. Another very popular device used for structural response control of bridges, especially in the United States, is the friction pendulum system (FPS) as shown in Fig. 1.38. This ingenious sliding friction-based isolation system works on the principle of the swinging pendulum, where the structure mounted on top of the FPS oscillates with a designed time period with a tendency always to return to its original equilibrium position. During the oscillation, energy is dissipated in the sliding friction phenomenon. The design of the FPS includes curvature, which governs the time period of the isolator (more the curvature less is the time period and vice-versa), and the coefficient of friction at the articulated surface. Some other isolation systems used in practice are the resilient friction base isolator (RFBI) and the electricite-de-France system (EDF), shown in

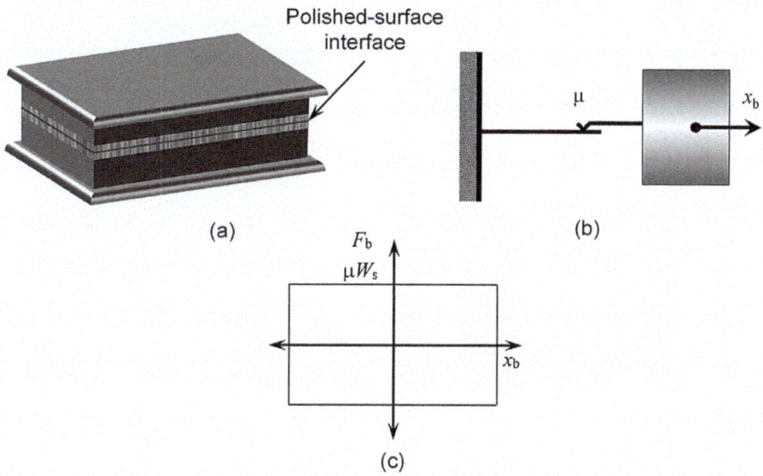

Fig. 1.37 **a** Pure friction isolation system (PF), **b** schematic diagram of PF, **c** force–deformation behavior of PF

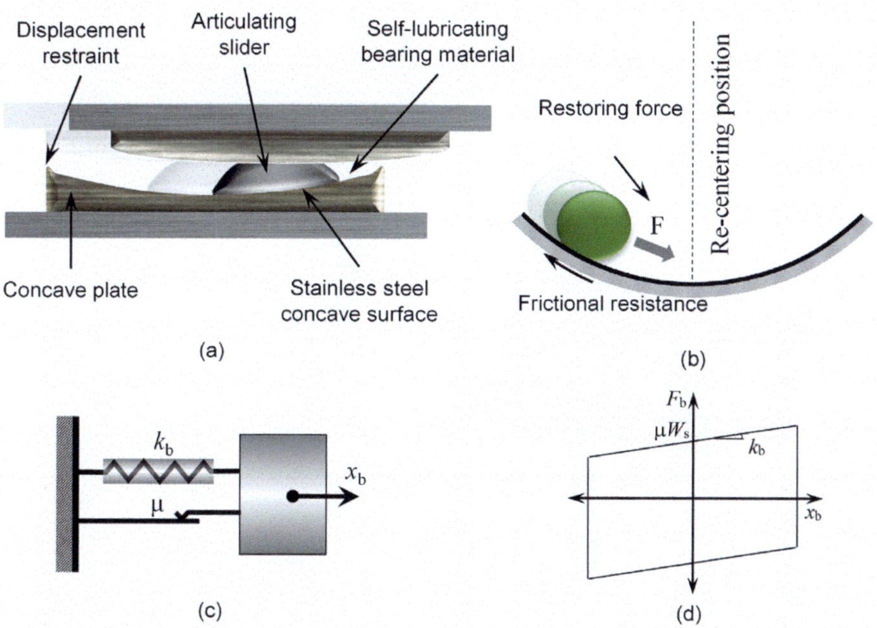

Fig. 1.38 **a** Friction pendulum system (FPS), **b** pendulum action, **c** schematic diagram of FPS, **d** force–deformation behavior of FPS

Figs. 1.39 and 1.40, respectively. The RFBI provides advantages of elastomeric and sliding isolation systems in one unit. The EDF is quite popular in France for seismic isolation of nuclear-containment structures.

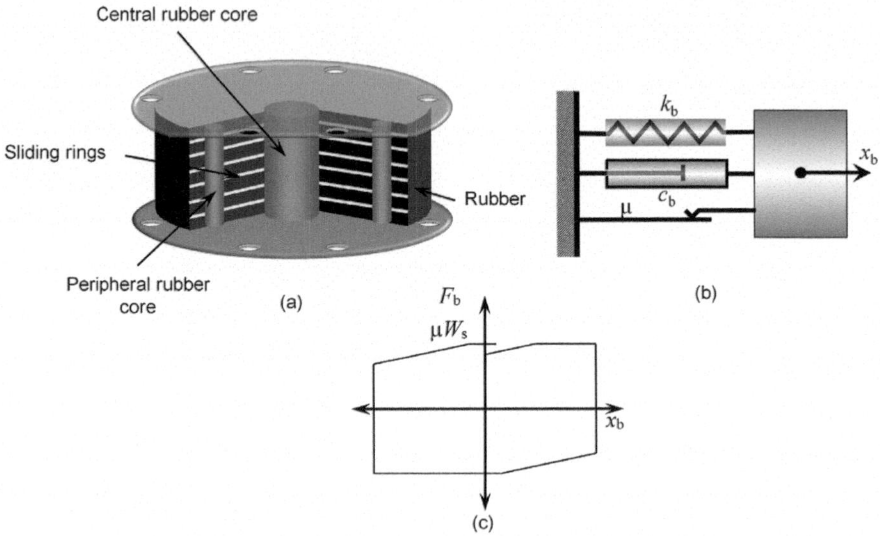

Fig. 1.39 **a** Resilient friction base-isolation system (RFBI), **b** schematic diagram of RFBI, **c** force–deformation behavior of RFBI

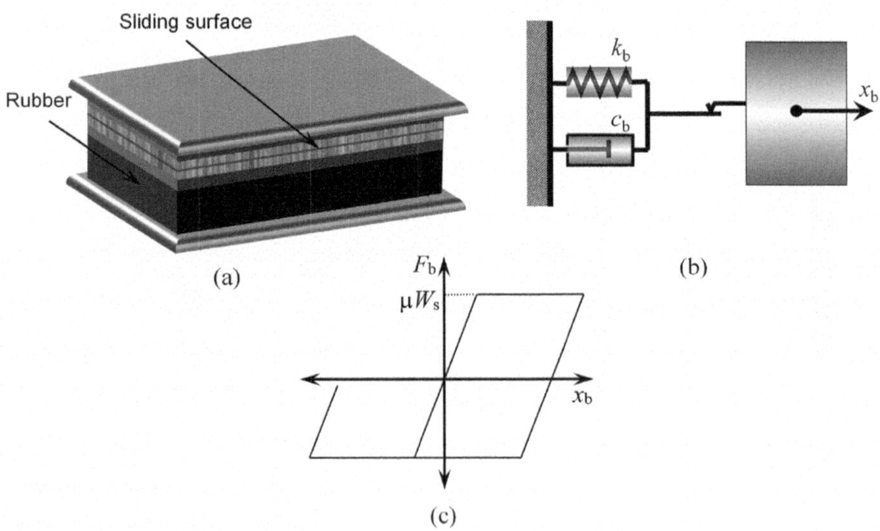

Fig. 1.40 **a** Electricite-de-France system (EDF), **b** schematic diagram of EDF, **c** force–deformation behavior of EDF

1.9.1 New Seismic Protection Devices—Isolation Systems

The response control devices discussed in the previous section were passive control devices, i.e., they function by virtue of their inherent dynamic properties which remain unaltered during the shaking. However, some seismic protection devices are designed to change their dynamic properties suitably during the shaking, which are called active or semi-active control devices. Especially, the semi-active type of control devices like magnetorheological (MR) dampers are employed to modify the structural response, as shown in Fig. 1.41. Here, a semi-active vibration control device in the form of an MR damper is tested at the Multi-Hazard Protective Structures (MHPS) Laboratory, Indian Institute of Technology (IIT) Delhi. The MR damper is mathematically modeled typically using the Bouc-Wen hysteresis model, for the purpose of simulating the dynamic response of the structures equipped with the MR dampers (Fig. 1.41). Experimentally obtained responses are used for validating the numerical simulation results so that the numerical models can be further used for modeling real-life buildings provided with the MR dampers and the seismic performance of the controlled structure could be estimated.

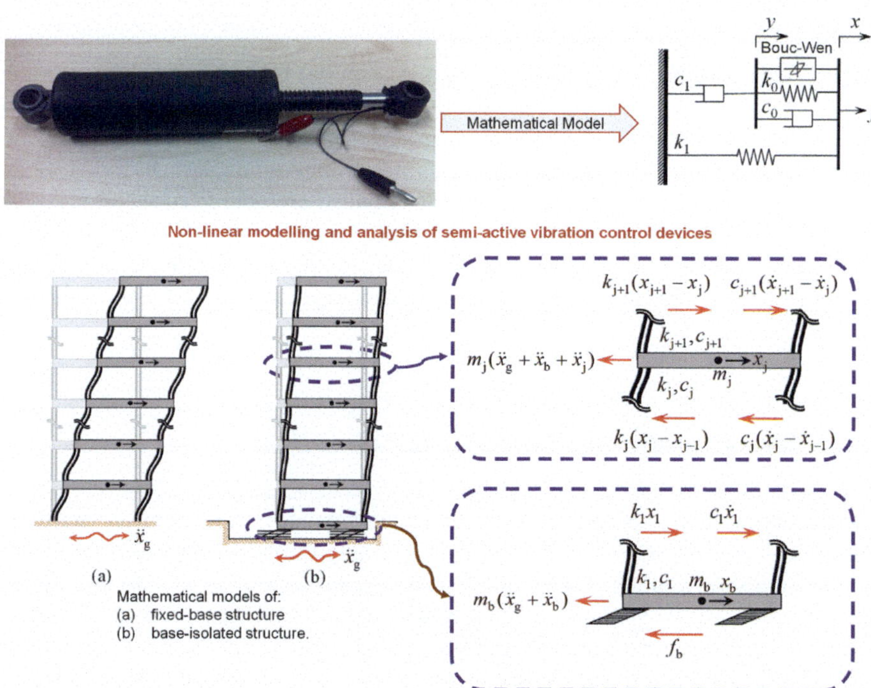

$$\begin{bmatrix} m_b & \mathbf{0}^T \\ \mathbf{0} & \mathbf{M}_s \end{bmatrix}\begin{Bmatrix} \ddot{x}_b \\ \ddot{\mathbf{X}}_s \end{Bmatrix}+\begin{bmatrix} 0 & -c_1\mathbf{r}^b \\ \mathbf{0} & \mathbf{C}_s \end{bmatrix}\begin{Bmatrix} \dot{x}_b \\ \dot{\mathbf{X}}_s \end{Bmatrix}+\begin{bmatrix} 0 & -k_1\mathbf{r}^b \\ \mathbf{0} & \mathbf{K}_s \end{bmatrix}\begin{Bmatrix} x_b \\ \mathbf{X}_s \end{Bmatrix}=-\begin{Bmatrix} m_b\ddot{x}_g + f_b \\ \mathbf{M}_s\mathbf{r}(\ddot{x}_g+\ddot{x}_b) \end{Bmatrix}$$

Fig. 1.41 Nonlinear modeling and analysis of semi-active vibration control devices

Fig. 1.42 Base-isolated showcase building (a collaborative project between Resistoflex and IIT Delhi)

It is crucial to take the investigated seismic response control technologies to the field for implementation in real-life structures. In a number of construction projects in India, the designers have shown keen interest in using some of these innovative technologies for the protection of structures from earthquakes. In Fig. 1.42, a base-isolated building constructed in Delhi is seen, where double curvature friction pendulum (DCFP) systems are used for isolation purposes. Here, the DCFP systems are installed at the first floor level of the structure, i.e., the isolators are placed on the column pedestals. The structure is designed for the operational (O) level of seismic performance.

1.9.2 Base-Isolated Structures at IIT Guwahati

Health monitoring of structures is carried out to understand how these novel vibration control devices or systems are performing, especially during an earthquake activity. The building shown in Fig. 1.43 is a demo base-isolated building constructed at the Indian Institute of Technology (IIT) Guwahati with the funding received from the Bhabha Atomic Research Centre (BARC). A health monitoring system is installed in this building to study the difference in the seismic performance of the non-isolated (fixed-base) and the base-isolated buildings during earthquakes. Moreover, a scientific study has been undertaken to study how the contents within the base-isolated building are protected as compared to the contents in the fixed-base building. For this purpose, Fig. 1.44 shows an algorithm developed and implemented for real-time

Fig. 1.43 Fixed-base and base-isolated structures at IIT Guwahati

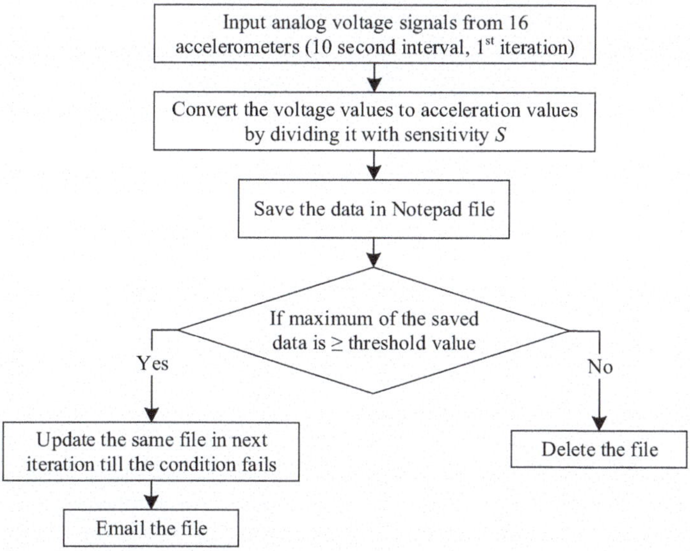

Fig. 1.44 Algorithm used for structural health monitoring

structural health monitoring of the two adjacent buildings (one base-isolated and another non-isolated) and secondary systems (SS) installed within them. The real-time data recorded in this process is also used for predicting the dynamic response of the structures during severe earthquakes. Further, data analytics tools are being used in this project to quantify the vulnerability of the base-isolated building under site-specific earthquake shaking of different intensities. Figure 1.45 shows the loca-

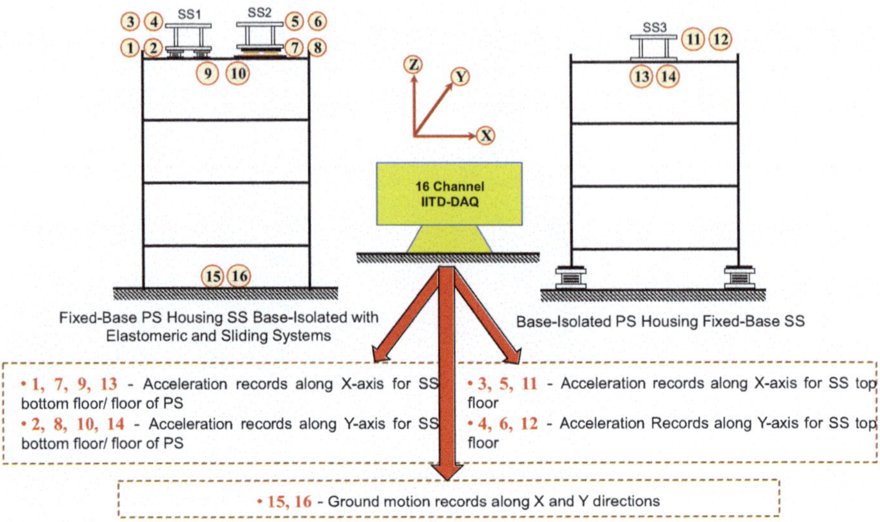

Fig. 1.45 Locations of sensor placement in the building

tions of the different sensors used for the real-time monitoring of the two buildings at the Indian Institute of Technology (IIT) Guwahati.

1.9.3 Tuned Mass Damper(s)

Apart from base isolation, another passive control device which is popularly used in real-life structures is known as the tuned mass damper (TMD), as introduced earlier. It is a classical dynamic response abatement device which uses an additional mass to dissipate the input seismic energy, thus reducing the response of the primary structure. Some recent modifications have been introduced to this device, by spatially distributing a number of TMD devices, which are called distributed multiple tuned mass dampers (d-MTMDs) as compared to the earlier single TMD (S-TMD) unit. Typically, such tuned mass damper(s) is/are used for controlling the vibration response in mid-rise to high-rise structures, as shown in Fig. 1.46. The TMDs are more popular for wind response control of structures, but they are quite beneficial for earthquake response control as well. The TMDs are used in bridges for both live load response control and earthquake response control (Fig. 1.46). In a classical problem of vehicle-bridge interaction (VBI), these devices are introduced to abate the dynamic response of the bridge under the vehicular loading. Moreover, the TMDs also find applications in tall structures such as reinforced concrete (RC) chimneys, provided in the form of annular rings in the cylindrical chimneys. Significant research contributions have been made in the application of the TMDs (passive, semi-active, and

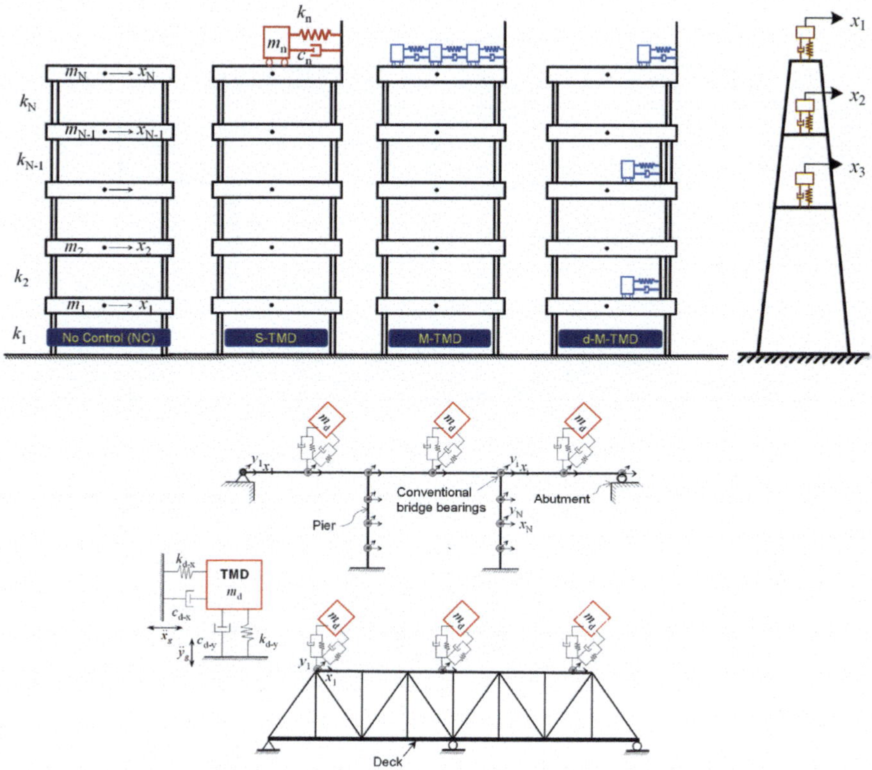

Fig. 1.46 Implementation of different types of tuned mass dampers in civil infrastructures [2, 14]

active) in dynamic response control of various structures under a variety of loading conditions, for which specialized literature can be referred.

1.9.4 Innovative Structural Control Algorithm

The latest technologies in the TMDs are meant to make them tuneable during real-time functioning, enhance semi-active response control by introducing some innovative control algorithms, and ensure their effectiveness under multi-hazard loading scenarios, by introducing artificial intelligence and machine learning (AI-ML) tools. Figure 1.47 shows the location of semi-active TMDs (SA-TMDs) provided in SDOF and MDOF structures, and an innovative response control algorithm for SA-TMD is shown in Fig. 1.48. In Figure 1.49, the acceleration and displacement response of the structure with the SA-TMD are found to be reduced significantly as compared to the uncontrolled structure and even to the structure with a single passive TMD. These are some of the ongoing investigations in dynamic response control devices.

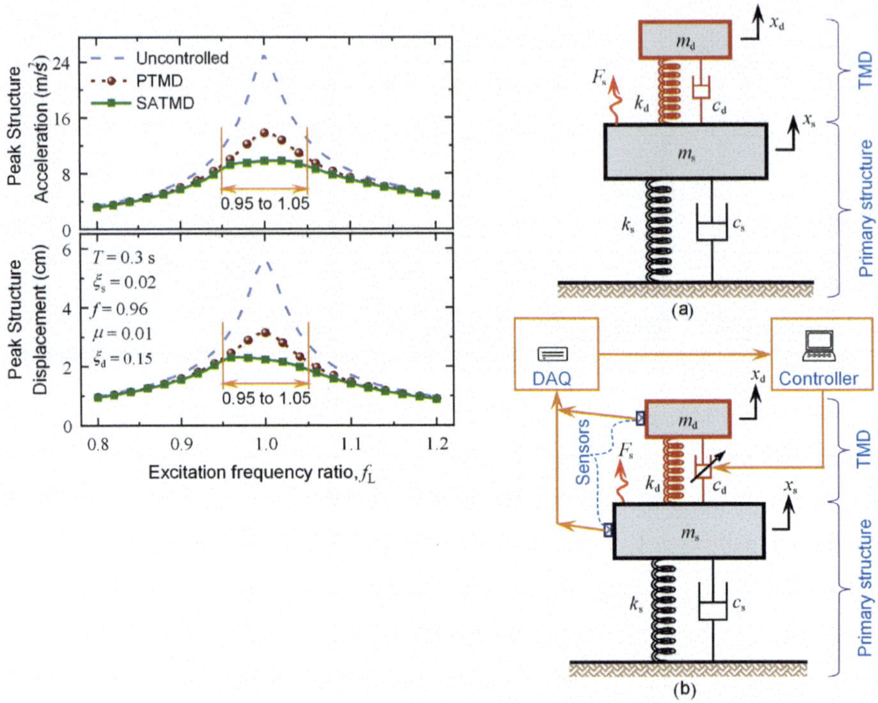

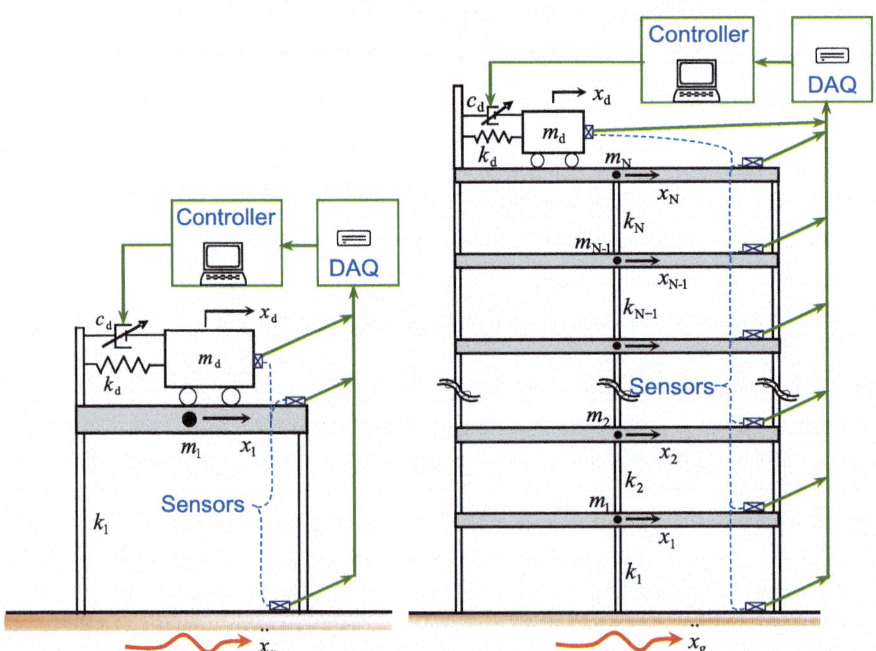

Fig. 1.47 Semi-active tuned mass damper (SA-TMD) installed in single and multistory buildings [26]

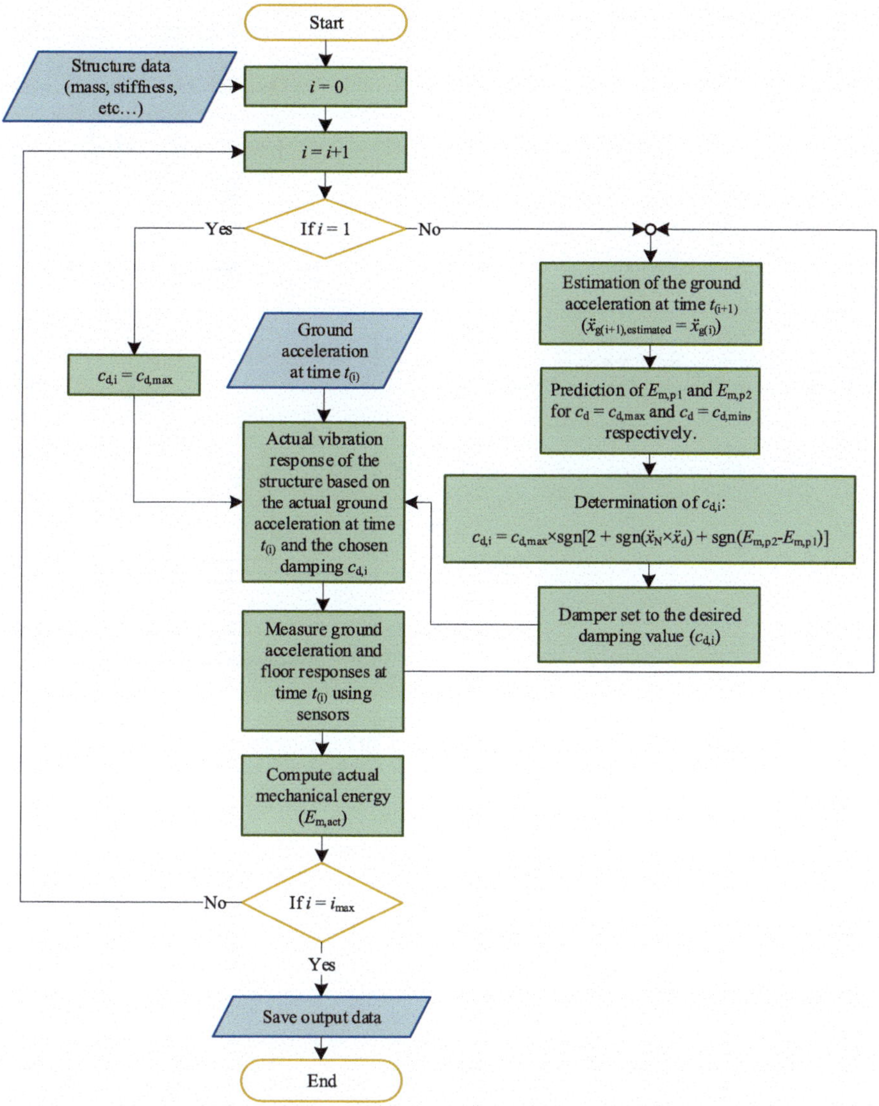

Fig. 1.48 A new energy-based predictive (EBP) control algorithm for the SA-TMD [26]

Subsequently, these devices will be further improved in terms of their mechanical performance by changing the way the energy is dissipating and the way the inertial forces are being added.

Upon conducting nonlinear time history analyses of bridges with distributed multiple tuned mass dampers (d-MTMDs), the response reduction (shown in Fig. 1.50) reveals the effectiveness of the passive d-MTMDs, as compared to the structure which is not controlled (NC) or structure controlled by a single TMD (STMD).

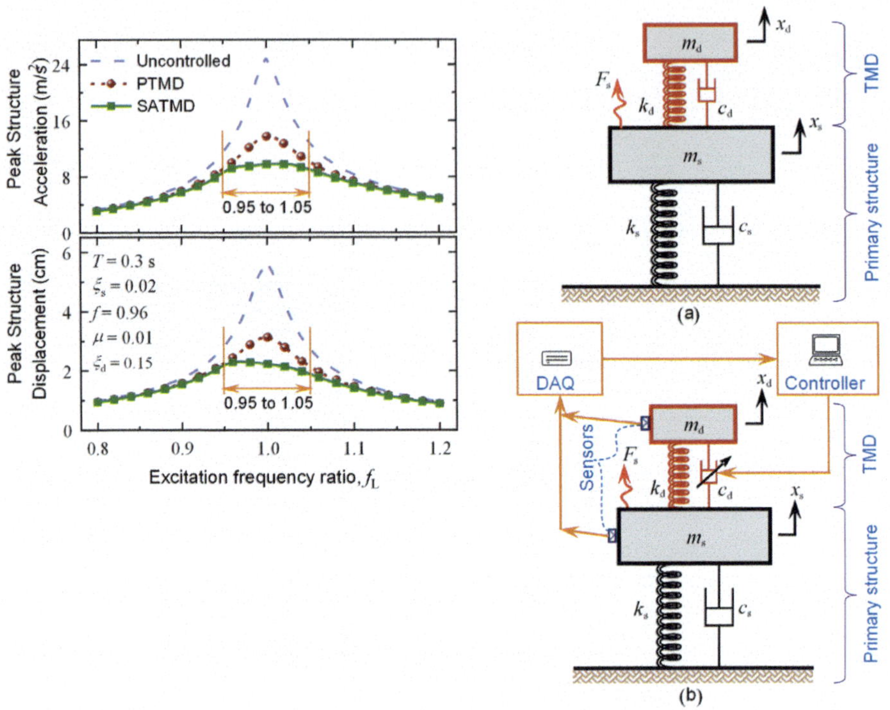

Fig. 1.49 Comparison of the effectiveness of the SA-TMD against the uncontrolled structure and the single passive TMD [27]

These have been some of the advanced technologies which are being used for controlling the seismic response of buildings and bridges. It gives merely a glimpse of the current state-of-the-art and the current practice, as well as opens new avenues for conducting research on emerging topics in earthquake engineering and technology.

1.9.5 Need for Future Research

In summary, various fundamental aspects of earthquake engineering and technology have been dealt with here at introductory level. The broad spectrum and various subdisciplines of earthquake engineering and technology have been presented systematically and succinctly. Upon putting the earthquake engineering and technology in a perspective, the future research needs are summarized below:

i. Improved strong ground motion prediction models are essential.
ii. Different materials are being innovated as compared to the conventional reinforced concrete (RC) or masonry structures. Such new materials are required to have high strength and still be lightweight as far as possible.

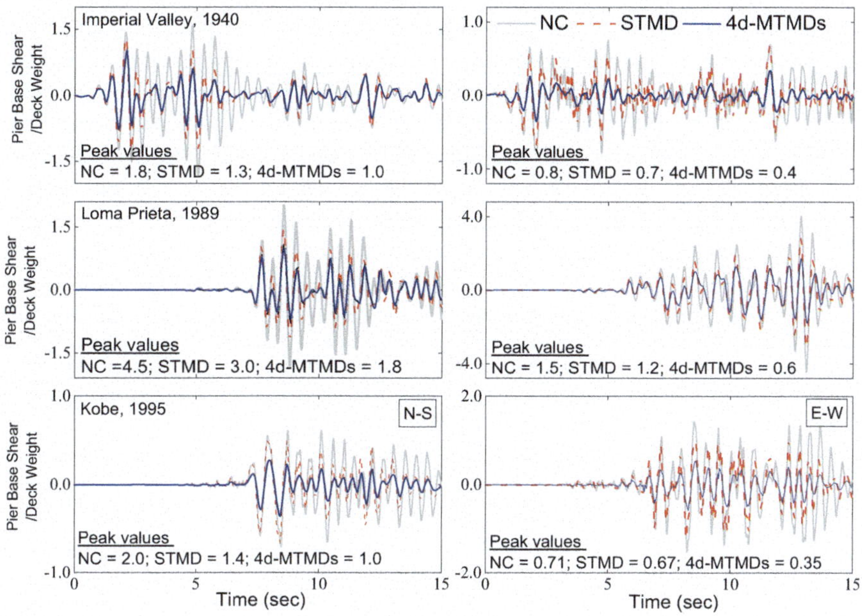

Fig. 1.50 Seismic response control of bridge with d-MTMDs [14]

iii. Nonlinear seismic analysis of structures should become a routine. In due course of time, it should be incorporated into the codal provisions as well.

iv. Computational structural dynamics will evolve much strongly with the advent of high-performance computing (HPC) facilities. Subsequently, not only for very important structures such as the nuclear power plant (NPPs) but also for other routinely designed important structures, conducting nonlinear time history analysis (NLTHA) should become a commonplace practice.

v. As compared to the conventional force-based seismic analysis, gradual transformation toward the displacement-based analysis is taking place, which is quite suitable for inelastic response evaluation as well as evaluating serviceability performance criteria.

vi. To quantify the seismic performance of structures even within the inelastic range, certain new approaches are being investigated as parts of development in performance-based earthquake engineering (PBEE). A need is felt to duly account for the probability of occurrence of earthquakes, randomness thereof, and uncertainties in the structural properties. Some of the current probabilistic approaches invented must be incorporated into the codes and standards in the future.

vii. Codes and standards which earlier were mainly following deterministic approaches should evolve to adopt more probabilistic approaches.

viii. Dynamic response control devices are being improved day by day. Drawing a parallel with the car industry, where years after years, the comfort level in riding the cars is improving with higher speeds and automation, there is a requirement to further improve the current structural response control devices to enhance their performance further.

ix. Structural health monitoring (SHM) will reveal how these new-generation engineered structures are performing during earthquakes as compared to the classical structures. The sensor-based data gathered from the continuous SHM should be used further to contribute to the development of new technologies in earthquake-resistant design structures with enhanced performance.

x. Some of the advanced techniques such as data analytics, machine learning, or deep learning should find their applications in structural analysis and response control.

References

1. BIS, IS 1893-Part 1 (2016) Criteria for earthquake resistant design of structures-part 1: General Provisions and Buildings. Bureau of Indian Standard (BIS), New Delhi, pp 1–29
2. Elias S, Matsagar V, Datta TK (2016) Effectiveness of distributed tuned mass dampers for multi-mode control of chimney under earthquakes. Eng Struct 124:1–16
3. Frequency of occurrence of earthquakes. https://www.usgs.gov. Accessed 21 Nov 2021
4. GovindaRaju L, Ramana GV, HanumanthaRao C, Sitharam TG (2004) Site-specific ground response analysis. Curr Sci 87(10):1354–1362
5. Henrych J, Abrahamson GR (1980) The dynamics of explosion and its use. J Appl Mech 47(1):218. https://doi.org/10.1115/1.3153619
6. Indo-Eurasian Plate. https://www.usgs.gov. Accessed 21 Nov 2021
7. Jangid RS. Introduction to earthquake engineering, IIT Bombay. https://nptel.ac.in/courses/105101004
8. Jayalakshmi S, Raghukanth STG (2017) Finite element models to represent seismic activity of the indian plate. Geosci Front 8(1):81–91
9. Kunde MC, Jangid RS (2006) Effects of Pier and deck flexibility on the seismic response of isolated bridges. J Bridge Eng ASCE 11(1):109–121
10. Lee D, Taylor DP (2001) Viscous damper development and future trends. Struct Design Tall Build 10(5):311–320
11. Major Tectonic Plates on Earth Surface. https://www.usgs.gov. Accessed 21 Nov 2021
12. Mander JB (2004) Beyond ductility: the quest goes on. Bull N Z Soc Earthq Eng 37(1):35–44
13. Marko J (2006) Influence of damping systems on building structures subject to seismic effects. Doctoral Dissertation, Queensland University of Technology, Brisbane, Australia
14. Matin A, Elias S, Matsagar V (2020) Distributed multiple tuned mass dampers for seismic response control in bridges. Proc Inst Civ Eng Struct Build 173(3):217–234
15. Matsagar VA, Jangid RS (2008) Base isolation for seismic retrofitting of structures. Pract Periodical Struct Des Constr 13(4):175. https://doi.org/10.1061/(ASCE)1084-0680
16. Murty CVR (2005). https://www.nicee.org/EQTips.php
17. Post-Earthquake Scenario. https://www.materialtree.com/bengaluru/blog/architectural-features-affect-buildings-during-earthquakes. Accessed 21 Nov 2021
18. Pujari NN, Ghosh S, Lala S (2016) Bayesian approach for the seismic fragility estimation of a containment shell based on the formation of through-wall cracks. ASCE-ASME J Risk Uncertain Eng Syst Pt A Civil Eng 2(3):1–13
19. Raghukanth STG, Iyengar RN (2007) Estimation of seismic spectral acceleration in peninsular India. J Earth Syst Sci 116(3):199–214
20. Rawat A, Ummer N, Matsagar V (2018) Performance of bi-directional elliptical rolling rods for base isolation of buildings under near-fault earthquakes. Adv Struct Eng 21(5):675–693

21. Saha SK, Sepahvand K, Matsagar VA, Jain AK, Marburg S (2016) Fragility analysis of base-isolated liquid storage tanks under random sinusoidal base excitation using generalized polynomial chaos expansion-based simulation. J Struct Eng ASCE 142(10):1–12
22. Secondary Systems and Piping in Nuclear Containment Structure. https://www.repsscan.com/wp-content/uploads/2015/11/05.jpg. Accessed 21 Nov 2021
23. Seed HB, Idriss IM (1982) Ground motions and soil liquefaction during earthquakes. In: Engineering monographs on earthquake criteria, structural design, and strong motion records. Earthquake Engineering Research Institute, Berkeley, California, CA, USA
24. Seismotectonic Features. https://pubs.usgs.gov/gip/dynamic/Vigil.html. Accessed 21 Nov 2021
25. Wolf JP (1997) Spring-dashpot-mass models for foundation vibrations. Earthq Eng Struct Dyn 26(9):931–949. https://doi.org/10.1002/(SICI)1096-9845(199709)26:9%3C931::AID-EQE686%3E3.0.CO;2-M
26. Zelleke DH, Matsagar VA (2019a) Energy-based predictive algorithm for semi-active tuned mass dampers. Struct Design Tall Spec Build 28(12):1–20
27. Zelleke DH, Matsagar VA (2019b) Semi-active algorithm for energy-based predictive structural control using tuned mass dampers. Comp Aid Civil Infrastruct Eng 34(11):1010–1025
28. Zelleke DH, Saha SK, Matsagar VA (2020) Multihazard response control of base-isolated buildings under bidirectional dynamic excitation. Shock Vib 2020:1–24

Chapter 2
Site Response Studies Application in Seismic Hazard Microzonation and Ground Characterization

Atindra Kumar Shukla

Dedicated to son Vishesh Shukla who left us to heavenly abode.

Abstract Site response study is one of the most powerful tools in engineering seismology as it models the influence of the near surface layers of soil on earthquake ground motions and is an important component of Seismic Hazard Microzonation. Site response parameters, viz. peak frequency (f_o), amplification (A_m) and nature of response curve defining the transfer function at site, are one of the most important inputs in hazard evaluation and ground characterization for seismic hazard and risk microzonation. There are several methods each having merits and demerits are available in literature for the estimation of site response and used world over. In the present chapter, therefore, the national and international scenarios of ground response studies are briefly reviewed, and choice of the methodologies is discussed in the Indian context.

The techniques of data capturing for 'Nakamura Type' and 'Event mode' studies are enumerated. The analysis of the data for a few Indian cities available in publication and reports is presented in the chapter with an objective to identify the pattern of ground response. The pattern defined in terms of 'type of response curves' is interpreted in conjunction with geologically classified ground conditions to identify predominant frequencies of different ground typologies.

A. Kumar Shukla (✉)
Meteorological Department, Ministry of Earth Sciences, Government of India, New Delhi, India
e-mail: atindrakumarshukla@gmail.com

T. G. Sitharam et al. (eds.), *Theory and Practice in Earthquake Engineering and Technology*, Springer Tracts in Civil Engineering,
https://doi.org/10.1007/978-981-19-2324-1_2

2.1 Introduction

Earthquakes caused by sudden movement of the fault release a great deal of energy, which then travels through the earth's crust in the form of different types of seismic waves setting earth in motion. The seismic waves travel to a great distance before finally losing most of their energy-generating different ground motion at different sites, owing to the distance from the source and attenuation relations of the area under study. This ground motion then subjected to the soil profile at particular site, gets modified according to the nature of soil profile. Thus, the travel of seismic waves is subjected to two complex processes. The first is the source effect with the influence of a propagation path that gives the input wave to the interface (Seismic/engineering bedrock). Another is the influence, i.e., amplification in the shallower sedimentary layers. Thus, the study of the effect of seismic waves excited due to an earthquake, i.e., seismic hazard at a site can be divided into two processes. (i) The first process is assessment of ground motion at the interface, i.e., at engineering bedrock from engineering point of view. (ii) The second process is the study of reaction of a specific point on the earth to this ground motion, which is influenced by the physical properties of the soil layers and is known as site response.

The effect of the site was initially recognized through pioneering works by Sezawa and Ishimotao in 1930s [60]. Subsequently, detailed observations and analysis of earthquakes such as great Japan earthquake, 1891; San Francisco earthquake of 1906; Long beach earthquake of 1933; well-studied Mexico earthquake of 1985; Loma Prieta earthquake, 1989; Northridge earthquake of 1994, over the past few decades have highlighted the importance of local geological conditions, such as characteristics of the lithological attributes, geometry and topography of soil on propagated ground motion. It has been observed that local geological conditions have significantly influenced the characteristics of ground shaking during earthquakes. It is now well recognized that each soil type responds differently, when subjected to the ground motions, imposed due to earthquake loading. Usually, the younger softer soil amplifies ground motion relative to older, more compact soils or bedrock. Local amplification of the ground is often controlled by the soft surface layer, which leads to the trapping of the seismic energy, due to the impedance contrast between the soft surface soils and the underlying bedrock. Moreover, the relatively simple onset of vertical resonances can be transformed into a complex pattern of resonances, strongly dependent on the characteristics of the lithological attributes, geometry and topography [5, 24, 30, 31].

The trapping of seismic waves due to the impedance contrast between sediments and the underlying bedrock is the fundamental phenomenon responsible for the amplification of motion over soft sediments. In previous years, there have been numerous studies about wave propagation in complex media [16, 90, 97]. When the structure is horizontally layered (1D structures), this trapping affects only body waves traveling up and down in the surface layers. When the surface sediments from a 2D or 3D structure, i.e., when lateral heterogeneities such as thickness variations are present, this trapping also affects the surface waves, which develop on these heterogeneities, and thus reverberate back and forth. The interference between these trapped waves leads to resonance patterns, the shape and the frequency of which are related with the geometrical and mechanical characteristics of the structure. While these resonance patterns are very simple in the case of 1D media (vertical resonance of body waves), they become more complex in the case of 2D and a fortiori 3D structures. The characteristics and depth of surface layers of the ground have also a great influence on the intensity at ground level. These soil layers over the firm bedrock may attenuate or amplify the base rock earthquake motion depending upon geotechnical characteristics, their depth and arrangement of layering. Thus, there are many factors that influence the way a site will respond to earthquake ground motion. These include:

(i) The source location.
(ii) The prevalence of energy focused or scattered from lateral heterogeneity and degree to which sediments behave non-linearly, which causes the response to depend on the level of input motion.
(iii) Amplification thus also depends on the match between the frequency content of the earthquake motion and natural period(s) of the ground.

Resonance frequency of each soil type also differs depending upon the physical property, and depth to bedrock. Site response parameters, viz. peak frequency (f_o) and amplification (A_0), form the most important input in hazard evaluation and ground characterization for Seismic Hazard and Risk Microzonation. Quantification of this amplification of ground motion throughout the area of interest and determination of natural resonance frequency is the main objective of site response study. This frequency-dependent amplification forms an important factor for seismic hazard analysis. The site response parameters are also used to distinguish regions where the seismic hazard is high due to amplification from the surface geology and match of natural frequency of the soil with the construction. The mapping of the soil behavior before a seismic wavefield propagates also provides an overview of the possible damage to individual structure or a set of buildings.

2.2 Damage Pattern Associated with Local Geological Condition During Earthquake

Many past and recent earthquakes in the world amply demonstrated large concentration of damage in specific areas due to site-dependent factors related to surface geologic conditions and local soil altering seismic motion and hence the engineering designs. These include the great Japan earthquake, 1891 [77], the San Francisco earthquake of 1906 [133], Long beach earthquake of 1933 [134]. Much of the damage caused by the 1989 Loma Prieta earthquake was in areas where the site response caused enhanced ground shaking [48]. The destructive effects of the 1994 Northridge earthquake, across the Los Angeles region, show pockets of severely damaged buildings within 1 km of largely undamaged area. More recently, the Jabalpur earthquake of 1997 and Bhuj earthquake of 2001 showed irregular damage patterns and buildings within 100 m of severely damaged structures are found to be almost unaffected [42, 52]. On the other hand, significant damage occurred in the Ahmedabad city in spite of a large epicentral distance of more than 300 km. This has been ascribed to site response characteristics of the Ahmedabad city area and natural period of ground structure. Much of this variability during the Bhuj earthquake can be attributed to local site response [83, 73, 121]. This is reported to be a similar situation as Mexico City during the Michoacan earthquake in 1985. The Wenchuan Ms8.0 earthquake in 2008 and the Minxian–Zhangxian Ms 6.6 earthquake in 2013 caused enormous buildings and houses to collapse due to the site amplification effect [128, 132, 135]. Numerous other studies have also demonstrated the ability of surface geologic conditions to alter seismic motion [4, 18, 34]. It has also been documented that the damaging effects associated with such soft deposits may lead to local intensity increments as large as 2–3° in MM scale, [4, 7, 29, 30, 38]. It is, therefore, a realistic definition of site response that has become one of the most relevant tasks in the seismic engineering analysis.

2.3 Spectral Characteristics of Ground Motion

In the frequency domain spectra of ground motion, the resonance pattern shows up as spectral peaks. The frequencies of these peaks are related both to the thickness and shear wave velocities in the surface layers. For embanked 2D or 3D structures, they also depend on their width. For one layer structures, this relation is very simple and given as:

$$\text{fo} = \beta_1/4 \quad \text{(Fundamental)}$$

$$\text{fn} = (-2n + 1)\,\text{fo} \quad \text{(Harmonics)}$$

where β_1 is the S wave velocity in the surface layer, 'h' its thickness and 'n' is an integer number characterizing the harmonic order. The fundamental frequency has been found ranging between 0.2 Hz (for very thick deposits, such as in Los Angeles or Tokyo or for extremely soft materials such as in Mexico City) and 10 Hz or more (for very thin layers such as diluvial deposits or weathered rocks).

The amplitude of these peaks is related mainly to the impedance contrast between the very surface and underlying bedrock, to the material damping in sediments, and to a somewhat lesser extent to the characteristics of the incident wavefield (type of wave, incident angle, near-field or far-field). In the case of 2D or 3D structures, the lateral geometry may also play an important role, especially in the case of small material damping.

For one layer 1D structures impinged by vertical plane S waves, the amplification at the fundamental frequency 'fo' is given by

$$Ao = 1./(1./C + 0.5\pi\zeta)$$

where $C = (\rho 2\beta 2/\ \rho 1\beta 1)$; is the impedance contrast; ρi is the density of medium ($i =$ 1for sediments and $i = 2$ for bed rock), and ζ is the material damping in sediments.

In the case of very small damping ($\zeta = 0$), the maximum amplification is simply the impedance C.

Such a simple simultaneous estimation of these two parameters is indeed possible only for sites that can be approximated as a one-layer over bed-rock structure.

Both experimental and theoretical studies have shown that peak spectral amplitude often varies between 6 and 10, and in extreme cases such as in Mexico City or in San Francisco, it exceeded 20.

These formulae show that the estimation of peak frequency fo is relatively easy, since only the S-wave velocity and thickness of the surface layer are needed, while the estimation of peak amplification Ao requires the additional knowledge of the bedrock velocity and the sediment damping. However, such a simple simultaneous estimation of these two parameters is indeed possible only for sites that can be approximated as a one-layer over bed-rock structure.

2.4 Site Effect on Different Earthquake Ground Motion Parameters

2.4.1 Site Effect on Peak Values (Peak Ground Acceleration and Peak Ground Velocity)

Before the Mexico earthquake, it was thought that peak ground accelerations (PGA) are not generally affected significantly by sediments, while peak ground velocities (PGV) are reaching higher values on soft soils. However, the Mexico City records exhibited PGA values 4 times higher in the lake-bed zone than in the hill zone, and this observation was later confirmed in many other earthquakes (Loma Prieta, San Salvador, Kobe, etc.). The aftershocks of the Wenchuan earthquake Ms8.0 recorded by the temporary strong motion array showed that the peak ground acceleration (PGA) at the top station was nearly 1.5 times larger than that at the bottom station [122]. On the other hand, it was also observed in a few recent earthquakes (Kobe earthquake) that liquefied sandy deposits induce drastic reductions of peak accelerations. Thus, it appears that PGA values on sediments cannot be predicted in a straightforward way from PGA values on rock. However, an increasing number of engineering seismologists expect an amplification of PGA on soft sediments, at least for moderate acceleration level (below 0.4 g) [49].

2.4.2 Site Effect on Duration

Local geological conditions can substantially alter the characteristics of seismic waves [20]. In particular, it has been shown that for unconsolidated deposits, resonant phenomena often appear. For these deposit sites, ground motion amplitude and duration over certain period bands may be several times larger than levels at sites located on rock. Several other Trifunac studies also reported a significant increase of duration on sediments, especially at long periods [40, 41, 66, 125, 137]. This issue is closely connected to the effects of lateral heterogeneities and the associated diffraction of surface waves.

2.4.3 Site Effect on Spatial Distance

Lateral heterogeneities within sediments have shown, both experimentally and theoretically, to induce significant differential motion, over spatial distances comparable to seismic wavelengths, which may be as short as a few tens of meters. Such spatial variability, therefore, bears a significant engineering interest, especially for lifeline structures such as bridges or pipes.

2.5 Methods of Estimating Site Response

There are plenty of ways to estimate site response. The simplest way is to characterize them in terms of soil type classification. Such soil type-specific amplification factors were implemented in the first version of Japanese building code as early as the 1950s. later on it was found that the conventional broad classification of soil type is not effective for characterizing site effects, because site amplification factors are strongly frequency and site dependent, any average values for different sites with same category yield relatively small and flat frequency characteristics, which is for from the reality at any sites in that category [4]. Based on further studies, several review papers have been published on different methods used for soil site characterization and microzonation [11, 13, 64, 95]. Broadly, these methods can be categorized as follows:

(i) Empirical Methods
(ii) Experimental Methods
(iii) Numerical Methods

The site response parameters to be evaluated from site response study are as follows:

(i) Peak frequency of soil column above bedrock
(ii) Peak amplification of soil column above bedrock

(iii) Peak ground acceleration (PGA) at surface for different periods of exceedance and damping
(iv) Amplification factor of soil column
(v) Spectral acceleration at different periods
(vi) Site specific response spectra

2.5.1 Empirical Methods

2.5.1.1 Using Direct Earthquake Observations

Empirical methods estimating site response by using direct earthquake observations and applying existing empirical attenuation relations generalize inversion spectral ratios for specific site class, and a regional earthquake source model. This technique has been adopted in the Southern California Earthquake Center (SCEC) and detailed in Phase II report [51]. However, the problem with this is that to reduce the uncertainty in the attenuation relation. Borcherdt [18] introduced sediment-to-bedrock spectral-ratio technique. This technique was used by Rogers et al. [96] and presented site response estimation using 1971 San Francisco earthquake data and underground nuclear explosions data from Nevada test. After averaging results over three frequency bands, they applied clustering techniques to determine which of several geotechnical parameters were influential on site response estimations. At frequencies above 2 Hz, they found near-surface void ratio (a proxy for shear wave velocity) to be most influential with Holocane thickness and/or depth to basement influence as well. Below 2 Hz, they found Quaternary thickness and/or depth to basement rock to be controlling factors. They noted, however, that some of the factors are highly interdependent [96] making it difficult to separate near surface effects from deeper basins.

2.5.1.2 Using Coda Wave

S-wave coda methodology was developed by Aki and Chouset [6], Tsujiura [127] and Phillips and Aki [3] to take advantage of the relatively abundant vertical component network data for which the corresponding direct S-wave arrivals are generally clipped. This methodology was used by Su and Aki [120] for the generation of empirical site amplification maps for Southern California. This approach is perhaps most appealing because the coda is thought to be composed of scattered energy coming in from a variety of directions [2, 6], so the site response estimates might naturally reflect an average over various source locations. Unfortunately, there is some disagreement on whether coda site response estimates are consistent with the direct S-wave response: while some have found agreement between the two [59, 127], others have not [75] Seekins et al. [100], Bonilla et al. [22], Field [33]. Another concern is that amplification factors based on vertical-component network data may not be applicable to horizontal component motion [120].

2.5.1.3 Using Ambient Seismic Noise

In order to make up the dearth of direct earthquake data, the use of ambient seismic noise (microseisms and/or microtremors) was also proposed. Kagami et al. [54] examined microtremor spectral ratios in Los Angeles basin and found the amplitudes to correlate with depth to basement rock, nuclear explosions spectral ratio amplitudes of Borcherdt et al. [19], and ratios obtained for the 1971 San Fernando earthquake. Kagami et al. [54] found similar results from a more extensive set of observations carried out in the San Fernando Valley, and a two-dimensional, theoretical model was constructed by Navaro et al., which shows a remarkable coincidence with the microtremor experiment data. In spite of these and other encouraging results [136], questions still prevail with respect to the applicability of ambient noise measurements [11]. As such, no microtremor-based amplification map has been forthcoming regardless of whether such a map can be generated, array analyses of ambient noise data may still prove useful in determining surface structure [1, 89].

2.5.1.4 Using Aftershock Data

Earthquake sequences at the region and aftershocks have also been used for site amplification estimation by inverting shear wave recordings and obtaining spectral-ratio estimates [22, 44]. The potential problem with such site effect estimates is that any difference between the true and assumed path effect will be mapped into the site response. Because all such earthquake sequences emanate from the main shock region and any regional focusing effects will contaminate the site response estimates. To avoid the focusing effect, Harmsen [43] performed similar inversion for site response at 281 locations in the Los Angeles region using strong motion records from four earthquakes of 1997, San Fernando, 1987; Whitter Narrow, 1991; Sierra Madre, and 1994 Northridge main shock emanating from different directions. Hartzell et al. [45] prepared a series of "First-Generation site-response Maps for the Los Angeles region" by combining Northridge aftershock results with Harmsen's [43] results of the above four earthquakes. They identified several isolated locations with relatively high amplification values and this remains to answer whether these will remain high for earthquakes other than those represented in the inversions, especially since no formal uncertainty estimates are provided.

2.5.2 Experimental Methods

The experimental methods can be used to evaluate the local seismic response using the records of seismic signals, microtremors that can be generated by artificial seismic sources, or ambient noise and earthquakes data collected either by conducting special experiments or available data. The Table 2.1 summarizes the most frequently used techniques for the experimental estimation of site response.

2.5.2.1 Nakamura Type Study (H/V of Microtremor)

The influence of local geological structure on the spectral characteristics of ambient noise of relatively distant sources has long been recognized. Many authors have studied the nature of microtremor source, propagation, effects at continental margins and interactions with continental structures [1, 70, 91]. Thus, atmospheric disturbances and meteorological phenomena over the land or the sea, as well as distant human activity, have been identified as generators of microtremors propagating as Rg and Lg phases over the continents. Most of the authors agree that microtremor spectral characteristics are associated with local geological structure, especially with the density and thickness of the surface layers.

The use of microtremors to determine dynamic site response characteristics of sites for microzonation purposes was first introduced by Kannai and his co-workers [55–57]. Udwadia and Trifunac [129] investigated whether microtremors could be used to determine site amplification characteristics for earthquake shaking by comparing data from strong motion earthquake records and microtremor measurements at EL Centro, California. They found little correlation between the ground

Table 2.1 Experimental methods of Site Response, i.e. evaluation of Peak frequency (f_0) and Peak Amplification (A_{0}) of soil column above bedrock

Technique	Earthquake(E) or Microtremors(M)	Reference site (R) or not (NR)	Example References
Sediment/bedrock(S/B) (receiver function type)	E	R	Brorcherdt [18]
Generalized inversion scheme	E	R	Andrew [9]
Parameterized source path inversion	E	NR	Boatwright et al. [21]
Coda waves analysis	E	NR	Philips and Aki [3]
Horizontal/vertical(H/V) (receiver function type)	E	NR	Lermo and Chavez-Garcia [69]
Sediment/bedrock (S/B)	M	R	Yamanaka et al. [136]
Horizontal/vertical(H/V)	M	NR	Nakamura [80]
Array analysis	M	NR	Malagnini et al. [74]

motions due to earthquake shaking and microtremor excitation. A major problem in interpreting the microtremor data was the nature of the unknown input. Furthermore, the microtremor motions were found to be stationary over a duration of 5–10 min, but nonstationary over a period of day. Therefore, they suggested that repeated measurements at the site would be necessary to establish some common features of response that could be attributed to site conditions.

Kobayashi and Midorikawa [61], Kobayashi et al. [62] measured microtremors at many sites in Mexico City. Subsequently, Lermo et al. measured microtremors at 97 additional sites and processed all the data to obtain site periods and spectral ratios. They found marked differences in the amplitudes of the spectral ratios and the period of peak response for the low traffic interval 02–04 h compared with busier intervals 06–22 h. These results stress the major impact that source effects have on the frequency and amplitudes of microtremors. Similarly, Seo et al. [101] show the clear differences in dynamic response characteristics derived from earthquake motion and microtremors using data from ground motion studies in the Ashigara valley Japan. The microtremors overestimate the spectral ratios in this region by factors up to 5. They concluded that microtremors cannot be used reliably to determine amplification ratios for strong shaking. Furthermore, the inference about site period may also be quite different from microtremors and strong data.

A very extended technique based on the H/V ratio (i.e., the ratio between the Fourier spectra of the horizontal and vertical components of microtremors) was introduced by Nakamura [80], which uses microtremors to estimate the amplification for horizontal displacements at the free surface during the occurrence of earthquakes. Nakamura [80] carried out extensive microtremor studies and demonstrated capability of the tool for site reference studies. He applied the technique at sites in Kanonomiya and Tabata, Japan, recording microtremors on an hourly basis over a 30-h period at each site. He proposed that the H/V ratio is a reliable estimation of site response of the S wave and provides reliable estimates not only about resonant frequency but also about the corresponding amplification. This method considers that spectral amplification of a surface layer could be obtained by evaluating the horizontal to vertical spectral ratio of the microtremors recorded at the site. This approach is being used by an increasing number of researchers with the aim of characterizing the seismic hazard in a small scale, and to provide detailed information for seismic microzonation in urban areas. This method is popularly known as Nakamura technique. Lermo and Garcia [70], Bard [11–14] reviewed the problems of site studies and have shown that the technique would be of characterization based on microtremor immense help in providing time and cost-effective method of response studies.

Horizontal to Vertical Spectral ratio (HVSR) technique can also be used to study the structural characteristics of a building besides the geological characteristics of a region and also the correlation of both geology and structure by recording noise data on the ground of the building and inside the building. HVSR method can also be applied in both ground and buildings in order to study the natural frequency, the index of vulnerability, the amplification factor of the ground as well as indexes of vulnerability of buildings and ground Triwulan et al. [126]. Ambient noise in concrete reinforced buildings, affected by subway trains, is also studied in Beijing by Luo et al. [72], who present the resonance frequency of the building on each floor. Mucciarelli et al. [79] claim that the study of microtremors and weak motion provides fast and reliable data for site amplification and structure vulnerability compared with other traditional methods (like geological study by drilling), and the correlation of damage and structural integrity is with physical parameters and not normalized dimensional indexes. Liu et al. [71] applied ambient noise survey on a seven-floor building using seismometers in order to assess the site amplification in an urban area and to study the ability of HVSR to assess building vulnerability under strong motion excitation. Further, by studying HVSR in buildings, there is the ability to study the fundamental frequency of the building excited by environmental noise of seismic activity, and compare it with the fundamental frequency of the surface around the building. The study of HVSR outside and inside the building potentially can provide two factors: initially, how close the fundamental frequency of the building is to the fundamental frequency of the surface layer, and also how high the amplitude of this frequency is? In an article Nakamura added the remark that «the Nakamura method to estimate not only the resonance frequency and amplification factor of a surface layer using HVSR at ground surface but also the resonance frequency and the amplification factor of both buildings and ground using HVSR at the top of a building» Nakamura [82].

Due to time and cost-effective method, microtremor-based Nakamura technique has been used for the estimation of site response particularly Peak Frequency and Peak Amplification of soil in all most all Seismic Hazard Microzonation studies conducted by India Meteorological Department (IMD)/Department of Science and Technology (DST)/Ministry of Earth Sciences for Indian cities such as Jabalpur, National Capital Territory (NCT) Delhi in 1:50,000 Scale, NCT Delhi on 1:10,000 scale, Guwahati, Bangalore, etc. [76, 78, 84, 85, 92, 93, 103–117, 119]. Site response studies alone have also been conducted for a few cities [119].

2.5.2.2 Theoretical Explanation of Horizontal/Vertical (H/V) Ratio

The main challenge to determine site amplification characteristics from microtremors is to remove source and path effect. Nakamura [80, 81] proposed an approximate procedure for removing source effects from microtremor records based on a modification of the conventional transfer function of the site.

The approach and hypothesis of Nakamura are summarized as follows:

(i) The horizontal tremor may be considered, to certain accuracy, to be amplified through multi-reflection of the S wave while the vertical tremor is amplified through multi-reflection of the P wave.
(ii) The effect of Rayleigh waves remarkably appears in the vertical tremor.

Accordingly, the degree of its effect may be known by determining the ratio between the vertical tremor recorded at the surface and at the substrate. Namely, the effect of Rayleigh waves is nearly zero when the ratio is approximately '1'. With an increasing ratio, the effect of Rayleigh waves may become more critical. Elimination of the effect of Rayleigh waves is obtained by using this ratio.

In other words, the surface sources of the microtremors were assumed to generate Rayleigh waves which affected equally the horizontal and vertical motion in the surface layer. Under these assumptions, the amplitude effects of sources were estimated by the ratio Es with

$$\mathrm{Es} = \mathrm{Svs}/\mathrm{Svb}$$

where, Svs and Svb are the frequency-dependent spectral ordinates of the vertical motions at the surface and at the base layer, respectively.

The transfer function of the site St is defined by

$$\mathrm{St} = \mathrm{Shs}/\mathrm{Shb}$$

where, Shs and Shb are the spectral ordinates of the horizontal motions at the surface and at base layer, respectively. The source effects are compensated by dividing St by Es giving the modified transfer function, Stt for the site as

$$\begin{aligned}\mathrm{Stt} &= (\mathrm{Shs}/\mathrm{Shb})/(\mathrm{Svs}/\mathrm{Svb}) \\ &= (\mathrm{Shs}/\mathrm{Svs})/(\mathrm{Shb}/\mathrm{Svs})\end{aligned}$$

This may be written as
$= \mathrm{Rs}/\mathrm{Rb}$

where

$$\begin{aligned}\mathrm{Rs} &= (\mathrm{Shs}/\mathrm{Svs}) \text{ and} \\ \mathrm{Rb} &= (\mathrm{Shb}/\mathrm{Svs})\end{aligned}$$

Nakamura assumed that Rb = 1.0 over a significant period of range of engineering interest based on data from the detailed studies at Kanonomiya and Tabata site. The transfer function is given by Rs, which can be determined from surface measurements of microtremors alone.

Although Nakamura's qualitative explanation looked questionable (as indicated in Lachet and Bard [10], and again in Kudio), various sets of experimental data [3, 27, 28, 34, 37, 39, 66, 68, 69, 88, 95, 100] confirmed that these ratios are much more stable than the raw noise spectra. In addition, on soft sites, they exhibit a clear peak that is well correlated with the fundamental resonant frequency. These observations are supported by several theoretical investigations [36]; Lachet and Bard [65]; Lermo and Chavez-Garcia [70]; Cornou [25]), showing that synthetics obtained with randomly distributed, near surface sources lead horizontal-to-vertical ratios sharply peaked around the fundamental S-wave frequency, whenever the surface layers exhibit a sharp impedance contrast with underlying stiffer formations.

Further, Lachet and Bard [65] proposed that the good match at fundamental frequency is due to the horizontal-vertical polarization of the Rayleigh waves, an interpretation that is in good agreement with the early Japanese studies [64]. According to this view, no straightforward relation exists between the H/V peak amplitude and the site amplification. However, this opinion is not shared unanimously. For instance, a one to one average correlation is claimed by Kanno and Ohmachi [58] on the basis of comparison between observed H/V peaks and numerical estimates of 1D transfer function. On the other hand, an empirical relationship between observed H/V peaks amplitude and local intensity increment (MM Scale) was argued in Toshinawa et al. [124] from a strictly experimental viewpoint.

A thorough comparison [13] between observed amplifications derived from earthquake records and observed H/V peak amplitudes at more than 30 sites demonstrates that the latter is almost always smaller than the former. Such an experimental result, although not yet explained by theoretical or numerical work, despite the new computations performed by Cornou [25], could be very useful indeed. It would mean that the Nogashi Nakamura technique could provide a lower-bound estimation to the actual amplification. This view needs to be confirmed, however, by a larger set of experimental data. Comparison with other techniques by different investigators, it shows that Nakamura technique allows obtaining, very simply, the fundamental resonant frequency, but fails for higher harmonics and that peak amplitude is somewhat different from amplification measured on spectral ratios. However, this Nakamura version of the microtremor method has already proved to be one of the most inexpensive and convenient techniques to reliably estimate fundamental frequencies of soft deposits.

Inexpensive and convenient techniques to reliably estimate fundamental frequencies of soft deposits are given in Table 2.2. It certainly deserves more work so as to elucidate the factors influencing peak amplitudes.

Table 2.2 Results of response studies: comments

Authors	Site	Frequency	Amplitude
Bard et al. [15]	Many sites (France, Greece)	Good	Generally smaller
Bindi et al. [17]	Umbria—Marche (Italy)	Good	Smaller
Duval [26]	Nice, Monaco, Venezuela	Good	Smaller or equal
Field and Jacob [37]	Oakland (California)	Good	Much smaller
Field et al. [35]	Gyumri (Armenia)	Good	Much smaller
Lachet et al. [66]	Thessaloniki (Greece)	Good	Smaller
Labrun [68]	Grenoble Annecy (France)	Good	Much smaller
Malagnini et al. [74]	Tuscany (Italy)	Good	Much smaller
Riepl et al. [95]	Vilvi	Good	Smaller
Seekins et al. [100]	San Francisco (USA)	Good	Comparable
Toshinawa et al. [124]	Christchurch (New-Zealand)	Good	Smaller
Volant et al. [131]	Garner Valley (USA)	Slightly larger	Smaller
Field [33]	Coachella Valley (USA)	Good	Much smaller

2.5.2.3 Receiver Function Type Technique (H/V of Earthquake Event)

The other extremely simple technique, which is combination of Langston [67], receiver-function method for determining the velocity structure of the crust from the horizontal to vertical spectral ratio (HVSR) of teleseismic P wave and the Nakamura technique [80, 81], using weak motion data recorded from small to moderate, natural or artificial seismic events. This method involves taking the spectral ratio between the horizontal and the vertical components of the shear wave part of the small events. This technique was first used by Lermo and Chavez Garcia [69] on different sites in Mexico City. Field and Jacob [37] also applied this technique in their systematic comparison and found that the method reproduces very well the shape of spectra; but, this technique underestimates the amplification level.

2.5.2.4 Receiver-Function Type Studies in Reference Site Dependent Mode

The greatest challenge in estimating site response from earthquake data is removing the source and path effect. In this respect, Borcherdt [18] proposed a procedure similar to receiver function type study as above, using bedrock reference site. The reference site is defined as a site that has a flat transfer function with an amplitude of 1. The spectral ratio gives a reliable estimate of site response, if the reference site fulfills basic criteria viz (i) it should be located near to the site to ensure that the differences between each site are only due to site conditions and not to difference in source radiation or travel path and (ii) it should also be unaffected by any kind of weathering or fracturing [118]. This method requires rigorous exercises for reference site selection.

2.5.2.5 Merits of Nakamura H/V Ratio-Based Techniques

Although theoretical aspects of the Nakamura's qualitative explanation are questionable [26, 32, 36, 65], the method applies a quick and cost-effective technique for ascertaining peak frequency and is found more stable for characterizing conditions based on nature of curve. As regards amplifications, the Nakamura Technique is a poor discriminator as compared to the H/V ratio of earthquake records; the technique gives lower bound estimates to the actual amplification [13] and shows wide variability in low frequency range. Therefore, as regards to determination of amplification, Nakamura technique is to be used with caution. However, it gives relative parametric evaluation of ground amplification.

Based on the results of site response study conducted for Seismic Hazard Microzonation of National Capital Territory (NCT) Delhi on 1:10,000 scale and earlier studies [85, 93, 106, 109, 111, 115, 116] and "moes.Gov.in") and their validation, it is found that Nakamura technique provides a rapid scanning and cost-effective

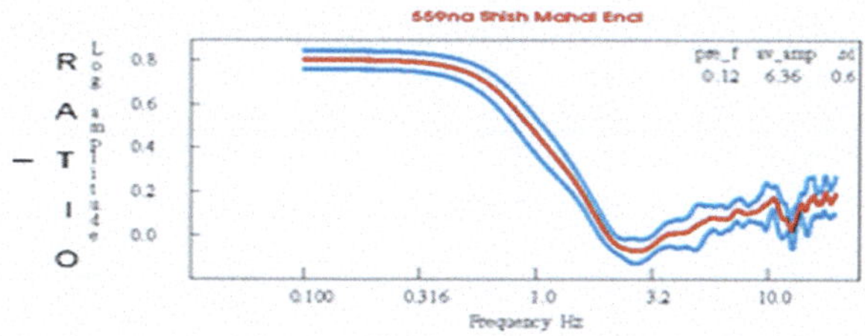

Fig. 2.1 Showing flat spectral ratio curves of horizontal to vertical component (QTS: Quasi-Transfer Spectrum) of microtremor below 0.3 Hz

method for identifying predominant frequency. Moreover, the response curves based on Nakamura method are stable enough to characterize a site [116].

2.5.2.6 Use of Sensors for the Generation of Microtremor Data for Site Response Study

Inference drawn on the basis of previous studies, Bard in his lecture note [13] recommended that for site response analysis at periods longer than 1 s., velocity sensors of period 5 or 10 s should be preferred. Some investigators recommended that a 4.5 Hz seismometer provides reliable results down to 0.5 Hz. On comparing different sensors, it has been suggested that both short period and broadband sensors yield very similar results in the frequency range 0.1–20 Hz and in particular, if the peaks related to the fundamental frequencies of the site occur at same frequencies. However, while undertaking study for Indian cities particularly of NCT Delhi, noted that estimation of peak frequency and amplification is difficult below 0.3 Hz due to flat response curves and alternate sensor characteristics need to be evaluated (SHM report of NCT Delhi on 1:10,000 scale available in MoES website). Figure 2.1 shows a typical example of flat spectral ratio curves of horizontal to vertical component (QTS: Quasi-Transfer Spectrum) of microtremor obtained in NCT Delhi below 0.3 Hz.

2.5.2.7 Precautions Regarding Experimental Setup

- Long external wiring that may bring in mechanical and electronic interference needs to be avoided.
- However, it is difficult to avoid recording close to roads in the urban area like Delhi but efforts should be made to put the sensor as far as possible from the road with heavy vehicles.
- Local weather conditions vis-à-vis rain may alter the site conditions and wind may cause artificial vibration of the sensor, therefore recording in rainy and windy days needs to be avoided.
- To acquire enough uniform waveform train and data representative to different local conditions, microtremor data of long duration need to be collected.

- Sensor and ground need to be coupled properly and efforts need to be made to put the sensor on the natural soil, remaining artificial overburden (SHM report NCT Delhi on 1:10,000 Scale).

2.5.3 Numerical Method of Ground Response Analysis

The ultimate objective of Seismic Hazard Microzonation study is to predict response of soil on ground surface motion for development of design response spectra, which can be used for design of building code; to evaluate dynamic stresses and strain for the evaluation of liquefaction hazard and to determine the earthquake-induced forces that can lead to instability of earth and earth-retaining structures. Thus, the study requires the evaluation of several other parameters than peak frequency and peak amplification evaluated using experimental techniques. This includes evaluation of the combined influence of amplitude of ground motion accelerations, their frequency components on different structures at different sites, which is represented by means of a ground response spectrum; that is plot showing the maximum response induced by the ground motion in a single-degree-of-freedom (SDOF) oscillators of different fundamental periods, but having the same degree of internal damping. These parameters can be evaluated through rigorous exercise of site response study based on numerical techniques.

Moreover, the experimental technique provides peak frequency of total soil column above firm bedrock (shear wave velocity about >1500 m/s). In engineering practices, the reference rock is considered as engineering bedrock (shear wave velocity about 760 m/s). Thus, it is imperative to study the effect of the soil column above this rock and derive all parameters with reference to engineering bedrock. Therefore, for Seismic Hazard Microzonation of urban agglomeration detailed study of site response based on numeral technique is also called for.

Several numerical techniques have been developed for ground response analysis, which allow computation of the seismic response of a given site, since the pioneer work by Thomson [123], Haskell [46], and applied in microzonation studies. Ground motion is dependent on near-surface conditions, [53] and if losses due to reflection, scattering and inelastic attenuation are negligible, then the energy along a tube is conserved and amplitude is proportional to $1/\sqrt{\rho\beta}$, where 'ρ' is density and 'β' is shear-wave velocity [8, 23]. This implies that response of sites can theoretically be estimated through numerical analysis using physical properties of the soil and rock.

The methods of 'ground response analysis' differ in the simplifying assumptions that are made, in the representation of stress–strain relations of soil and in the methods used to integrate the equation of motion. The one most widely used method is based on the multiple reflection theory of S waves in horizontally layered deposits and very often referred to as '1D analysis of soil columns'. According to this theory when a fault ruptures below the earth's surface, body waves travel away from the source in all directions. As they reach boundaries between different geologic materials, they are reflected and refracted. Since the wave propagation velocities of shallower materials are generally lower than the materials beneath them, inclined rays that strike horizontal layer boundaries are usually reflected to a more vertical direction. By the time, the rays reach the ground surface, multiple refractions have often bent them to a nearly vertical direction.

The principle of one-dimensional ground response analysis is therefore based on the assumption that all boundaries are horizontal and soil and bedrock are assumed to extend infinitely in the horizontal direction (half-sphere). The second assumption is that inclined incoming seismic rays are reflected to a near-vertical direction, because of decrease in velocities of surface deposits. Therefore, the response of the soil deposit is caused by shear waves propagating vertically from the underlying bedrock. The methods of one-dimensional ground response analysis are useful for level or gently sloping sites with parallel material boundaries. Such conditions are not uncommon and one-dimensional analyses are widely used in geotechnical earthquake practice and have been known to give reasonable estimates of ground vibration under a seismic event [49, 63].

The numerical method can be broadly grouped into the following three categories: (i) Linear Analysis (ii) Equivalent Linear Analysis, and (iii) Non-Linear Analysis.

2.5.3.1 Linear Approach

Linear approach is a widely used technique for ground response analysis and is based on the use of transfer functions, which is the ratio of displacement amplitudes at two points at the top and bottom of the soil layers, respectively. In this procedure, a known time history of bedrock (input) motion is represented as Fourier series, usually using the FFT. Each term in the Fourier series of the bedrock (input) motion is then multiplied by a transfer function to produce the Fourier series of the ground surface (output) motion. The ground surface (output) motion can then be expressed in the time domain using the inverse FFT. Thus, the transfer function determines how each frequency in the bedrock (input) motion is amplified or diamplified, by the soil deposit. Thus, the key to the linear approach is the evaluation of transfer function of the soil layers, which is the ratio of displacement amplitudes at any two points at the top and bottom of the soil layer. The modulus of the transfer function is the amplification factor, which is the ratio of the free surface motion amplitude to the bedrock motion amplitude. The soil layer is seldom homogeneous and heterogeneity of soil layers can be modeled by inclusion of more soil layers. Because it relies on the principle of superposition, this approach is limited to the analysis of linear systems.

2.5.3.2 Nonlinear Approach

The nonlinear stress–strain behavior of soil can be represented more accurately by cyclic non-linear models that follow the actual stress–strain path during cyclic loading. Such models are able to represent the shear strength of the soil, and with an appropriate pore pressure generation model, changes in effective stress during undrained cyclic loading. In the nonlinear approach, the actual nonlinear response of a soil deposit is analyzed using direct numerical integration in the time domain. In this approach by integrating the equation of motion in small time steps any linear or nonlinear stress–strain model or advance constitutive model can be used. At the beginning of each time step, the stress–strain relationship is referred to obtain the approximate soil properties to be used in that time step. By this method, a nonlinear inelastic stress–strain relationship can be followed in a set of small incrementally linear steps. Nonlinear methods require a reliable stress–strain or constitutive model. The parameters that describe such models are not as well established as those of the equivalent linear model (described in following section). A substantial field and laboratory testing program may be required to evaluate nonlinear model parameters.

2.5.3.3 Equivalent Linear Analysis: Equivalent Approximation of Nonlinear Response

The linear approach assumes that shear strength (G) and damping (ξ) are constant. However, soil exhibits nonlinear, inelastic, stress–strain behavior under cyclic loading conditions, and non-linear behavior of soils can be determined very well in a laboratory environment. At low strain levels, the stiffness of the soil is the greatest and damping is the smallest. At higher strain levels, the effect of nonlinearity and inelasticity increases, producing lower stiffness and greater damping. These relationships can be tested and plotted in curves, called shear modulus reduction curves and damping curves, respectively. The linear approach, therefore, needs to be modified to provide reasonable estimates of ground response. Equivalent linear models treat soil as linear viscoelastic material. Nonlinear behavior is accounted for by the use of strain-dependent stiffness and damping parameters. The stiffness of the soil is usually characterized by the maximum shear modulus, which mobilizes at low strain, and a modulus reduction curve, which shows how the shear modulus decreases at larger strain. Damping behavior is characteristic by the damping ratio, which increases with increasing strain amplitude. The shape of the modulus reduction and damping curves are influenced by soil plasticity and, for soil of very low plasticity, by effective confining pressure. The non-linear analyses, therefore, require a quantitative knowledge of the actual nonlinear material behavior. The actual nonlinear hysteretic stress–strain behavior of cyclically loaded soils can be approximated by equivalent linear soil properties. The equivalent linear shear modulus 'G' is generally taken as a secant shear modulus (G_{sec}) and the equivalent linear damping ratio, ξ, as the damping ratio that produces the same energy loss in single cycle as the actual hysteresis loop. The non-linearity is very often approximated by a 'linear-equivalent' method that uses an iterative procedure to adapt the soil parameters (i.e., rigidity and damping) to the actual strain it undergoes.

In an equivalent linear approach, the non-linearity of the shear modulus and damping is accounted for by the use of equivalent linear soil properties using an iterative procedure to obtain value for modulus and damping compatible with the effective strains in each layer [50]. In this approach, the following steps are involved:

First step: A known time history of bedrock motion is represented as a Fourier series, usually using the Fast Fourier Transform (FFT).

Second step: The Transfer Functions for the different layers are determined using the current properties of the soil profile. Transfer functions give the amplification factor in terms of frequency for a given profile.

Third step: The Fourier spectrum is multiplied by the soil profile transfer function to obtain an amplification spectrum transferred to the specified layer.

Fourth step: The acceleration time history is determined for that layer by the Inverse Fourier Transformation.

Fifth step: Peak acceleration is obtained from the acceleration time history and with properties of the soil layer, the shear stress, and strain time histories are determined.

Sixth step: New value of soil damping and shear modulus are obtained from the damping ratio and shear modulus degradation curves corresponding to the effective strain from the strain time history. With these new soil properties, new transfer functions are obtained and the process is repeated until there is not much variation from one iteration to the next and this is called convergent of the solution.

This is although the linear equivalent of nonlinear phenomenon, it is still a linear analysis. However, this approach has been shown to provide a reasonable estimate of soil response and use widely in seismic microzonation studies. Site response using above algorithm has been coded in many software SHAKE [98], PROSHAKE (User's Manual), SHAKE 91 [50], DYNEQ [86, 87].

Advance Methods

There are certain advanced methods, however, that have the same base, i.e., the wave equation. Many different models have been proposed to investigate several of the various aspects of site effects, which involve complex phenomena. For example, one has to consider various types of incident wave fields (near-field, far-field, body waves, surface waves); the structure geometry may be 1D, 2D or 3D; or the mechanical behavior of earth materials may encompass a very wide range (visco-elasticity, non-linear behavior, water-saturated media, liquid domains, etc.). Typically, these advanced methods may be classified into four groups:

(i) Analytical methods, which can be used only for very simple geometries, are extremely valuable as benchmarks.
(ii) Ray methods, which are basically high-frequency techniques. It is uneasy to use them when wavelengths are comparable to the size of heterogeneity, a situation that is generally the most interesting one.
(iii) Boundary-based techniques (including all kinds of boundary integral techniques, or those based on wave function expansions), which are the most efficient when the site of interest consists of a limited number of homogeneous geological units.

(iv) Domain-based techniques (such as finite-difference or finite-element methods), which allow to account for very complex structures and material behavior, but are very heavy from a computational viewpoint.

As stressed by Aki and Irikura [7], these methods have their advantages and disadvantages, and, in general, those that can deal with more realistic models are less accurate, while those achieving a higher accuracy are more time-consuming. Most of these methods are still actually developed.

2.6 Methodology Adopted for Site Response Study in Indian Cities

The empirical and experimental methods are useful for the evaluation of peak frequency (f_0) and to some extent peak amplification (A_0) of soil above firm bedrock. Numerical methods are used to predict response of soil on ground surface motion for the development of design response spectra, which can be used for design of building code; to evaluate dynamic stresses and strain for evaluation of liquefaction hazard, and to determine the earthquake-induced forces that can lead to instability of earth and earth-retaining structures. Therefore, in order to meet the requirement of Seismic Hazard Microzonation, a combination of experimental and numerical techniques has been adopted.

Experimental methods based on microtremors are cost and time effective as a rapid screening tool for the evaluation of peak frequency and approximate estimation of peak amplification of soil column above base/firm rock (Shear wave velocity >1500 m/s) and therefore used for the evaluation of peak frequencies (f_0) and peak amplifications (A_0) at different sites. The studies when conducted in conjunction with other complementary techniques, such as shear wave velocity measurements (Aki's technique), receiver-function type technique based on weak motion data and specially in reference site-dependent mode, would form an effective tool for site characterization ascertaining peak/resonance frequency and amplification.

Site-specific ground motion parameters and response spectra etc. have been evaluated using numerical technique. The study based on numerical technique also permits the evaluation of Peak Frequency and Peak amplification of soil column above engineering bedrock (Shear wave velocity 760 m/s) and therefore used for the validation of Peak frequencies obtained from experimental technique for different equivalent depths of soil column.

Author of the chapter was associated in the Seismic Hazard and Risk Microzonation (SHRM) study of Indian cities since the formulation of methodology in the Indian contest. Site response study is an important component of SHRM. Based on this formulated methodology, the first SHRM study in India was conducted in 2001 for a small city of Jabalpur, where an earthquake had occurred in 1997, and data were available for verification. Since then Seismic Hazard Microzonation studies were undertaken for several Indian cities, and the author was involved in almost all studies. An Earthquake Risk Evaluation Center (EREC), mandated with such studies was established in 2004 in India Meteorological Department, (IMD), under the controlling Ministry of the Department of Science and Technology (DST), which subsequently shifted to the newly formulated Ministry, the Ministry of Earth Sciences (MoES) in 2006. A major project on Seismic Hazard Microzonation for a large area of 1485 km^2 and in a large mapping scale of 1:10,000 of National Capital Territory (NCT) Delhi was initiated in 2007 by Earthquake Risk Evaluation Center (EREC), which completed in 2014. The report along with maps was made public in February 2016 and is also available in the website of Ministry of Earth Sciences (MoES), Government of India (moes.gov.in) for public consumption. This study was supervised by a multidisciplinary Advisory Committee consisting of engineers, seismologists, geologists, geophysicists associated with institutions of national repute.

As a case study, site response study conducted at NCT Delhi is briefly discussed in the following para.

2.6.1 Site Response Studies of NCT Delhi

National Capital Territory (NCT) of Delhi occupies an area of 1482 sq. km spreading between Lat 28°24′01″ & 28°53′00″ N and Long 76°50′24″ & 77°20′37″ E (Toposheet53 D/14 & 53H/1,2,3,6, and 7) lies in Seismic Zone IV of the Seismic Zoning Map of India (BIS IS 1893, Part 1:2002). Geographically, the region is located on folded crustal ramp represented by basement rocks of Delhi Supergroup, bounded by two regional faults viz Mahendragarh-Dehradun Sub Surface Fault (MDSSF) in the west and Great Boundary Fault (GBF) in the East Delhi. The ramp trending NNE-SSW across the 'fore deep' is juxtaposed to Himalayan thrust belt. Another important structural element of the belt is NW–SE trending Delhi-Sargodha Ridge (DSR), which passes through Delhi and is flanked by basins on either side, viz Sahaspur Basin in the north and Bikaner Basin in southwest.

The geoscientific constitution of the area provides a highly variable domain with a complex scenario having scope of rapid changes in seismic accentuations. This complexity is further aggravated by long history of the evolution of Delhi as a mega polis and its further growth with new emerging centers of suburban agglomeration under the new concept of National Capital Region (NCR). Thus, the complex interplay of seismotectonics, lithological assemblage in terms of soft sediment accentuation, and co-existence of structures of medieval period as well as the most modern skyscrapers and well-laid modern residential buildings set unique conditions, which need closer evaluation for Seismic Hazard.

The objective of Seismic Hazard Microzonation on 1:10,000 scale of NCT Delhi was to evaluate all these hazard parameters such as (i) Peak frequency of soil column above bedrock, (ii) Peak Amplification Factor/ratio of soil column above bedrock, (iii) Peak Ground Acceleration (PGA) at surface for different periods of exceedance and damping, (iv) Amplification factor of soil column (v) Spectral acceleration at different periods, (vi) Site-specific response spectra and, therefore, a combination of experimental and numerical techniques have been adopted.

Experimental methods based on microtremors are cost and time effective as a rapid screening tool for the evaluation of peak frequency and approximate estimation of peak amplification of soil column above base/firm rock (Shear wave velocity >1500 m/s) and therefore used in the present study for the evaluation of peak frequencies (f_0) and peak amplifications (A_0) at different sites spread over NCT Delhi. Site-specific ground motion parameters and response spectra etc. have been evaluated using numerical technique.

The present study based on numerical technique also permits the evaluation of peak frequency and peak amplification of soil column above engineering bedrock (shear wave velocity 760 m/s) and, therefore, used for the validation of peak frequencies obtained from experimental technique for different equivalent depths of soil column.

2.6.1.1 Site Response Studies of NCT Delhi Based on Experimental Methods

Site response studies at NCT Delhi were conducted basically resorting to Nakamura-type studies based on microtremor data. Receiver function type study based on earthquake events in non-reference mode has also been resorted with a few events.

Fieldwork/data acquisition for experimental techniques

There are several small amplitude vibrations, which appear on the surrounding ground surface. The period range of such vibrations is from 0.1 to 10 s. Vibrations that have small periods, less than 1 s, are called microtremors or Kanai's microtremors [102], and those with larger periods are called microseism. The origin of microtremors is probably due to traffic vehicles, heavy machinery facilities, household appliances and so on that are not related to earthquakes; however, small waves propagate from artificial sources surrounding daily life.

At NCT Delhi the experimental approach for site response studies has been adopted, resorting to techniques of Nakamura type studies on H/V ratio based on microtremor record. These studies have been conducted in five station shifting array mode, deploying Digital Triaxial Portable velocity sensor (L-4 3D) of 1 Hz. The array covered about 600 sites in NCT and adjoining terrain of NCR. Continuous waveform data at each site were collected for 1 h. Smooth common wave form of 120 s was sampled from a 1-h plot, and 30-s data were processed after smoothening for 300 cycles.

Analysis of data

As regards to analysis, Fourier transform spectra for each event (micro tremor as well as earthquake events) were computed using the SPEC program tagged with SEISAN-the earthquake analysis software [47]. The SPEC program has been customized to present a data set for generating parameter files (default: spec. par) and event file (default spec. inp.); the program produces one output file ('spec.out'). The FT spectra were generated for a time window of 30 s for each component of the individual microtremor for Nakamura type study; relative spectra of horizontal versus vertical components were produced for receiver function type study. The adopted program has facility of diagnosing Peak Frequency (fo), Peak Amplification (Ao) and obtaining frequency and amplification values at desired points on spectral ratio curves of horizontal to vertical component (QTS: Quasi-Transfer Spectrum) of microtremor along with error messages.

Thus, in mathematical terms, the seismic signal acquired in time series A(t) is converted into frequency domain A(f) using Fast Fourier Transform(FFT) technique. This process is applied to all the three component data, i.e., Vertical, EW and NS. The amplitude of horizontal component $A_E(f)$ and $A_N(f)$ is divided by vertical component $A_Z(f)$ to get the relative spectra of horizontal (EW & NS), and the root mean square average amplitude A(f) is calculated as

$$A(f) = \sqrt{\frac{\{A_E(f)/A_Z(f)\}^2 + \{A_N(f)/A_Z(f)\}^2}{2}}$$

From the average relative spectra, the amplification factor at various peak frequencies within 0.3–10 Hz is calculated with standard deviations. The maximum site amplification factors with corresponding peak frequencies are estimated for each site. On the basis of results of the study, Peak amplification and Peak frequency map of NCT Delhi have been generated. Seismic Hazard Microzonation report of NCT Delhi on 1:10,000 scale available in public domain in MoES website may be referred for detailed methodology and results.

A few typical spectral ratio curves of horizontal to vertical components (QTS: Quasi-Transfer Spectrum) of microtremor along with error messages obtained at few sites with different geological formations are shown in Fig. 2.2.

2.6.1.2 Ground Response Analysis of NCT Delhi Based on Numerical Methods

Analytical method as discussed in previous sections based on multiple reflection theory of S waves in horizontally layered deposits, referred to as 1D analysis of soil columns has been adopted. To consider nonlinear behavior of soil equivalent linear method as detailed in previous section, which uses an iterative procedure to adapt the soil parameters (i.e., rigidity and damping) to the actual strain it undergoes has been adopted. This technique allows computation of the seismic response of a given site providing spectral acceleration at different periods. Peak frequency and peak amplification can also be ascertained for the soil above reference rock (seismic/engineering rock).

The specific parameters required for such an analysis are (i) time history (input motion), (ii) index and dynamic properties of soil (soil type, unit weight, density, shear wave velocity), (iii) shear modulus reduction curves and damping curves, (iv)

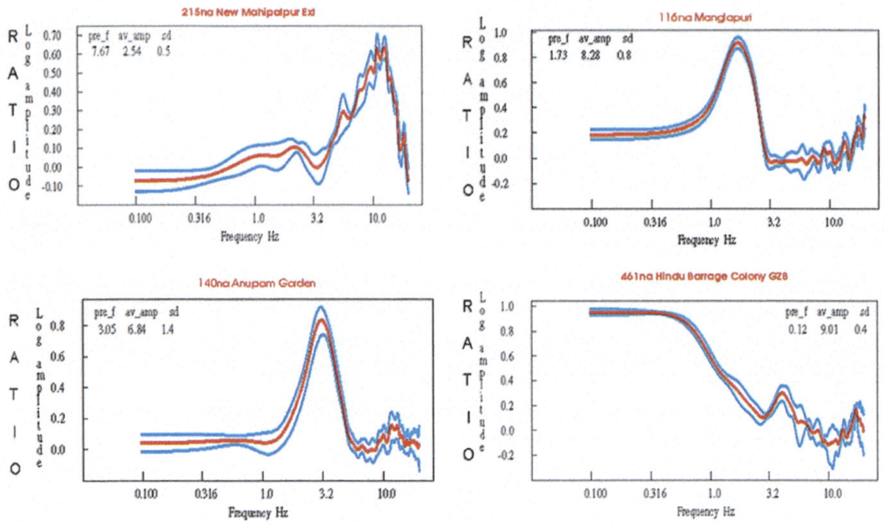

Fig. 2.2 Shows a few typical Spectral ratio curves of horizontal to vertical component (QTS: Quasi-Transfer Spectrum) of microtremor along with error messages for different types of soils of NCT Delhi

delineation of engineering bedrock and generation of subsurface soil model and (v) groundwater level maps.

To achieve these required input parameters, time history has been generated at 499 sites through Probabilistic Seismic Hazard analysis (PSHA), Geotechnical geophysical investigation has been carried out at these 499 sites spread over NCT Delhi for Index property of soil and delineation of engineering bedrock. Resonant column tests on representative samples of Sand, Silt and Clay of Delhi at different depths to determine shear modulus and damping ratio of soil under different confining pressure etc. have also been conducted.

Ground response analysis has been carried out at these 499 sites using a recently developed computer program DYNEQ, a computer Program for Dynamic response analysis of level ground by Equivalent linear method, version 3.25 [86], which is similar to SHAKE [98] and also incorporate the latest development such as frequency dependent characteristics, as damping due to scattering etc. has been used. Such nonlinear analysis requires a quantitative knowledge of the actual nonlinear material behavior, which can only be obtained by means of sophisticated laboratory tests. Some generic average curves have been proposed for different types of material as sand or clay [98, 99] and available in software databases, but the actual behavior of a given soil at a given site may strongly depart from these averages. Due to this Resonant Column test on representative samples of Sand, Silt and Clay of Delhi at different depths to determine shear modulus and damping ratio of soil under different confining pressure etc. have been conducted and used in place of generic average curves provided in the database of the software.

In DYNEQ program, some routine has been added at Earthquake Risk Evaluation Center (EREC) now merged with National Center of Seismology (NCS), Ministry of Earth Sciences, to plot products, such as response spectra, Frequency dependent amplification plot, etc. using single command, to make product generation more user friendly.

Products from Ground Response Study

(i) Peak ground acceleration at surface

Based on geotechnical/geophysical investigations and subsequently generated soil model and earthquake acceleration time histories at engineering bedrock, ground response analysis has been performed using DYNEQ software at 449 sites spread over NCT Delhi and Peak Ground Acceleration (PGA) at these sites have been obtained for:

(a) 2% probability of exceedance in 50 years (based MCE) for 5% damping and
(b) 10% probability of exceedance in 50 years (based on DBE) for 5% damping.

(ii) Site Amplification factor

At a particular site, when earthquake ground motion is subjected to, the soil column gets amplified due to site effects, which depends upon the material properties of the subsurface sediments layers, distribution soil type, surface topography and strength of the incoming earthquake ground motion. Amplification factor of the soil column is therefore the ratio of the strength of ground motion at the surface (top layer) to the strength of incoming earthquake ground motion actually applied to the bottom of the soil column (last layer).

Amplification factor has been evaluated for both input motions based on MCE and DBE at all 449 sites. Obviously, as amplification of soil column is the characteristic of soil at a particular site, the amplification factor for both periods of exceedances is the same. The amplification factors for NCT Delhi soil are less than 4.0 except at one site where amplification factor is 4.6.

(iii) Acceleration Response Spectra for different sites of NCT Delhi

Acceleration response spectra have been generated at 449 sites for 0 to 3 s, spread over NCT Delhi, represented by 75 topo sheets on 1: 10,000 scale covering all nine districts of NCT Delhi (i) North-West, Delhi (ii) North, Delhi (iii) North-East, Delhi (iv) East, Delhi (v) West Delhi (vi) South-West Delhi (vii) Central, Delhi (viii) New Delhi and (ix) South Delhi. Considering earthquakes for (a) 2% probability of exceedance in 50 years (MCE) for 5% damping and (b) 10% probability of exceedance in 50 years (DBE) for 5% damping. A typical response spectrum obtained from ground response study of NCT Delhi at a site is shown in Fig. 2.3 for MCE and Fig. 2.4 for DBE.

(iv) Variation of Peak Ground Acceleration Soil response with depth

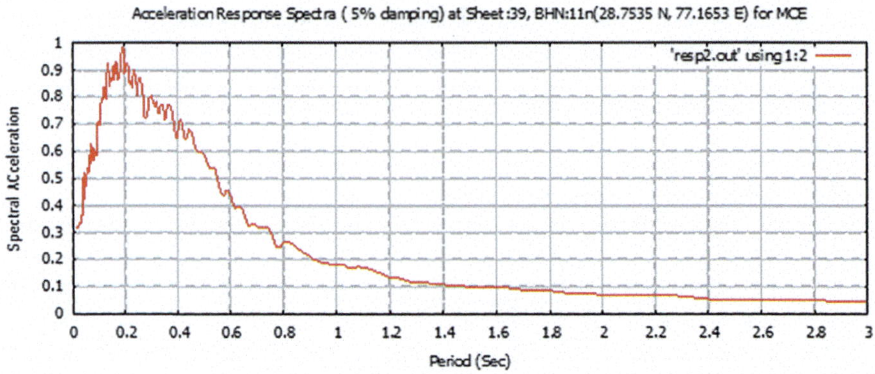

Fig. 2.3 Typical response spectra obtained from ground response study of NCT Delhi at a site, considering Maximum Considered Earthquake (MCE)

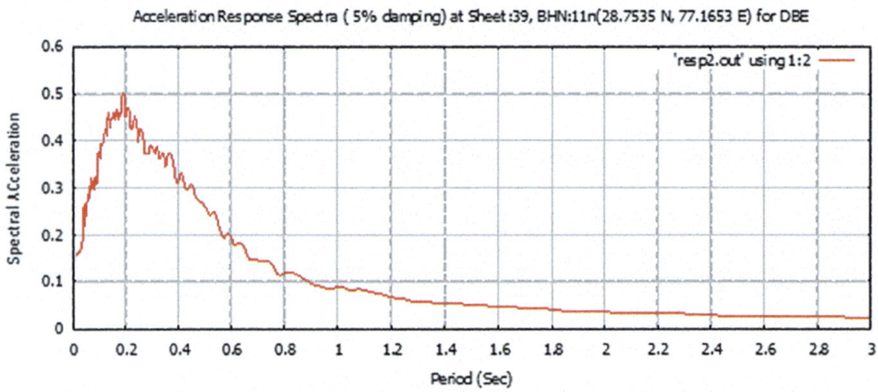

Fig. 2.4 A typical response spectra obtained from ground response study of NCT Delhi at a site considering Design Basis Earthquake (DBE)

Local geological conditions, characteristics of the lithological attributes and depth play a significant role in ground response and variation of PGA. The significant variation in PGA is observed close to the surface below a few meter depths from ground level. Variation of PGA with depth at a site considering MCE and DBE are shown in Fig. 2.5. It is seen from the figure that at this site significant variation of ground motion (PGA) is due to the top about 9 m of soil. At this site in case of MCE, PGA at surface is around 0.3 g, which reduces to 0.15 g at about 9 m depth below ground level.

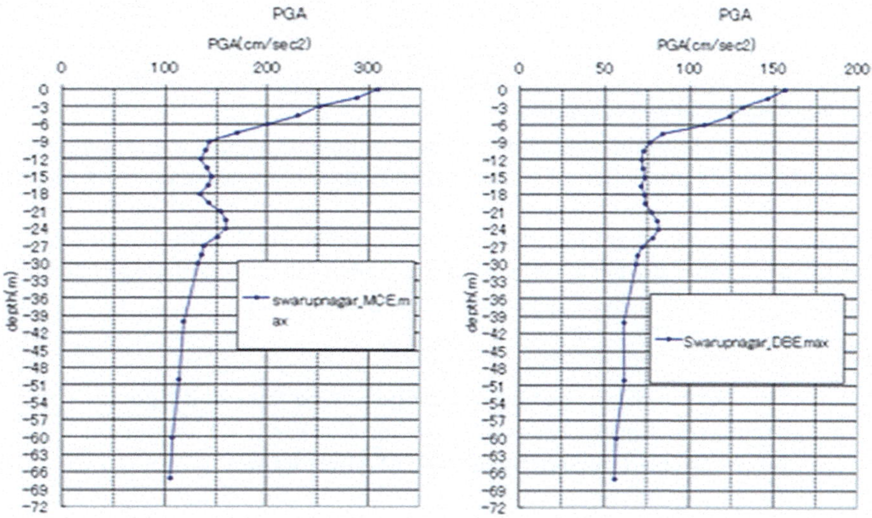

Fig. 2.5 Vertical distribution PGA at a site of NCT Delhi, considering MCE and DBE

2.7 Geological Dependence of Transfer Function Vis-a-Vis Ground Characterization

Geological dependence of ground characterization has been studied based on the results of site response study of NCT Delhi. Results of site response study have been separately laid on geological condition maps and reviewed in reference to shear wave characteristics of the sites. On-site spectral ratio curves (QTS: Quasi-Transfer Spectrum), pattern at different conditions is evaluated in terms of (a) nature/type of response curves; (b) variations in predominant frequency (o_f) (c) peak amplification; a strong geological dependence of transfer functions is suggestive. The details of these are enumerated in the following paragraphs.

2.7.1 Ground Characterization Based on Nature of Spectral Ratio Curves of Horizontal to Vertical Component (QTS: Quasi-Transfer Spectrum) of Microtremor

The microtremors to determine the dynamic characterization for microzonation purpose was used by Kanai et al. [57], Kanai and Tanaka [56]. Based on the microtremors survey, they have tried to characterize Japanese Building Code on the basis of a distinctive microtremor spectral pattern. Kanai & Tanaka used two methods for microzonation based on microtremor record; one was based on the largest amplitude in microns and the predominant period. However, precise differentiation between different soil typologies could not be done as (I) Site Periods overlap for different typologies. In fact, absolute amplitude is considered to be a poor discriminant of soil conditions.

However, now with the application of new techniques of spectral rationing and better scope of determination of site period and site-amplification, the microtremor studies can form an important input in microzonation and hazard assessment. Hence, an attempt has been made to characterize ground condition in NCT Delhi area in terms of nature of site response curve instead of only predominant period and amplification [78, 116]. Similar efforts have also been made in Jabalpur area.

2.7.1.1 Ground Characterization of Jabalpur Area Based on Nature of Spectral Ratio Curves of Horizontal to Vertical Component (QTS: Quasi-Transfer Spectrum) of Microtremor

During the process of Seismic Hazard Microzonation of Jabalpur, site response study based on H/V ratio was conducted jointly by India Meteorological Department (IMD) New Delhi, National Geophysical Research Institute (NGRI) Hyderabad, Central Building Research Institute (CBRI) Roorkee and Geological Survey of India (GSI) Nagpur [94]. Spectral ratio curves of horizontal to vertical component (QTS: Quasi-Transfer Spectrum) of microtremor obtained by site response study can broadly be classified in total 11 types considering the (a) Modal characteristics/Nature of mode/modes (unimodal, bimodal etc.), (b) resonant frequency window, (c) Peak frequency and (d) amplification ranges. This helps in characterizing the ground attributes and also understanding the transfer functions in a defined manner. Among these 11 types of curves, seven types are the basic clearly identifiable curves. Type nature and associated geological characteristics, typology of these seven basic curves are described in the following para and shown in Fig. 2.6.

Type ‘A’ curves (Fig. 2.6) are flat curves with low amplification up to 5 Hz followed by steep rise to higher frequencies. These characterize hard, compact granite and meta-volcano sediment areas. These denote undamped ground motions with high impedance and low amplification. This curve typology is found in granite forming exfoliation dome and tor and exposed Mahakoshals in structural ridges.

Type ‘B’ curves (Fig. 2.6) are essentially a unimodal response curve with flat response in initial frequencies as in Type A curves and well-defined frequencies as in Type A curves followed by a well-defined resonant window between 7 and 12 Hz. The amplification values are comparatively higher than the type A curves. This type of curve is associated with ground domains having thin (<10 m) alluvial cover on Mahakoshal meta-volcano-sediments. The response pattern characterizes the domain around Bhedaghat and in the southern part of the area.

Type ‘C’ curves (Fig. 2.6) are unimodal response curves with a broad resonant frequency window between 3 and 6 Hz with wide amplification variation ranging from 1 to 10. This curve characterizes the ranging thickness of alluvial cover

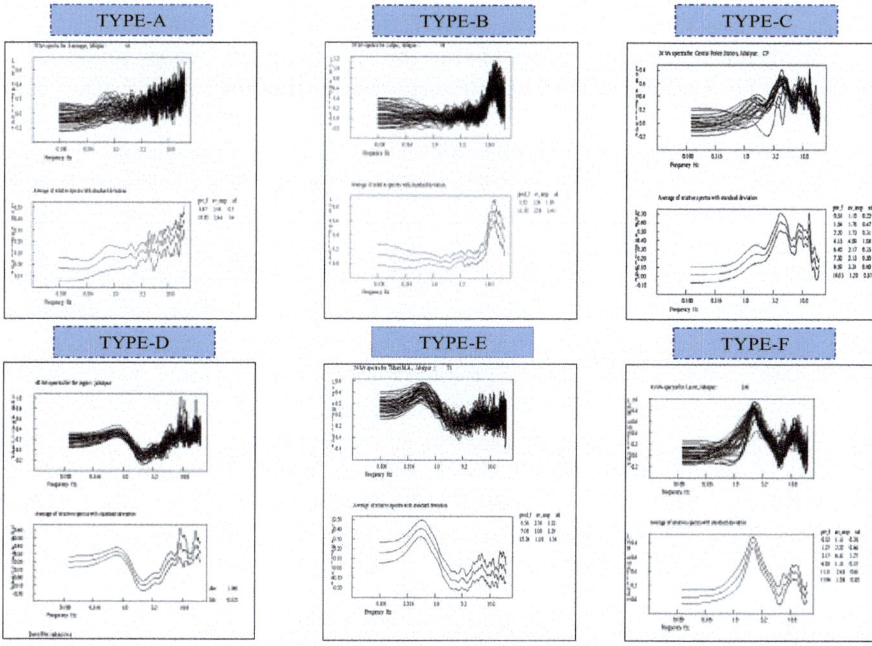

Fig. 2.6 The response curve types denoting different ground conditions of hard rock and alluvium fill terrains

on Mahakoshal meta-volcano sediments, Lameta formation and soft sediments of Gondwana formation.

Type 'D' and 'E' response curves (Fig. 2.6) are bimodal curves with two peak resonant windows at lower frequency range (<3 Hz) and the other peak frequency at about 6 Hz, separated by a trough of deamplification with its lowest at 3.2 Hz.

Type 'F' curves (Fig. 2.6) are unimodal with resonant frequency between 1 and 3 Hz with pronounced peak. The amplification ranges up to 5. This curve is very well constrained, based on the Nakamura type method. This type of curve is associated with thick alluvium cover of the order of 30 m as Mahakoshal meta-sediments and deeply weathered soft sediments of Gondwana formation.

An analysis and comparative studies of response curve type and associated ground conditions demonstrate that basically the curves depicting hard bed rocks are modified with their weathering, fracturing pedogenetic modifications and soil cover generation and extent of alluvial cover on the bed rock. On superposition of sediment curve of varying damping and impedance values, the Type A response curves get modified with a cognizable trend and pattern and reach to type F curves. Modified patterns are shown in Fig. 2.6.

2.7.1.2 Ground Characterization of NCT Delhi Area Based on Nature of Spectral Ratio Curves of Horizontal to Vertical Component (QTS: Quasi-Transfer Spectrum) of Microtremor

During the process of Seismic Hazard Microzonation of NCT Delhi, site response study based on H/V ratio was conducted by Earthquake Risk Evaluation Center (EREC), India Meteorological Department(IMD) New Delhi at about 600 sites spread over Delhi and environ. About 30 types of spectral ratio curves (QTS: Quasi-Transfer Spectrum) obtained by this study at NCT Delhi can broadly be classified in seven families of curves. A brief description of these is as follows. A few representative curves of each family are shown in Fig. 2.7. Figure 2.7 also shows, how nature of spectral ratio curves and associated ground conditions demonstrate that basically the curves depicting hard bed rocks (1 and 2 of first row, left to right) are modified with their weathering, fracturing pedogenetic modifications and soil cover generation and extent of alluvial cover on the bed rock. On superposition of sediment curve of varying damping and impedance values, the response curves get modified with a cognizable trend and pattern and reach to the curve 3rd of 4th row (right to left) showing deeper soil cover of the area.

Family I curves are characterized by flat/low amplification curves characterizing exposed rocks with thin weathered/sediment cover. Family I consists of three types of curves.

(i) Flat curves with no pronounced resonant window. Peak amplification (<2) at mid-level frequencies (>5 Hz)

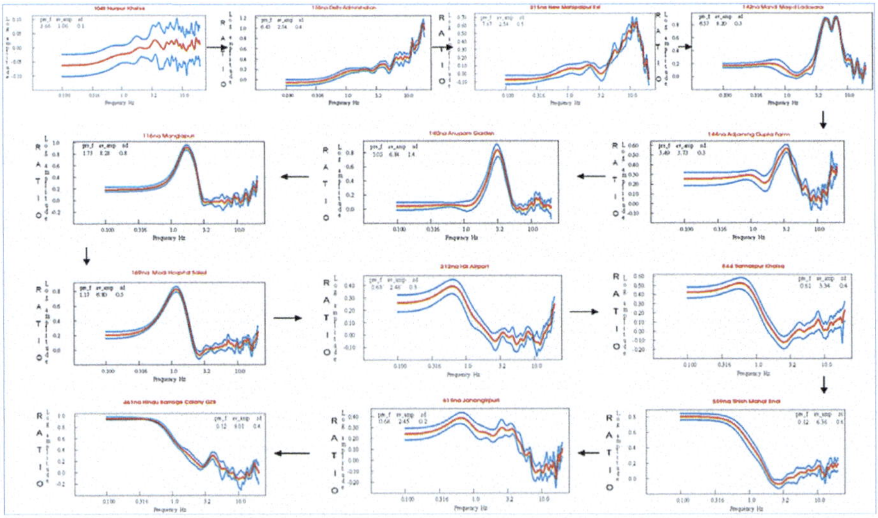

Fig. 2.7 Showing a few representative curves of each family and trend in modification of Spectral ratio curves of horizontal to vertical components of microtremor with change in nature of sediment cover

(ii) Flat response up to 3 Hz followed by low amplification level of 3 beyond 8 Hz
(iii) Flat response up to 3 Hz followed by gradual accentuation of amplification with poorly defined mode between 6 and 10 Hz.

Family II curves are characterized by strongly unimodal curves characterizing thin (<20 m) sediment cover proximal to Ridge and Central Delhi, with strong resonant mode between 3.2 and 10 Hz, showing amplification range between 2 and 6 with high Impedance contrast.

Family III curves are characterized by strongly unimodal, with well-defined resonant windows characterizing moderately thick (>30 m) sediment cover on bedrock with strong impedance contrast at the base. This family of curves consists of five types of curves.

(i) Unimodal curve with resonant window between 2.5 and 5 Hz, showing variable amplification (3–4) in the central Delhi area and more than 8 in north Delhi, west of the ridge. Accentuation in amplification at higher frequencies is noted.
(ii) Unimodal curve with well-defined resonant window between 2.5 and 5 Hz with preceding and succeeding flat response. The curve is typical of domain around ridge and central Delhi area with sediment cover thickness 30–100 m.
(iii) Unimodal curve with moderately wide resonant band between 2 and 4 Hz followed by a de-amplification window between 4 and 8 Hz and strong accentuation in amplification at higher frequencies. The curve co-occurs with other curves of the family and characterizes a domain around Ridge and central Delhi with sediment cover thickness 30–100 m.
(iv) Unimodal curve with 1 Hz with resonant band between 2 and 4 Hz. The curve is typical of domain around ridge and central Delhi area with sediment cover thickness 30–100 m.
(v) Strong mode around 1 Hz with resonant band between 0.3 and 2 Hz. The amplification is mostly around 3–4 units with exceptional up to 6. The curve characterizes moderately thick (>30 m) sediment cover on quartzite with strong impedance contrast at the base.

Family IV curves are characterized by moderate resonant windows at low frequency (<1 Hz) characterizing thick sediment cover to the west of Ridge. This family consists of two types of curves.

(i) Moderate resonant window at low frequencies (<1 Hz) followed by de-amplification around 2.5–3.0 Hz. The curve characterizes thick sediment cover mostly in the Southwestern and northeastern domain.
(ii) Moderate resonant window at low frequencies (<1 Hz) followed by de-amplification around 2.5–3.0 Hz. The curve characterizes thick sediment cover mostly in the southwestern and northeastern domain.

Family V curves are characterized by mostly unimodal with subdued mode at <1 Hz followed by de-amplification trough around 3 Hz frequencies; occasionally bi-modal with low secondary mode. The curves characterize deep sediment cover containing strong damping layer/layers. This family of curves consists of seven types of curves.

(i) Weak resonant window at <1 Hz frequency followed by steep attenuation in amplification dropping down to <1 (de-amplification) at 3 Hz followed by flat response at higher frequencies. The curve characterizes sediment cover with strong damping at moderate frequency.
(ii) Weak resonant window at <1 Hz frequency followed by steep attenuation in amplification dropping down to <1 (de-amplification) at 3 Hz followed by flat response at higher frequencies. The curve characterizes thick sediment cover with strong damping at moderate frequency.
(iii) Weak resonant mode with peak frequency between 0.56 and 0.76 Hz followed by a prominent de-amplification through around 3 Hz and subsequently secondary mode around 6 Hz. The curve characterizes thick sediment cover with high damping at frequencies around 3 Hz.
(iv) Weak resonant mode with peak frequency between 0.56 and 0.76 Hz followed by a prominent de-amplification through around 3 Hz and subsequently secondary mode around 6 Hz. The curve characterizes thick sediment cover with high damping at frequencies around 3 Hz.
(v) A weak resonant window with peak frequency <0.71 Hz followed by wide de-amplification window around 3 Hz; the amplification gradually accentuates at further higher frequencies attaining high values (~2) only beyond 10 Hz. The curve has typical Z shape and characterizes thick sediment cover around Najafgarh and in trans Yamuna.
(vi) A weak resonant window with peak frequency <0.71 Hz followed by wide de-amplification window around 3 Hz; the amplification gradually accentuates at further higher frequencies attaining high values (2) only beyond 10 Hz. The curve has typical Z shape and characterizes thick sediment cover south Najafgarh domain and locally in trans Yamuna area.
(vii) A weak resonant window with peak frequency <0.71 Hz followed by wide de-amplification window around 3 Hz; the amplification gradually accentuates at further higher frequencies attaining high values (>4) only beyond 10 Hz. The curve has typical Z shape and characterizes thick sediment cover around Najafgarh and in trans Yamuna.

Family VI curves are characterized by Flat curves at low frequencies, well-defined de-amplification window; occasionally show bi-modality. These curves characterize deep sediment cover containing clay-rich damping layer/layers. This family of curve consists of eight types of curves.

(i) Plateau of amplification at low frequencies (<1 Hz) followed by rapid attenuation to a low (<1) at 3 Hz; flat or weak attenuation in amplification showing V-shaped response curve
(ii) Plateau of amplification at low frequencies (<1 Hz) followed by rapid attenuation to a low (<1) at 3 Hz; flat or weak attenuation in amplification showing V-shaped response curve

(iii) Bimodal with a prominent amplification plateau at low frequencies (<1 Hz) and secondary window of moderate amplification (~2) at mid-level frequencies around 6 Hz
(iv) Bimodal with a prominent amplification plateau at low frequencies (<1 Hz) and secondary window of moderate amplification (~2) at mid-level frequencies around 6 Hz
(v) Sickle-shaped curve with high amplification plateau of frequencies up to 1 Hz followed by attenuation in amplification dropping to <1 around 3 Hz. A prominent U shaped low amplification window between 3 and 7 Hz followed by recovery of amplification at higher frequencies
(vi) Sickle-shaped curve with high amplification plateau of frequencies up to 1 Hz followed by attenuation in amplification dropping to <1 around 3 Hz. A prominent U shaped low amplification window between 3 and 7 Hz followed by recovery of amplification at higher frequencies
(vii) Sickle-shaped curve characterized by flat plateau amplification at low frequencies (<1 Hz) followed by V or U shaped low amplification trough between 2.5 and 5.0 Hz; occasionally secondary window around 7.5 Hz followed by strong recovery in amplification at higher frequencies
(viii) Sickle-shaped curve characterized by flat plateau amplification at low frequencies (<1 Hz) followed by V- or U-shaped low amplification trough between 2.5 and 5.0 Hz; occasionally secondary window around 7.5 Hz followed by strong recovery in amplification at higher frequencies.

Family VII curves are characterized by polymodal curves characterizing Khadar and Bela areas proximal to river Yamuna. This family consists of four types of curve.

(i) Polymodal curves with moderate resonant window at low frequencies (<1.4 Hz) followed by cyclic attenuation in amplification dropping to de-amplification beyond 7–8 Hz followed by poor recovery of amplification. These curves characterize Khadar area with likely late Pleistocene sediment.
(ii) Polymodal curves with moderate resonant window at low frequencies (<1.4 Hz) followed by cyclic attenuation in amplification dropping to de-amplification beyond 7–8 Hz followed by poor recovery of amplification. These curves characterize Khadar area with likely late pleistocene sediment.
(iii) Flat plateau of amplitude (2.13–3.98) followed by cyclic attenuation with de-amplification around 9–10 Hz followed by poor recovery in amplification. These curves characterize Bela area either in palaeochannel or active floodplains.
(iv) Flat plateau of amplitude (2.13–3.98) followed by cyclic attenuation with de-amplification around 9–10 Hz followed by poor recovery in amplification. These curves characterize Bela area either in palaeochannel or active flood plains.

2.7.2 Ground Characterization Based on Peak Frequency (f_o), Peak Amplification and Nature of Spectral Ratio Curves of Horizontal to Vertical Component (QTS: Quasi-Transfer Spectrum) of Microtremor

During the process of Seismic Hazard Microzonation of NCT Delhi, site response study based on H/V ratio was conducted by Earthquake Risk Evaluation Center (EREC), India Meteorological Department (IMD) New Delhi at about 600 sites spread over Delhi and environ. Based on peak amplification, Peak frequency and nature of Spectral ratio curves derived at these sites, NCT Delhi can be classified in three structural blocks (I) ST-I Central Delhi Block, (ii) ST-II East Delhi and (iii) ST-III Western Delhi blocks. Striking difference is seen in the pattern of resonant frequency in these three structural blocks. A brief description of these blocks is presented as follows.

The ST-I (Central Delhi Block) can be divided into sub-blocks IA (Central Delhi ridge triangle) and IB (Chhatarpur basin) have highly variable peak frequency scenarios.

The sub-block IA can further be divided into two geomorphological zones, i.e., structural hills (Kohi) and alluvial plains (Khandarant). The structural hills (Kohi) consist of the units of ridge ambience with positive relief (+60 m max) exposing hard, fractured and massive quartzite; CS > 20Mpa. Locally weathered and reclaimed with soft soil. Characterized by flat response curves of high fundamental frequency (>5 Hz) and low amplification (<1) and very low vulnerability; anticipated $V_S{}^{30}$—about 1000 m/s. The alluvial plains (Khandarant) consist of **(a)** Shallow rock sites with alluvial and colluvial cover consisting sporadic boulders with sandy matrix at the foothills, on quartzite with high impedance contrast at shallow depth. Highly dissected characterized by well-defined uni-modal response curve with high frequency (>3), moderate (<2) amplification and very low vulnerability and no liquefaction**.** Extensive ground modifications –Khandarat; **a**nticipated $V_S{}^{30}$—about 700 m/s. **(b)** Shallow soil sites with level ground of alluvial fill (sand-silt–clay) as above with base rock quartzite of high impedance placed at depths (>30 m). Characterized by uni-modal curves with high amplification at fundamental mode, around 1 Hz with amplification ranging up to 7. Extensive ground modifications—Khandarat. $V_S{}^{30}$ about 350 m/s.

The Sub-block IB can further be divided in two Geomorphological zones (i) Chhatarpur Basin with shallow to deep soil sites: structurally controlled oval-shaped basin (width/depth ratio 0.30) with fill consisting of semi-consolidated sand-silt–clay of alluvial—lacustrine-aeolian origin; dissected topography at margin and level ground in basin spread. Characterized by a uni-modal response curve with strong peak at around 1 Hz; strong 'Basin effect' anticipated. Variable $V_S{}^{30}$: 300 m–700 m/s. Bedrock PGA (ii) Weathered rock sites: Fractured Quartzite associated with argillites and structurally controlled weathering. Characterized by bimodal response curve with fundamental frequency around 3 Hz; anticipated V_S^{30}—about 900 m/s.

The ST-II (East Delhi Block) consists of two sub-blocks. (i) Sub-block IIA characterizes by deep/very soft soil sites—active regime of Jamuna river: Bela domain bound by guide bunds confining active river regime, consists of cohesionless fine micaceous sand forming bars and channel deposits. Characterized by distinct curve type with low frequency (<0.6 Hz) and high amplification (>4); Very high vulnerability and Liquefaction susceptibility. Anticipated $V_S30 < 200$ m/s. (ii) Sub-block IIA characterizes deep soil sites (Bangar Plains): level ground of alluvial plains with signatures of river avulsion—palaeo channels, meander scroll, oxbow lakes etc.; lately under extensive reclamation and urbanization. Characterized by distinct curve typology, with low frequency, high amplification and variable vulnerability ranging from low to very high. Liquefaction susceptibility is high. V_S^{30} −300 m/s.

The ST-III (Western Delhi Block) consists of two sub-blocks. (I) Sub-block IIIA characterized by shallow to deep soil sites with steeply sloping bedrock regime. The plains west of Delhi ridge, consisting of variable thickness of pervious, collegial and lacustrine sediments over quartzite with many locations of bedrock inliers; characterized by steeply sloping bedrock regime with a maximum of 300 m. The zone has a unique bimodal curve typology of low frequency (<1) and moderate (<4) amplifications and moderate vulnerability. The unit is susceptible to basin margin effects demanding 2D response analysis. Variable anticipated V_S^{30}: 300 m–800 m/s. (ii) Sub-block IIIA, on the basis of distinct geomorphological features can further be classified (a) SW Debar plain with very deep soil sites (Debar Plains): level grounds of Debar plains that partially emerged after draining of Najafgarh lake consist of lacustrine sediment-silt–clay with Kankar beds with local aeolian sediment cover. Characterized by distinct curve typology with low frequency (0.6–0.8 Hz) and moderate to high amplifications and moderate to high vulnerability. V_S^{30} −300 m. (b) NW Dabar plain with deep soil sites (Bangar Plains): The upland plains under intensive irrigation, consist of very deep sequence of older alluvium with local aeolian sands characterized by distinct curve typology of very low frequency (0.2–0.6 Hz), high amplification and high vulnerability. V_S^{30} −250 m/s.

2.8 Application of H/V Ratio-Based Techniques for the Delineation of Geological Attributes

It is seen from the discussion of the previous sections that Peak frequency, Peak amplification and particularly nature of spectral ratio curves of horizontal to vertical component (QTS: Quasi-Transfer Spectrum) obtained from microtremor-based Nakamura type study is quiet efficient to identify geological variations and their characteristics. Thus, this study may be very helpful to delineate boundaries of variable geological attributes of sites. However, for variable domains extensive observations by deploying equipment in small grid spacing may be required. It is, therefore, suggested that:

(i) As a first step Nakamura type study by collecting one-hour data at minimum possible grid needs to be attempted for delineating broad outlines of ground condition of domains and their boundaries.
(ii) On identifying domains of similar characteristics and their boundaries, in the second phase of study, a network for long duration as per the seismicity of the domain needs to be set up for collecting event data originating from all possible directions for further receiver function type study.

Acknowledgements Thanks to all team members associated with fieldwork of site response study of Indian cities, conducted by Earthquake Risk Evaluation Center (EREC), India Meteorological Department (IMD), Ministry of Earth Sciences, Govt. of India, particularly with Jabalpur, Guwahati, NCT Delhi on 1:10000 scale.

Dedication This little Chapter is affectionately dedicated to my dear Son 'Vishesh' Who left us for his heavenly abode on 11 May 2021 due to pandemic COVID-19. His brief life of 38 years on this planet has been fully devoted to the welfare of the society, as he preferred to work in the social sector. He was always encouraging me to keep engaged in social services through his recently created platform of KODE foundation. Recently, we jointly submitted a research paper for the coming Fifth World Congress on Disaster Management (WCDM-5) November 24–27, 2021. Acceptance of writing this chapter of the book was received around the same time and I would like to dedicate him.

References

1. Aki K (1957) Space and time spectra of stationery stochastic waves special reference to microtremors. Bull Earthq Res Inst, 415–456
2. Aki K (1969) (1969) Analysis of the seismic coda of local earthquakes as scattered waves. J Geophys Res 74(2):615–631
3. Phillips WS, Aki K (1986) Site amplification of coda waves from local earthquakes in central California. Bull Seismol Soc Am 76(3):627–648
4. Aki K (1988) Local site effects on Strong Ground motion. In: Earthquake engineering and soil dynamics II—recent advances in ground motion evaluation, June 27–30, Park City, Utah
5. Aki K (1993) Local site effects on weak and strong ground motion. Tectonophysics 218:93–111
6. Aji K, Chouet B (1975) Origin of coda waves: source attenuation and scattering effects. J Geophys Resa 80:3322–3342
7. Aki K, Irikura K (1991) Charecterization and mapping of earthquake shaking for seismic zonation. Proceedings of the fourth international conference on seismic zonation, August 25–29, Stanford, California, E.E.R.I (editor), Oakland CA, 1, 61–110
8. Aki K, Richards PG (1980) Quantitative seismology, vol I, W.H. Freeman, San Fransisco
9. Andrews DJ (1986) 'Objective determination of source parameters and similarity of earthquakes of different size. In: Das S, Boatwright J, Scholz CH (eds) Earthquake Source Mechanics'. American Geophysical Union, Washington, D.C., pp 259–268

10. Bard PY (1994) Effect of surface geology on strong ground motion: recent results and remaining issues. Proceedings 10th European conference Earthquick Engineering, Vienna, Austria, 1, pp 305–323
11. Bard P-Y (1998) Microtremor measurements: a tool for site effect estimation?, State-of-the art paper. In: Irikura K, Kudo K, Okada H, Satasini T (eds), Effects of surface geology on seismic motion, Yokohama, Rotterdam, Balkema, vol 3, pp 1251–1279
12. Bard PY (1999) Local effects on strong ground motion: physical basis and estimation methods in view of microzoning studies. Proceedings of the advances study course "Seismotectonic and Microzonation techniques in earthquake engineering: integrated training in earthquake risk reduction practices", pp 75–124
13. Bard PY (2000) Lecture note and Exercise note on seismology; seismic data analysis, hazard assessment and risk analysis
14. Bard PY (2004) Guidelines for the implementation of the H/V spectral ratio technique on ambient vibrations: measurements, processing and interpretation. SESAME European Research Project (WP12—Deliverable D23, 12), EVG1-CT-2000–00026, pp. 1–62, Sesame-Team
15. Bard PY, Duval AM, Lebrun B, Lachet C, Riepl J, Hatzfeld (1997) Reliability of the H/V technique for site effects measurements: an experimental assessment. Seventh international conference on soil dynamic and earthquake engineering, Istanbul, July 19–24, 1997
16. Benites R, Olsen KB (2005) Modeling strong ground motion in the wellington metropolitan area, New Zealand. Bull Seism Soc Am 97:2180–2196
17. Bindi D, Parolai S, Spallarossa S, Cattaneo M (1998) Site effects and H/V methods: comparison of different approaches, Fourth EEGS Meeting
18. Borcherdt RD (1970) Effects of local geology on ground motion near San Fransisco Bay. Bull Seism Soc Am 60:29–61
19. Borcherdt RD, Gibbs JF, Fumal TE (1979) Progress on ground motion predictions for the San Francisco Bay region, California, in U.S. Geol Surv Cir 807:13–35
20. Borcherdt RD, Glassmoyer G, Andrews M, Cranswick E (1989) Effect of site conditions on ground motion and damage, First Published August 1, 1989 Research Article https://doi.org/10.1193/1.1585233
21. Boatwright J, Fletcher JB, Fumal TE (1991) A general inversion scheme for source, site, and propagation characteristics using multiply recorded sets of moderate-size earthquakes. Bull Seism Soc Am 81:1754–1782
22. Bonilla LF, Steidl JH, Lindley GT, Tumarkin AG, Archuleta RJ (1997) Site amplification in the San Fernando valley, California: variability of site effect estimation using S-wave, coda, and H/V methods. BSSA 87–3:710–730
23. Bullen KE (1965) An Introduction to the theory of seismology. Cambridge University Press, Cambridge
24. Chavez-Garcia FJ, Sanchez LR, Hatzefld D (1996) Topographic Site Effects and HVSR. A comparison between observations and Theory. Bull Seismol Soc Am 89(3):1559–1573
25. Cornou C (1998) Etudes theoriques et numeriques sur la methode de Nakamura-Nogoshi, Memoire de Diplome d'Ingenieur, EOST Strasbourg, LGIT Grenoble, 132 pages (in French)
26. Duval A-M (1994) Détermination de la réponse d'un site aux séismes à l'aide du bruit de fond: Evaluation expérimentale. Thèse de Doctorat, Paris VI, 265 pp. (in French)
27. Duval AM, Bard PY, Meneroud JP, Vidal S (1994) Usefulness of microtremor measurements for site effect studies, Proceedings of the 10th European conference on earthquake engineering, Vienna, Austria, Balkema, Duma Ed., I, pp 521–528
28. Duval AM, Bard PY, Meneroud JP, Vidal S (1995) Mapping site effects with microtremors Proceedings of the fifth international conference on seismic zonation, October 17–19, 1995, Nice France II, pp 1522–1529
29. ESG98 (1998) Proceedings of the ESG'98 international symposium, The effects of surface geology on ground motion, A.A. Balkema Publishers, (Irikua K, Kudo KK, Okada H, Sasatani T eds)

30. Faccioli E (1991) Seismic amplification in the presence of geological and topographic irregularities. Proceedings of II international conference on recent advances in geotechnical earthquake engineering and soil dynamics, St. Louis, Missouri, vol II, pp 1779–1797
31. Faccioli E (1996) Site effects in the Eurocode 8. Proceedings 11th World conference earthquake engineering, Acapulco, CDROM, paper 2043
32. Fäh D, Rüttener E, Noack T, Kruspan P (1997) Microzonation of the City of Basel. J Seismolog 1:87–102
33. Field EH (1996) Spectral amplification in sediment filled valley exhibiting clear basin edge-induced waves. Bull Seismol Soc Amer 86:991–1005
34. Field EH, Jacob KH, Hough SE (1992) Earthquake weak motion estimation: a weak motion case study. Bull Seismo Soc Am 82:2283–2307
35. Field EH, Clement AC, Jacob KH, Aharonian V, Hough SE, Frieberg PA, Babaian TO, Kara-Petian SS, Hovanessian SM, Abramian HA (1995) Earthquake site response in Giumri (Formerly Leninakan), Armenia using ambient noise observations. Bull Seism Soc 85:349–353
36. Field EH, Jacob KH (1993) The theoretical response of sedimentary layers to ambient seismic noise. Geophys Res. Letters 20–24:2925–2928
37. Field EH, Jacob KH (1995) A comparison and test of various site response estimation techniques, including three that are non reference-site dependent. Bull Seism Soc Am 85:1127–1143
38. Finn WD, Liam (1991) Geotechnical engineering aspects of seismic microzonation, Proceedings of the fourth international conference on seismic zonation, August 25–29, Stanford, California, vol 1, pp 199–250
39. Gitterman Y, Zaslavsky Y, Shapira A, Shtivelman V (1996) Empirical site response evaluations: case studies in Israel. Soil Dyn Earthq Eng 15:447–463
40. Gutenberg (1956) Effects of ground on earthquake motion. Trans Am Geoph Union, vol Tl, 1956, pp. 757–760
41. Gutenberg B (1957) Effects of ground on earthquake motion 11. Soc Am 1 47:221–250
42. Harsh KG, Chandha RK, Rai MN, Narayana BL, Mandal P, Ravi Kumar M (1997) The Jabalpur earthquake of May 22, 1997. J Geolog Soc India 50(1):85–91
43. Harmsen SC (1997) Determination of Site amplification in the Los Angeles urban area from inversion of strong motion records. Bull Seism Soc Am 87:866–887
44. Hartzell S, Leeds A, Frankel A, Michael J (1996) Site response for urban Los Angeles using aftershock of the Northridge Earthquake. Bull Seism Soc Am 86:S168–S192
45. Hartzell S, Cranswick E, Frankel A, Carver D, Meremonte M (1997)Variability of site response in the Los Angeles urban area. Bull Seism Soc Am 87:1377–1400
46. Haskell NA (1953) The dispersion of surface waves on multilayered media. Bull Seismol Soc Am 43(1):17–34
47. Havskov J, Ottemoeller L (2001) Seisan software manual. University of Bergen, Norway
48. Holzer Thomas L (1994) Loma Prieta damage largely attributed to enhanced ground shaking. EoS 75(26):299–301
49. Idriss IM (1990) Response of soft soil sites during earthquakes. Proceedings H. Bolton seed memorial symposium, pp 273–290
50. Idriss IM, Sun JI (1992) SHAKE91: A computer program for conducting equivalent linear seismic response analyses of horizontally layered soil deposits. University of California, Davis, California, User's Guide, p 13
51. Jackson DD, Aki K, Cornell CA, Dieterich JH, Henyey TL, Mahdyiar M, Schwartz D, Ward SN (1995) Seismic hazards in Southern California: probable earthquakes. Bull Seis Soc Amer 85:379–439, 60 p
52. Jain SK, Murty CVR, Arlekar JN, Sinha R, Goyal A, Jain CK (1997) some observations on engineering aspects of the jabalpur earthquake of 22 May (1997). EERI special earthquake report, EERI Newsletter, vol 32, No 2
53. Joyner WB, Fumal TE (1985) Predictive mapping of earthquake ground motion. U.S. Geol Sur Profess Pap 1360:203–220

54. Kagami H, Duke CM, Liang GC, Ohata Y (1986) Observation of 1 to 5 second micro tremors and their application to earthquake engineeringIII-a two dimensional study of site effects in the San Fernando Valley. Bull Seism Soc Am 76:1801–1812
55. Kanai, K. (1957) Semi-emperical formula for the seismic characterstics of the ground. Bull Earthq Res Inst Tokyo Univ 35(308):325
56. Kanai K, Tanakai T (1961) On Microtremor VIII. Bull Earthq Res Inst Univ Tokyo 39
57. Kanai K, Tanaka T, Oada K (1954) On Microtremors. Bull Earthq Res Inst Unv Tokyo 32
58. Kanno K, Ohmachi T (1998) Ground motion characteristics estimated from spectral ratio between horizontal and vertcal components of micrtremors. Bull Seism Soc Am 88:228–241
59. Kato K, K. aki and M. Takemura, (1995) Site amplification from coda waves; Validation and application to S-wave site response. Bull Seism Soc Am 85:467–477
60. Kawase H, Keiiti A (1989) A study of the response of a soft basin for incident S,P, and Rayleige waves with special reference to the long duration observed in Maxico city. Bull Seismol Soc Amer 79(5):1361–1382
61. Kobayashi H, Midorikawa S (1986) Study of Site effect in Mexico city using microtremors. Proceedings of national symposium on earthquake engineering, vol 7, pp 355–360
62. Kobayashi H, Seo K, Miclorikawa A, Katakoa S (1996) " Measurenment of Microtremors in and around Mexico, DF" Part I in Report of Seismic Microzonation studies of the Mexico earthquake September (1995); Tokyo, The graduate school of Nagatsuta Inst. of Technology
63. Kramer SL (2003) Geotechnical earthquake engineering. Pearson Education Publication
64. Kudo K (1995) Practical estimates of site response, state-of-the-art report. Proceedings of the fifth international conference on seismic zonation, October 17–19, Nice, France, Ouest Editions Nantes, vol 3, pp 1878–1907
65. Lanchet C, Bard PY (1994) Numerical and theoretical investigations on the possibilities and limitations of Nakamura's technique. J Phys Earth 42:377–397
66. Lachet C, Hatzfeld D, Bard PY, Theodulidis N, Papaioannou C, Savvaidis A (1996) Site effects and microzonation in the city of Thessalonki (Greece): comparision of different approaches. Bull Seism Soc Am 86:1692–1703
67. Langston CA (1979) Structure under Mount Rainier, Washington, inferred from teleseismic body waves. J Geophys Res 84:4749–4762
68. Lebrun B (1997) Les Effects de site: etude experimentale et simulation de tros configurations. These de Doctorat de l' Universite Joseph Fourier – Grenoble I, November 27, 1997, 208 pp (in French)
69. Lermo J, Chavez-Garcia FJ (1993) Site effect evaluation using spectral ratios with only one station. Bull Seismo Soc Am 83:1574–1594
70. Lermo J, Chavez-Garcia FJ (1994) Are microtremors useful in site response evaluation? Bull Seism Soc Am 83:1574–1594
71. Liu L, Chen QF, Wang W, Rohrbach E (2014) Ambient noise as the new source for urban engineering seismology and earthquake engineering: a case study from Beijing metropolitan area. Earthq Sci 27:89–100
72. Luo GC, Liu LB, Qi C, Chen QF, Chen YP (2011) (2011); Structural response analysis of a reinforced concrete building based on excitation of microtremors and passing subway trains. Chin J Geophys 54:2708–2715
73. Madabhushi SPG, Haigh SK (2005) The Bhuj, India Earthquake of 26th January 2001 A Field Report, EEFIT 2005 January 2005 ISBN 0 901297 37 2, University of Cambridge Earthquake Engineering Chapter 3 Geotechnical Aspects of the Bhuj Earthquake Dr S P G Madabhushi, Mr Dinesh Patel Dr S K Haigh
74. Malagnini L, Tricarico P, Rovelli A, Herrmann RB, Opice S, Biella G, de Franco R (1996) Explosion, earthquake, and ambient noise recording in a Pliocene sediment-filled valley: Inferences on seismic response properties by reference- and non-reference-site techniques. Bull Seism Soc Am 86(3):670–682
75. Margheriti L, Wennerberg L, Boatwright J (1994) A comparision of coda and S-wave spectral ratios as estimates of site response in the southern San Francisco Bay Area. Bull Seism Soc Am 84:1815–1830

76. Mandal HS, Khan PK, Shukla AK (2012) Soil responses near Delhi ridge and adjacent region in Greater Delhi during incidence of a local earthquake. J Nat Hazards; published on line 1–26, April 04, 2012. Reference No: Nat. Hazards. https://doi.org/10.1007/s11069-012-0098-4
77. Milne J (1898) Seismology 1ˢt Ed., Kegan Paul, Trench, Trube, London
78. Mishra PS, Shukla AK, Prakash R, Singh RK, Pandey AP, Ray TK, Bhatnagar AK (2006) Issues and problems of seismic Microzonation of national capital territory of Delhi 2006; Proceedings on" Himalaya Earthquake: A fresh appraisal", pp 63–66
79. Mucciarelli M, Contri P, Monachesi G, Calvano G, Gallipoli (2001) An empirical method to assess the seismic vulnerability of existing buildings using the HVSR technique. Pure Appl Geophys 158(2635–2647):2001
80. Nakamura Y (1989) A method for dynamic characteristics estimation of subsurface using microtremor on the ground surface. Quart Reptort of R.T.R.I., Japan, 30, 25–33
81. Nakamura Y (1996) Real time information systems for hazard mitigation, X World Conference Eathquake Engineering. Acapulco, #2134, Elsevier Science Ltd
82. Nakamura Y (2019) What is the Nakamuara method? Seismol Res Lett 90(4):1437–1443. https://doi.org/10.1785/0220180376
83. Narayan JP, Sharma ML (2004) Effect of local Geology on damage severity during Bhuj India Earthquake.13 World Conference on Earthquake Engineering , Vancouver, B.C, Canada, August 1–6, 200 Paper No. 2042
84. Nath SK, thingbaijam KKS, Raj A, Shukla K, Pal I, Nandi DR, Yadav MK, Bansal BK, Dasgupta S, Majumdar K, Kayal JR, Shukla AK, Dev SK, Pathak J, Hazarika PJ, Paul DK (2007) Seismic scenario of Guwahati city (2007); Presented in international workshop on earthquake hazard mitigation "EHAM-2007, IIT Guwahati. 7–8 December 2007
85. Nath SK, Raj A, Sharma J, Thingbaijam KKS, Kumar A, Nandi DR, Yadav MK, Dasgupta S, Majumdar K, Kayal JR, Shukla AK, Deb SK, Pathak J, Hazarika PJ, Paul DK, Bansal BK (2008) Site amplification, Qs, and source parameterization in guwahati region from seismic and geotechnical analysis. Seismol Res Lett 79:526–539
86. Nozomu Y, Suetomi I (2004) DYNEQ A computer program for dynamic analysis of level ground based on equivalent linear method; See discussions, stats, and author profiles for this publication at: https://www.researchgate.net/publication/281466287
87. Nozomu Y (2014) DYNEQ A computer program for Dynamic response analysis of level ground by Equivalent linear method Version 3.35 (December, 2014) Revised from Original version (May 1995) Tohoku Gakuin University
88. Ohmachi T, Nakamura Y, Toshnawa T (1991) Ground motion characteristics in the San Francisco Bay area detected by micro tremor measurements. In: Proceedings of the second international conference on recent advances in geotechnical earthquake engineering and soil dynamics, March 11–15, St. Louis, Missouri, S. Prakash (Editor), Univ. of Missouri-Rolla, 1643–1648
89. Okada HT, Matsushima S, Hidaka E (1987) Comparision of spatial autocorrelation method and frequency-wave number spectral method of estimating the phase velocity of Rayleigh waves in long period micro remors. Geophys Bull Hokkaido Univ 49:53–62
90. Olsen KB, Archuleta RJ (1996) 3D simulation of earthquakes on the Los Angeles fault system. Bull Seism Soc Am 86:575–596
91. Omote S, Srivastava HN, Drakopoulos J, Tokumitsu T (1992) "Investigations of Microtremors in the Akita Plain in Japan", Pure Appl. Geophys 99:85–93
92. Pandey AP, Singh RK, Shukla AK, Prakash R, Singh D, Bhatnagar AK (2008) Site Characterization and Vulnerability Analysis of NCT Delhi (2008); Presented in a conference on 45th annual convention meeting on Seismic hazard and crustal earthquakes: Indian scenario 5–7 Nov 2008 organized by NGRI and Department of geophysics, BHU
93. Rao NP, Ravi Kumar M, Seshunarayana T, Shukla AK, Suresh G, Pandey Y, Raju D, Pimprikar SD, Das C, Gahalaut K, Mishra PS, Gupta H (2011) Site amplification studies towards seismic Microzonation in Jabalpur urban area, central India: Available online 21 January 2011. Journal of physics and Chemistry of the earth (2011)

94. Rao NP, Ravi Kumar M (2012); Seismic Microzonation of Jabalpur Area (Abridged report); No. 63 Geological Society of India , Bangalore
95. Riepl J, Bard PY, Hatzfeld D, Papaioannou C, Nechtschein S (1998) Detailed evaluation of Site Response estimation methods across and along the sedimentary valley of Volvi (EURO-SEISTEST). Bull Seism Soc Am, 88–2, 488–502
96. Rogers A, Tinsley JC, Borcherdt RD (1985) Predicting relative shaking response, in Evaluating earthquake hazards of the Los Angeles region -- an earth-science perspective, Ziony, J.I., (ed.), U.S. Geol. Survey Prof. Paper 1360, 221–248. https://doi.org/10.3133/pp1360
97. Satoh T, Kawase H, Sato T, Pitarka A (2001) Three-dimensional finite-difference waveform modeling of strong motions observed in the Sendai Basin, Japan. Bull Seism Soc Am 91:812–825
98. Schnabel PB, Lysmer J, Seed HB (1972) SHAKE: A computer program for earthquake response analysis of horizontally layered sites. Report No. EERC 72–12, Earthquake Engineering Research Center, University of California, Berkeley, California
99. Seed HB, Idriss IM (1970) Soil moduli and damping factors for dynamic response analysis, Report No. UCB/EERC-70/10, University of California, Berkeley
100. Seekins LC, Wennerberg L, Margheriti L, Liu HP (1996) Site amplification at five locations in San Francisco, California: a comparision of S waves, codas and micro tremors. Bull Seism Soc Am 86–2:627–635
101. Seo K, Samano T, Yamanaka H,. Hao H, Koyama S, Takeuchi M (1989) Comparative considerations about the effect of surface geology on seismic motion among different site conditions in Meeting of the IASPEI, Istanbul
102. Seo KS, Yamanaka T, Hao H, Koyama X, Takeuchi S, Fujioka M, Kishino K, Kawano Y, Asano K, Nakajima K, Murai N, Mualchin M,L, HisadaY (1991) Microtremor measurements in the San Francisco Bay area ,Part I; Fundamental Characteristics, -Proceeding 4th international conference of Seism. Zonation", Earthquake Engineering Res Inst Stanford, California, vol 2, pp 417–424
103. Shukla AK (2003) Presented in international workshop on earthquake risk assessment and mitigation, held at CBRI Roorkee on February 28, 2003
104. Shukla AK (2004), Site Response study of Jabalpur urban Area. 2004 Presented in Workshop on Seismic Hazard and Risk Microzonation of Jabalpur, held in NGRI, Hyderabad on 27th September 2004
105. Shukla AK (2006); Seismic monitoring, hazard and risk evaluation. Proceedings of Sixth international workshop on seismic analysis in the South Asia Region held in Chiang Mai, Thailand, 4–7 December 2006
106. Shukla AK, Parakash R, Singh D, Singh RK, Pandey AP, Mondal HS, Nayal BMS, Seismic Microzonation of NCT Delhi: (2007a); Presented in an International Workshop at IISC Bangalore. Interline Publishing Bangalore vol I, pp 39–42
107. Shukla AK, Seismic Microzonation and Risk Evaluation: A Concept (2007b) A workshop on Microzonation held in ISC Bangalore, Interline publishing Bangalore vol I, pp 90–93
108. Shukla AK, Parakash R, Singh RK (2007c) Seismic Microzonation: a tool for seismic hazard mitigation. Presented in Tropmet 2007 at Barktulla University Bhopal 19–20 Dec 2007
109. Shukla AK (2007d) Seismic Microzonation of NCT Delhi; National Seminar on Seismic Microzonation held at Institute of Seismological Research, Ghandhinagar, Ahmedabad on 26 Oct. 2007, Extended Abstract, pp 53–55
110. Shukla AK, Pandey AP, Prakash R, Singh D, Singh RK, Mandal HS, Jaryal JS, Rana BS, Gupta A, Bali S, Bhatnagar AK (2009) Application of GIS in Seismic Hazard Microzonation of NCT Delhi- A First Level Inferences, GEOMAP -2009, March 20–21, 2009. Conference on Urban Infrastructure and Geoinformatics, Bangalore
111. Shukla AK (2010) Invited talk on "seismic hazard mitigation efforts in indian context" (2010); Proceeding of 14th international symposium on earthquake engineering (ISEE), 17–19 Dec.2010 at IIT Roorke, vol I, pp 11–25
112. Shukla AK, Invited talk on Earthquake Hazard Mitigation Approaches vis-a-vis Seismic Hazard Microzonation (2011a) Conference on earthquake risk management for industries

(ERMI - 2011) July 21–22, 2011 ,organized by national disaster management authority (NDMA), Federation of Indian chambers of commerce & industry (FICCI) and global forum for disaster reduction (GFDR) at New Delhi on July 21–22, 2011
113. Shukla AK, Invited talk: Use of Seismic Microzonation for disaster management and methodology (2011b) Workshop on "Disaster preparedness and management of State of J&K" Srinagar , July 28, 2011.Organised by NDMA and J.K.Government at Srinagar J&K
114. Shukla AK, Invited talk Seismic Hazard and Risk Microzonation: (2011c) Proceeding of Indo-Norwegian Workshop on Geohazards; 12–14 September, 2011, New Delhi; Ministry of Earth Sciences, pp 67–70
115. Shukla AK, Mandal HS, Singh RK, Pandey AP, Sisodia HS (2013a) Ground Response Analysis of a few representative sites in NCT Delhi by Presented in international symposium: advances in earthquake science (AES-2013) held on 1–2 February, 2013 at Institute of Seismological Research Ghandhinagar
116. Shukla AK, Singh RK, Pandey AP, Mandal HS, Sisodia HS (2013b) Aplication of Site Response Spectral curves for the interpretation of local ground condition in NCT Delhi; Presented in international symposium: advances in earthquake science (AES-2013) held on 1–2 February, 2013 at Institute of Seismological Research Ghandhinagar
117. Shukla AK, Pandey AP, Singh RK, Mandal HS (2013c) Vulnerability Indices of Ground (K_g) in different part of NCT Delhi, Presented in international symposium: advances in earthquake science (AES-2013) held on 1–2 February, 2013 at Institute of Seismological Research Ghandhinagar
118. Steidl JH, Tumarkin A, Archuleta RJ (1996) What is a reference site?. Bull Seism Soc Am 86:1733–1748
119. Suresh G, Singh RK, Prakash R, Singh D, Jaryal JS, Rana BS, Shukla AK (2005) Site response study using microtremor and local earthquake data in seismically active region of Jind, (2005) Proceedings seismic hazard analysis and Microzonation, IIT Roorkee, pp 303–318
120. Su F, Aki K (1995) Site amplification factors in central and Southern California determined from Coda waves. Bull Seism Soc Am 85:452–466
121. Susan EH, Martin S, Bilham R, Atkinson GM (2002) The 26 January 2001, M 7.6 Bhuj, India, earthquake: observed and predicted ground motions. Bull Seismol Soc Am 92(6): 2061–2079
122. Tang LM, Liang, Wu ZJ (2010) Influence of site condition on seismic amplification effects during the Wenchuan earthquake. J Civil Archit Environ Eng 32(Supp. 2):175–178
123. Thomson WT (1950) Transmission of elastic waves through a stratified solid. J Appl Phys 21:89–93
124. Toshinawa T, Taber JJ, Berrill JJ (1997) Distribution of ground motion intensity inferred from questionnaire survey recordings, and microtremor measurements—a case study in Christchurch, New Zealand, during the 1994 Arthurs Pass Earthquake. Bull Seism Soc Am 87–2:356–369
125. Trifunac MD, Novikova EI (1994) State of the art review of strong motion duration. In: Duma G (ed) Proceedings of the tenth European conference on earthquake engineering, Vienna, Austria, Balkema, Rotterdam, I, pp 131–140
126. Triwulan W, Utama D, Warnana D, Sungkono (2010) Vulnerability ng index estimation for building and ground using microtremor. The second international seminar on applied technology, Science and Arts
127. Tsujiura M (1978) Spectral analysis of coda waves from Local earthquakes. Bull Earthq Res Inst Tokyo Univ 53:1–48
128. Tou C, Wu Z, Mu Y, Wang P, Zhu O (2020) numerical analysis of seismic site effects in loess region of western china under strong earthquake excitations. Hindawi Shock Vib, Article ID 3918352, 12 p https://doi.org/10.1155/2020/3918352
129. Udwadia FE, Trifunac MD (1978) Studies in earthquake motions and micro tremor process. Proceedings II international conference on Microzonation, San Francisco, California
130. User's Manual, EduPro Civil Systems, Inc. Sammamish, Washington PROSHAKE Computer Programme for Ground Response Analysis

131. Volant P, Cotton F, Gariel JC (1998) Estimation of site response using the H/V method. Applicability and limits of this technique on Garner Valley Downhole Array dataset (California). In: Proceedings of 11th Europe conference earthquake engineering, Balkema, Rotterdam, 13 pp., CD-ROM
132. Wang LM, Wu ZJ (2013) Earthquake damage characteristics of the MinXian-ZhangXian Ms6.6 earthquake and its lessons. China Earthq Eng J 35(3):401–412
133. Wood HD (1908) Distribution of apparent intensity in San Francisco in the California earthquake of April 18, 1906, report of the State Earthquake Investigation Commission, Carnegie Institute, Washington, D.C., Publ. vol 87, pp 220–245
134. Wood HD (1933) Preliminary report on the Long Beach earthquake. Bull Seism Soc Am 23:42–56
135. Wu ZJLM, Wang, Wang P, Chen T, Shi H, Yang X-P (2013) Influence of site conditions on ground motion at far field loess sites during strong earthquake. J Central South Univ 20(8):2333–2341
136. Yamananka H, Dravinski M, Kagami H (1993) Continuous measurements of microtremors on sediments and basements in Los Angeles, California. Bull Seism Soc Am 83:1595–1609
137. Zeevaert L (1962) Strong ground motions recorded during earthquakes of May the 11th and 19th, 1962 in Mexico City 11. Bull Seis Soc Am 54(1):209–231

Chapter 3
Seismic Design of Shallow Foundations: Principles, Design Methodologies and Current Indian Practices

Ravi Kiran Nandyala and Ravi S. Jakka

Abstract The purpose of a foundation/substructure is to transfer the loads from the superstructure safely to the underlying soil. Safe and economical design of foundations to withstand various probable loading conditions is the prime role of a geotechnical engineer. The design of an earthquake-resistant foundation is highly challenging, as seismic loads are quite complex in nature. Design of a foundation against seismic loading requires not only a thorough understanding of the behaviour of the founding soil, response of the structure and interaction of the founding soil and the structure under earthquake loading but also a proper understanding of the basis for the recommendations of the relevant design standards of the country. An attempt is made here to discuss the basis for the recommendations of the Indian geotechnical design standards for shallow foundation design which deal with fixing the plan area and the founding depth (location from the ground surface), the prime focus being demystifying the design philosophy of IS 1893 for the seismic design of shallow foundations. In summary, a clear step-by-step design procedure that is to be adopted while designing a shallow foundation as per Indian Standards is presented.

Keywords Shallow foundation · Seismic design · IS 1893 Part 1 · IS 1904

R. K. Nandyala
Centre of Excellence in Disaster Mitigation and Management, IIT Roorkee, Roorkee, India
e-mail: ravikiran@alumni.iitd.ac.in

R. S. Jakka (✉)
Department of Earthquake Engineering, IIT Roorkee, Roorkee, India
e-mail: ravi.jakka@eq.iitr.ac.in

R. K. Nandyala · R. S. Jakka
Indian Institute of Technology Roorkee, Roorkee 247667, Uttarakhand, India

T. G. Sitharam et al. (eds.), *Theory and Practice in Earthquake Engineering and Technology*, Springer Tracts in Civil Engineering,
https://doi.org/10.1007/978-981-19-2324-1_3

3.1 Introduction

Seismic risk is defined as the convolution of hazard, vulnerability and exposure [1–4]. Seismic risk reduction is possible only through either lowering exposure or decreasing the vulnerability, since minimising seismic hazard is not possible. Seismic risk reduction is done either through planning interventions such as avoiding high hazard areas for infrastructure development, or engineering interventions such as designing and constructing infrastructure meeting the performance levels for a given target seismic hazard or through the deployment of early warning systems to reduce human losses. Under the ambit of lowering exposure, economic losses can be minimised primarily through proper urban planning and human losses can be minimised by using reliable early warning systems, whereas decreasing vulnerability is done through designing and constructing the infrastructure to desired performance levels. However, to either lower exposure or decrease vulnerability, accurate estimation of seismic hazard is essential.

Seismic codes aim to accomplish seismic risk reduction through publishing hazard maps, codifying simplified procedures for the approximate estimation of hazard, giving generic guidelines to lower exposure and specific guidelines to reduce the vulnerability of the structures/foundations to be designed and constructed for the estimated hazard. Towards lowering exposure, seismic codes recommend avoiding the construction of structures/foundations near the vicinity of active seismic faults, unstable slopes, landfills of hazardous waste sites, dormant or active mine or cavernous lime stones, flood plains and loose to medium dense fine sands (SP) located adjacent to deep rivers/active seismic regions. As land is premium and avoiding sites is to be done judiciously, seismic codes recommend carrying out site-specific studies in high seismicity regions to assess the ground stability under the design seismic action, considering the effect of local soil and site conditions on site amplification. This chapter deals with the specific guidelines of the Indian geotechnical design standards to reduce the vulnerability of the shallow foundations to be designed and constructed.

Foundations are broadly classified into shallow and deep foundations. Selection of a suitable type of foundation depends on subsurface soil characteristics, magnitude of loads from the superstructure and the requirements of the superstructure. If a shallow foundation, i.e. a Strip/Isolated/Combined/Raft foundation is selected, then the length, width, and founding depth of the selected foundation need to be specified. In case of a deep foundation like a Pile/Pier/Well/Pile Group, the length, diameter, and the number of piles, as the case may be, must be specified. A foundation is usually considered as shallow if its founding depth is lower than its width. Shallow foundations derive resistance from the bearing soils. They are used when soil at shallow depths is strong enough to withstand the stresses imparted by the superstructure. Shallow foundations should usually be the first choice, as they are economical, do not need special construction techniques or special equipment for drilling and provide better quality control. However, isolated foundations are not recommended when excessive settlements and/or large lateral soil movements (lateral spreading) are expected under the design earthquake. Mat foundations are preferred on soft or loose soils. They perform better than isolated, continuous and tied foundations by minimising differential soil movements due to earthquakes by bridging over loose pockets of soil. Deep foundations are preferred when loads from the superstructure are too high, water table is shallow, there are restrictions over open excavation or to bypass liquefiable soils, if competent soil (dense sands, stiff clays or rocks) exists at shallow depths (say, 10–20 m).

The design of a shallow foundation involves fixing the plan area and founding depth of the foundation, which comes under the purview of the geotechnical design of foundations and the structural design of the foundation element which comes under the purview of the structural design of foundations. The discussion here is limited to the geotechnical design of foundations. Across the world, the geotechnical design of foundations is done through either the Working Stress Design (WSD)/Allowable Stress Design (ASD) or the Limit State Design (LSD)/Load and Resistance Factor Design (LRFD). The difference between LSD and LRFD design philosophies is graphically shown in Fig. 3.1 [5]. The partial safety factors of the LSD method which was adopted by the Eurocode in 1993 were calibrated against the global factors of safety of the ASD philosophy to produce comparable design results [6]. The differences in philosophies of geotechnical design of foundations are treated at length in [5, 6]. The North American building codes like the UBC/IBC and the loading standard ASCE 7 permit the use of both ASD and LRFD, whereas the Indian codes of practice permit the use of only the ASD philosophy for the geotechnical design of foundations. The geotechnical design problems are traditionally dealt with under two distinct categories, the stability and the elasticity problems. LSD deals with the stability problems under the ultimate limit state and the elasticity problems under the serviceability limit state [6].

Various stages in the seismic design and construction of a foundation are depicted in the flowchart shown in Fig. 3.2. Apart from 'Hazard Estimation' and 'Seismic Design', the rest of the stages are more or less similar to the design of foundations subjected to gravity loads and done the same way world over.

Shallow foundations are subjected to varying combinations of vertical load, V, horizontal load, H, and moment, M. Many design standards and codes of practice [8, 9], while estimating the ultimate bearing capacity of shallow foundations, consider the effect of H using correction factors and M using reduced foundation dimensions. These approaches to account for the influence of H and M on the ultimate bearing capacity were developed independently, without considering the interaction between the vertical load, horizontal load and the moment. Some design standards/codes/guidelines [10–13] do provide V–H–M capacity envelopes derived through experimental studies [14, 15–20] or analytical studies [21–27] to check the safety of shallow foundations [28, 29]. Apart from these standards, in recent decades, some literature was published on the interaction surface approach for the design of shallow foundations resting on cohesive [30–44] or cohesionless soils [14, 16, 45, 46]. However, all the above studies are focussed on the design of shallow foundations resting on level ground and consider either exclusively cohesive or cohesionless soils only. Also, only a few of them include seismic loading.

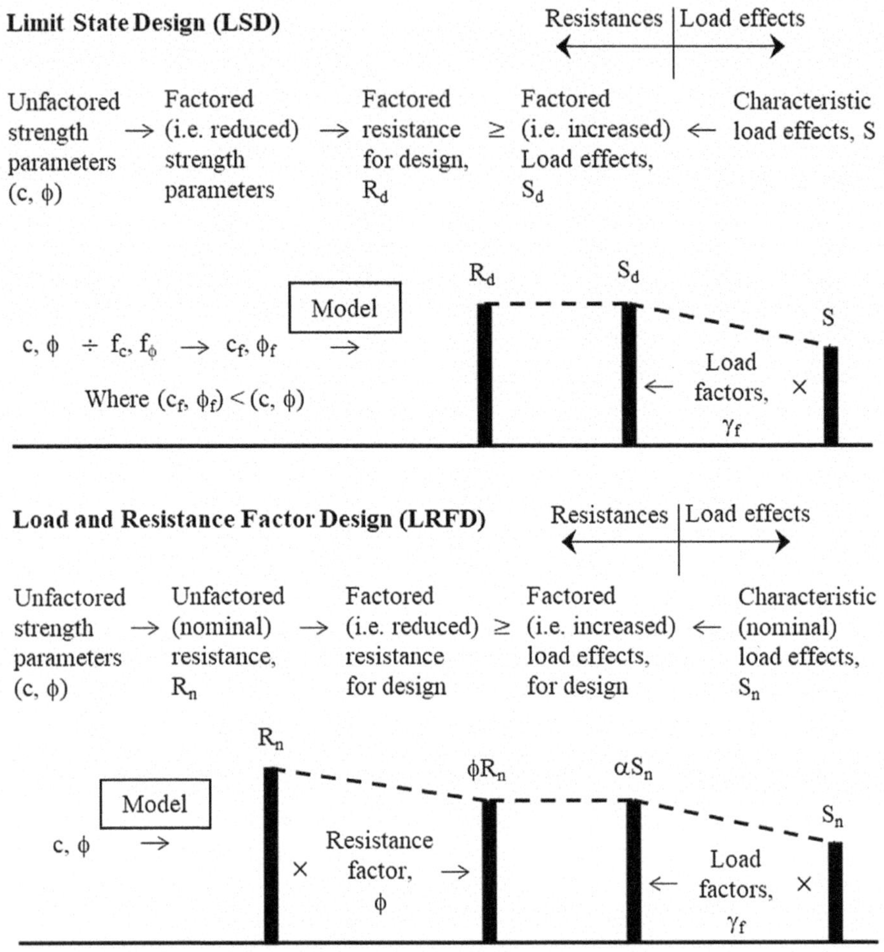

Fig. 3.1 Difference between LSD and LRFD [7]

3.2 Shallow Foundation Resisting Mechanisms, Earthquake Loading and Possible Failure Modes, Design Approaches for Various Loads

3.2.1 Resisting Mechanisms of the Founding Soil

The way a shallow foundation withstands stresses is different based on the nature of the applied loading. Figure 3.3 presents the load resisting mechanisms of the founding soil against different types of foundation loads.

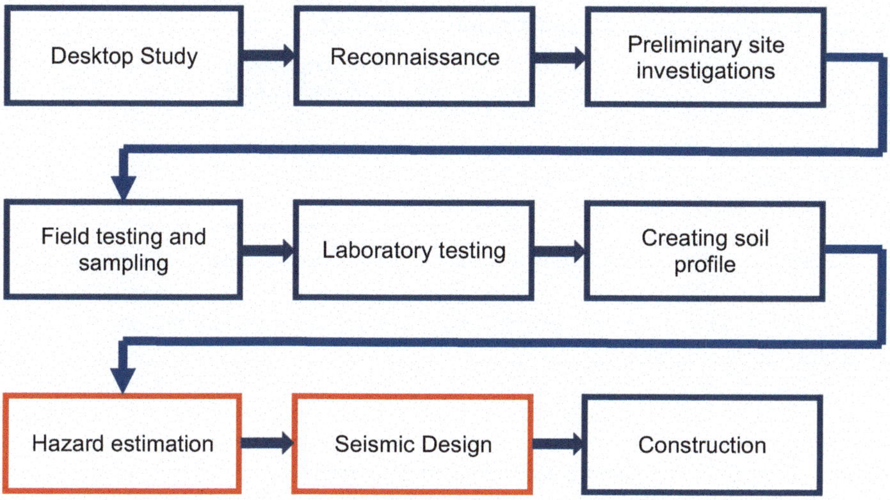

Fig. 3.2 Stages in the seismic design and construction of foundations

Fig. 3.3 Resisting mechanisms of the founding soil against different types of loads: **a** Soil bearing pressure; **b** Dead weight of the foundation and backfill; **c** Friction resistance or adhesion of soil and lateral earth pressure; and **d** Redistribution of bearing pressure

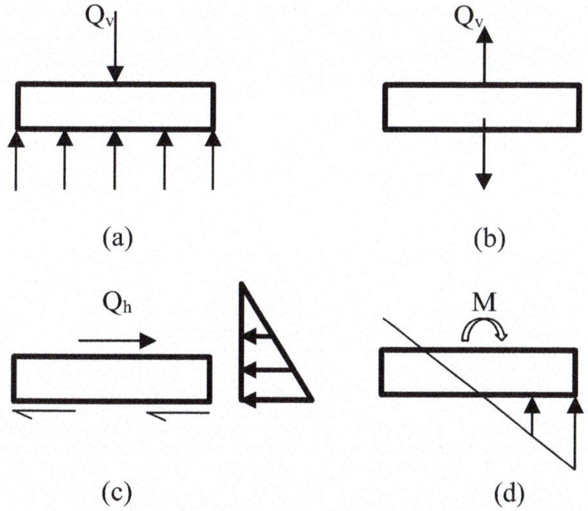

Earthquake loading is quite complex and the analysis of foundations under earthquake loading is highly challenging. Table 3.1 in Section 3.2.3 lists various possible modes of failures of shallow foundations under earthquake loading, the understanding of which is essential for the design of safe and economic foundations.

3.2.2 Components of Earthquake Loading

Earthquakes generate the following types of loads on foundations.

- Alternating Horizontal loads,
- Alternating Vertical loads and
- Alternating Moments.

In addition to the generation of the above type of loads, earthquakes reduce the strength of the soils due to the development of excess pore water pressures and/or liquefaction. Excess pore water pressures can also cause ground failures such as sand boils and lateral spreading.

3.2.3 Possible Shallow Foundation Failures

Shallow foundations experience different modes of failure based on the type and intensity of loading and the type of foundation soil. Broadly, foundation failures can be classified as shown in Table 3.1 [47].

Seismic design of a shallow foundation requires the consideration of all the above possible modes of failure, their causative loads and failure mechanisms. While calculating the design seismic loads, it is important to include the effects of site amplification. The design approach to be adopted to ensure safety under the action of the causative loads is described in brief in the following sections.

Table 3.1 Foundation failures, corresponding causative loads and soil failures

Failure mode (global)	Primary causative loading	Possible soil failure(s)
Tilting of foundation/overturning of the structure	Moment loading	• Bearing Capacity failure • Liquefaction • Sand boils and lateral spreading
Ground instability/Slope failures	Horizontal loading	• Flow slides and lateral spreading
Sliding of foundations	Horizontal loading	• Lateral spreading of underlying soils
Settlement of foundations	Vertical loading	• Consolidation of liquefied soil • Sand boils and lateral spreading

3.2.4 Design Against Vertical Loads

World over, the basic approach for the geotechnical design of foundations for a concentrically vertical load remains more or less same. However, the approach for design against the horizontal and the moment loads vary significantly among countries and between structural and geotechnical engineers.

3.2.5 Design Against Horizontal Loads

Table 1806.2 of the International Building Code 2018 [48] apart from containing the settlement-limited presumptive vertical foundation pressure values also includes presumptive values of lateral bearing pressure and lateral sliding resistance which are used in the check against lateral loading. The effect of horizontal loads on the vertical load carrying capacity is not considered. On the other hand, Eurocode 7 [8] and the Indian standard IS 6403 [9] have adopted different versions of the bearing capacity equation containing the inclination factors to consider the effect of horizontal loads on the vertical load carrying capacity of the foundation. Although the AASHTO LRFD Bridge Design specifications [49] includes a version of the bearing capacity equation containing the inclination factors, as stated in the commentary of the code, the usual practice is to ignore the inclination factors. The reason for this practice is attributed to the fact that the information on the design vertical and horizontal loads is missing at the time of geotechnical testing and report preparation. The code further suggests omitting the use of inclination factors in the case of foundations with modest embedment, citing two reasons. The first reason given is that the tests by Meyerhof which led to the inclusion of inclination factors have shown that for foundations with the depth of embedment ratio close to 1, the effect of load inclination on bearing resistance is relatively small. The second reason given is that the resistance factors provided in Article 10.5.5.2.2 of the code were derived for vertical loads alone without considering the effect of horizontal loads.

3.2.6 Design Against Moments

The geotechnical design for moments is done through either Peck's approach (elastic stress method) or Meyerhof's approach (equivalent area method). When using Peck's approach, the maximum value of the elastic bearing stress is compared with the value of the allowable bearing capacity value estimated for the case of a concentric vertical load without accounting for the effect of moment loading. Many methods and charts based on Peck's approach were proposed over the years. A comprehensive list of such methods is presented in [50]. While using the equivalent area method, codes prescribe to use reduced foundation dimensions while calculating the capacity as well as the demand. However, the usual practice is to calculate the demand using reduced dimensions, while the capacity is calculated using the original foundation dimensions since the values of design moments are unknown at the time of the preparation of the geotechnical design recommendations report.

3.3 Shallow Foundation Design Alternatives

Seismic design of a shallow foundation, in a broad sense, can be done in one of the following two ways as depicted in Fig. 3.4:

1. Allow inelastic deformations in the structural system and keep the founding soil elastic: Allowing inelastic deformations in the structural system alone can be done in one of the following two ways:
   (i) Capacity design of the foundation, such that the founding soil and the foundation (structural) remain elastic, while the column undergoes inelastic deformations.
   (ii) Not applying the capacity design of the foundation, resulting in the column and the founding soil remaining elastic, while the foundation (structural) undergoes inelastic deformations. Such a design alternative is not desirable.
2. Allow inelastic deformations in the founding soil and keep the structural system elastic: Allowing inelastic deformations in the founding soil alone can be achieved through relying upon either foundation rocking or/and foundation sliding. However, care should be taken to ensure that the inelastic deformations in the founding soil do not lead to the following kind of foundation failures.
   (i) Excessive permanent tilting of foundation,

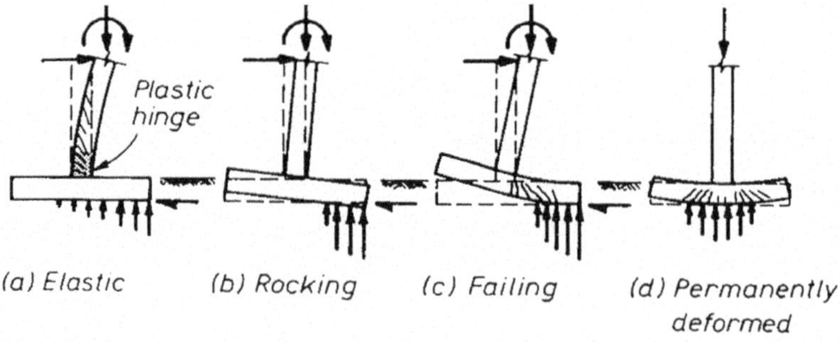

Fig. 3.4 Possible shallow foundation responses to seismic loading [51]—With permission from John Wiley & Sons, Inc. (**a**) Elastic Soil and Hinged Column; (**b**) Elastic Column and Inelastic Soil; (**c**) Elastic Column and Elastic Soil, Inelastic Foundation Element; and (**d**) Permanently deformed Foundation Element due to Cyclic loading

(ii) Excessive permanent sliding of foundation and
(iii) Excessive permanent settlement of foundation.

The mechanism shown in Fig. 3.5 needs to be ensured when capacity design is adopted for the seismic design of the structure. Such mechanisms are only possible when the foundation is capacity designed. Eurocode 8 recommends capacity design of foundation as the guiding principle for the seismic design of shallow foundations. The American and Indian codes currently do not recommend the capacity design of foundation. A foundation designed as per Eurocode 8 will likely have the plastic hinge in the column while the foundation (structural) and the founding soil remain elastic, whereas a foundation designed as per the American/Indian code is more likely to have the inelastic deformations either in the foundation (structural) or the founding soil since the foundation is not capacity designed. So, many foundations designed as per the current American/Indian code are likely to undergo rocking. However, the level of inelastic deformations in the founding soil cannot be estimated using the current code-based procedures. Increasingly, there is a call from the Geotechnical Earthquake Engineering fraternity to adopt unconventional seismic responses like foundation rocking, sliding and seismic isolation of surface foundations exploiting the properties of natural liquefiable soil, to minimise damage in the superstructure. By not adopting capacity design of foundation, as in the case of current American/Indian code, one is highly likely to end up designing a rocking/sliding foundation, although not intending to do so.

A simplified analysis, assuming a pinned base, shows that the founding soil is highly likely to enter the inelastic range when subjected to a rare event. It is important to note that estimation of seismic forces will have a bearing on the seismic design of shallow foundations. Hence, it is very important to understand the assumptions in computing the seismic forces. The computation of seismic forces depends on the period estimation, hazard estimation (seismic zone) and the shape of the design response spectra. Capacity design needs to be adopted for safety–critical structures. However, it would be uneconomical to adopt for normal structures. For all normal structures, designing using either the current practice or rocking principles

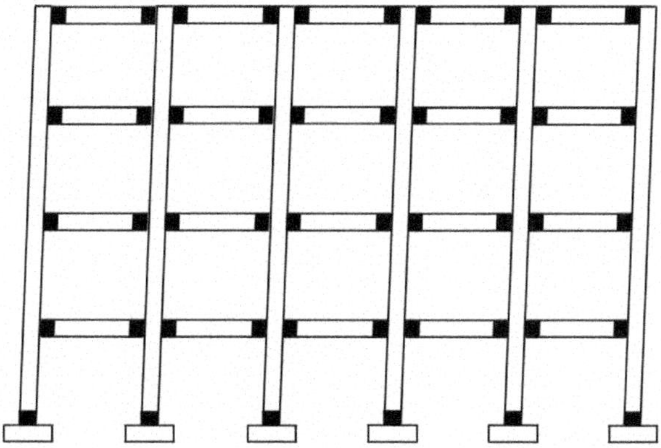

Fig. 3.5 Mechanism to be ensured through Capacity Design [52]

is desirable. The problem with the use of current practice is, though it allows rocking unconsciously, it may lead to foundation failures because of the lack of foundation design considering actual rocking. Hence, it is highly desirable to come up with guidelines/design procedures allowing rocking within acceptable limits.

3.4 Seismic Settlements

Settlement calculations are done using Terzaghi's Consolidation theory by which the compression of the foundation soils under the applied building loads is determined from laboratory consolidation curves. The time rate of consolidation is dependent upon the rate at which water and air can escape from the voids of the soil subjected to increased pressures from building loads. This rate is relatively fast for sands and slow for saturated clays.

If we consider the effect of seismic loads on the settlement behaviour of a typical building as shown in Fig. 3.6, we observe that the earthquake loads result in an increase in the vertical loads on the exterior column foundations. In considering how the foundations should be designed for the increased vertical loads due to these lateral forces, it might seem obvious that the exterior foundations should be made larger to accommodate the combined vertical and seismic loads. If the size of the exterior foundations is increased because of seismic design considerations while leaving the interior foundations at the same size as required for support of dead loads, the settlement of the exterior foundations under dead loads would be less than

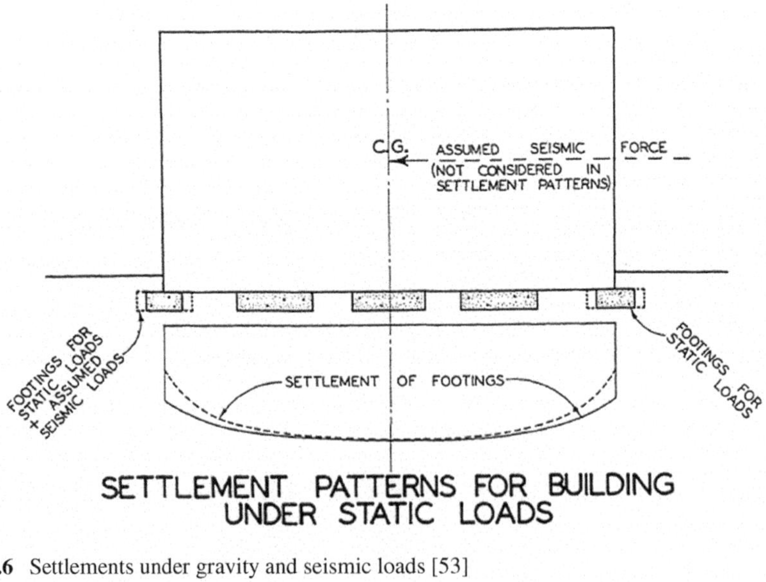

Fig. 3.6 Settlements under gravity and seismic loads [53]

previously considered for the smaller foundations and the differential settlements between exterior and interior columns would be increased.

Thus, an apparent conflict exists in the design of foundations for vertical static and lateral seismic loads. To resolve this conflict, it is necessary to consider the difference in loading conditions for the vertical and lateral loads, and the behaviour of the foundation soils under these different conditions. First, it should be realized that the time duration of loading is very important in predicting the response of foundation soils to loading. Design wind loads may be imposed frequently or for a sufficient length of time so that they can be considered similar to other live loads to which the structure is subjected. However, short time wind, seismic or blast loads require a different concept of foundation design than used for conventional live loads.

One approach to the design of shallow foundations for these transient loads is to permit an increase in the allowable bearing pressures. However, soil behaviour under transient loads is very much dependent on its physical properties. Generally speaking, sands are quite sensitive to the transient loads, while clays are relatively insensitive. To examine this difference between sandy and clayey soils and to evaluate its effect on foundation design, the load deflection characteristics of each soil type will be considered.

Figure 3.7 presents a typical load–settlement curve for an isolated foundation on sandy soils. The solid line represents the deflection of the foundation under permanently applied (dead) loads. Because of the relatively fast drainage characteristics of sand, these settlements will occur quite rapidly. If instrumentation were provided to record the load–settlement response of the foundation to transient loads, a pattern

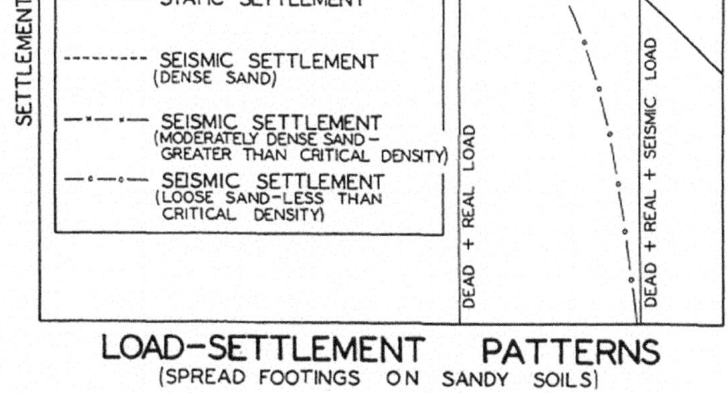

Fig. 3.7 Load–Settlement patterns in sandy soils due to Seismic Load [53]

similar to that indicated by the dashed lines in Fig. 3.7 would be obtained. The magnitude of the settlement is predominantly dependent upon the soil structure. For dense sandy soils, the settlement is comparatively small. As the density of the sand deposit decreases, the settlement increases. In the case of saturated sands, if the density is less than a certain critical value, the soil structure might collapse during an earthquake, transmitting the load to the pore water, resulting in a flow failure known as liquefaction. If this behaviour is anticipated, either the loose sandy soils should be stabilised or the building should be supported independently of this stratum.

In establishing allowable bearing pressures and in estimating foundation settlements for short-term loads, a general knowledge of the behaviour of the foundation soils must be relied upon. Foundations founded on loose sands may yield or fail during seismic disturbances, even though their behaviour under permanently applied vertical loads may have been satisfactory. Buildings founded on a dense sand formation may suffer no damage during an earthquake, even though the short time vertical loads may be much higher than those imposed permanently. Thus, some increase in bearing pressures for foundations founded on dense sandy soils may be permitted for transient seismic loads without forfeiting a significant safety margin against failure.

Usually, clays are less preferred as foundation materials than sands owing to their greater compressibility. However, clays do possess some advantages over sands vis-à-vis their behaviour under transient loads. Figure 3.8 presents a typical load–settlement curve for an isolated foundation on clayey soils. The settlement curve for permanent loads is similar to that of sandy soils, although the magnitude of the settlement and the time rate of settlement are quite different. Also, the curve

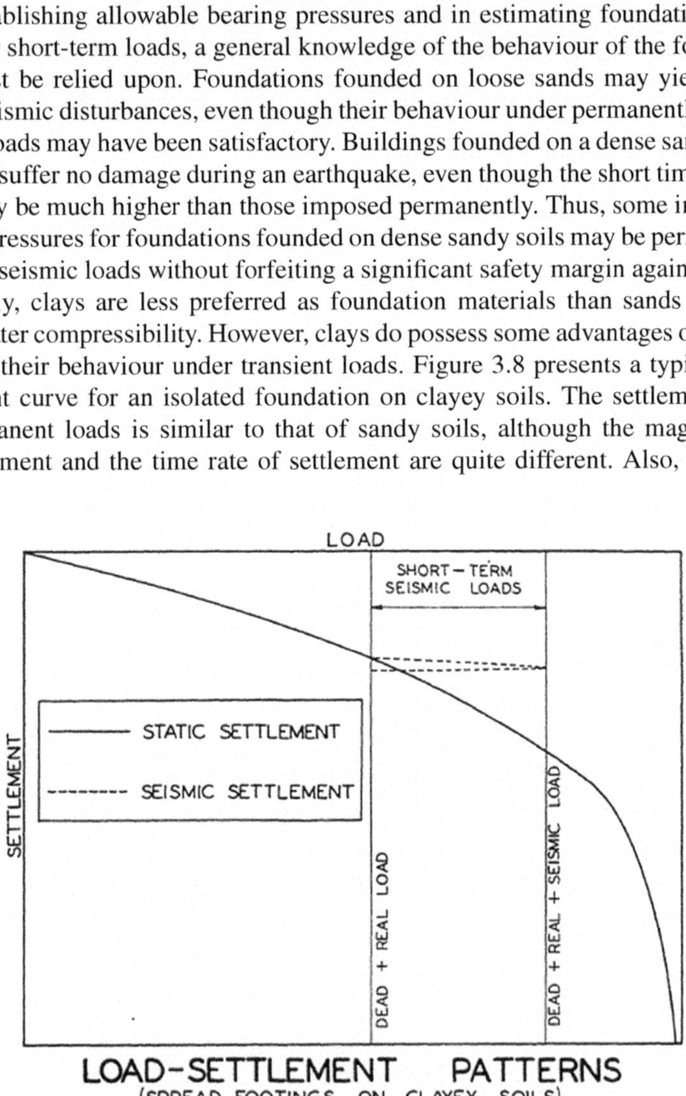

Fig. 3.8 Load–Settlement patterns in clayey soils due to Seismic Load [53]

indicates more abrupt soil yielding at loads beyond the design range. As illustrated in Fig. 3.8, a foundation resting on clay would not exhibit substantial settlement during an earthquake, because the intergranular structure of the clay cannot adjust to the change of load during this short time. Despite the development of high pore pressures in the clay by the increased loads imposed during the earthquake, the time lag for consolidation of the clay will prevent appreciable deflections from occurring. Thus, foundations on clayey soils may be designed for comparatively high, transient loads without excessive settlement of the foundations. The allowable bearing pressures depend largely upon the physical characteristics of the soil and to a certain extent upon the type of foundation and the structure.

3.5 Seismic Design of Shallow Foundations as per IS 1904

IS 1904 [54] considers seismic load as a transient load and due to the reasons mentioned in Sect. 3.4, the stresses in the founding soil due to seismic loads are allowed a certain percentage increase over the allowable values.

Shallow foundations shall be proportioned for the following load combinations:

(1) Dead load + Live load and
(2) Dead load + Live load + Seismic load,

where

(i) Dead load includes the weight of the superstructure, the substructure and the overlying fill.
(ii) Live loads from the superstructure, in accordance with IS 875 Part 2 1987 shall be taken for proportioning and designing the foundations.
(iii) If seismic load is less than 25 percent of the load combination 1, then load combination 2 may be neglected in design and the load combination 1 shall be compared with the safe bearing load to satisfy allowable bearing pressure.
(iv) If seismic load is more than 25% of the load combination 1, foundations shall be checked for load combination 2. The safe bearing capacity shall be increased in accordance with Table 1 of IS 1893 Part 1. In cohesionless soils, liquefaction analysis and computation of seismic settlements shall be made as well.

3.5.1 Design Against Settlements, as per IS 1904

For foundations resting on cohesionless soils, the settlements shall be computed corresponding to Dead load + live load + seismic load, since in such types of soils, settlements occur near instantaneously.

For cohesive soils, the settlements shall be computed corresponding to permanent loads alone. Permanent loads include dead load, weight of all fixed equipment and one half of the design live load.

The distinction between permanent and temporary loads largely depends upon the judgement of the engineer-of-record of the project. The design engineer should strive to minimise the differential settlements due to the live load variation by ensuring equal bearing pressure for all the foundations under the service load.

3.6 Seismic Effects to be Considered as Per IS 1893 Part 1

3.6.1 Influence of Soil Type on Intensity of Shaking:

(i) Influence of soil type on the intensity of shaking (horizontal) is accounted for, in the calculation of horizontal seismic coefficient.

$$\alpha_h = A_h = \frac{\left(\frac{Z}{2}\right)\left(\frac{S_a}{g}\right)}{\left(\frac{R}{I}\right)}$$

where Z = Zone Factor (Range: 0.36, 0.24, 0.16 and 0.10). An appropriate zone factor is to be selected based on the location of the project site.

I = Importance Factor (Range: 1.0–1.5)

R = Response Reduction Factor (Range: 3.0–5.0 for RC Buildings)

$\frac{S_a}{g}$ = Avg. spectral acceleration coefficient (Fig. 3.9), where the effect of soil type is considered (Maximum value is 2.5).

(ii) Influence of soil type on the vertical shaking is also considered indirectly.

$$\alpha_v = A_v = \frac{2}{3} \cdot \alpha_h = \frac{\left(\frac{2}{3} \times \frac{Z}{2}\right)\left(\frac{S_a}{g}\right)}{\left(\frac{R}{I}\right)}$$

Note: The value of $\frac{S_a}{g}$ in the above equation is taken as 2.5, for buildings governed by IS 1893 Part 1 and liquid retaining tanks governed by IS 1893 Part 2.

3.6.2 Increase in Allowable Bearing Pressures in Soils

IS 1893 Part 1 permits an increase in allowable bearing pressures in soils where earthquake forces are not expected to cause significant settlement, to avoid undesirable differential settlements that can take place prior to earthquake occurrence. This provision drew inspiration from [53] to have different (increasing) factors for different soils, unlike the UBC/IBC codes, which use the same increasing factor irrespective of the type of soil. When earthquake forces are included, the allowable bearing pressure in soils shall be increased as per Table 1 of IS 1893 Part 1, depending upon the type of soil.

3.6.3 Accounting for Liquefiable Soils

(i) In soil deposits consisting of submerged loose sands and soils falling under classification SP with corrected standard penetration N-values less than 15 in seismic Zones III, IV, V and less than 10 in seismic Zone II, within the top 5 m, the vibration caused by an earthquake may cause liquefaction or excessive total and differential settlements. Such sites should preferably be avoided while locating new structures and should be avoided for locating structures of important projects.

(ii) If N-values (corrected values) at the project site are lower than the desirable N-values listed in Table 3.2 and if there is no option to avoid the site, appropriate site improvement techniques (such as improving compaction or stabilisation) should be adopted to achieve suitable N-values.

(iii) Alternatively, a deep pile foundation may be provided and taken to depths well into the layer which is not likely to liquefy.

(iv) Marine clays and other sensitive clays are also known to liquefy due to the collapse of soil structure and will need special treatment according to site condition.

Table 3.2 Desirable N-values (corrected values)

Seismic zone	Depth (m) below ground level	N-values	Remarks
III, IV and V	≤5 ≥10	15 25	For values of depths between 5 and 10 m, linear interpolation is recommended
II	≤5 ≥10	10 20	

3.6.4 Other Guidelines for the Seismic Design of Shallow Foundations as per IS 1893 Part 1

(i) The allowable bearing pressure shall be determined in accordance with IS 6403 [47] or IS 1988 [55].
(ii) If any increase in bearing pressure has already been permitted for forces other than seismic forces, the total increase in allowable bearing pressure when seismic force is also included shall not exceed the limits specified in Table 1 of IS 1893 Part 1.
(iii) Isolated R.C.C. foundations without tie beams or unreinforced strip foundations shall not be permitted in soft soils with $N < 10$.

3.7 Seismic Design of Shallow Foundations Using Pseudostatic Method as per IS 1893 Part 1

Foundations subjected to earthquake loading can be analysed/checked for their safety by using pseudostatic analysis. Various steps involved in the application of pseudostatic analysis are given below:

I. Estimation of Earthquake Loading,
II. Conversion of the Earthquake Load into an Equivalent Static Load and
III. Apply conventional Bearing Capacity theories to obtain the Safe Bearing Capacity.

3.7.1 Estimation of Earthquake Loading

(i) ***Horizontal Dynamic Load Calculation***

Horizontal seismic coefficient for the Design Basis Earthquake (DBE) is estimated using

$$\alpha_h = A_h = \frac{\left(\frac{Z}{2}\right)\left(\frac{S_a}{g}\right)}{\left(\frac{R}{I}\right)}$$

Horizontal load due to earthquake

$$H_e = \alpha_h W$$

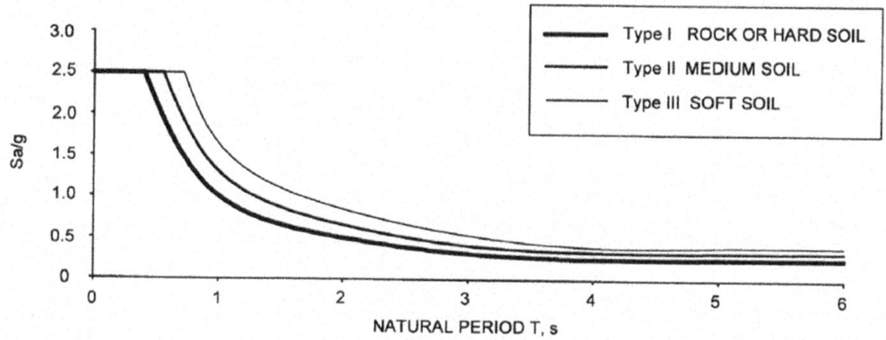

Fig. 3.9 Response spectra for obtaining (S_a/g) based on the fundamental period of the structure (IS 1893 Part 1) [56]

(ii) ***Vertical Dynamic Load Calculation***

Vertical seismic coefficient is taken as

$$\alpha_v = A_v = \frac{2}{3} \cdot \alpha_h = \frac{\left(\frac{2}{3} \times \frac{Z}{2}\right)\left(\frac{S_a}{g}\right)}{\left(\frac{R}{I}\right)}$$

Note: The value of $\frac{S_a}{g}$ in the above equation is taken as 2.5, for buildings governed by IS 1893 Part 1 and liquid retaining tanks governed by IS 1893 part 2.

Vertical load due to earthquake

$$V_e = \alpha_v W$$

(iii) ***Calculation of Dynamic Moment***

Moment may be calculated as the product of the maximum dynamic horizontal load estimated in step (i) and the centre of gravity of the structure or obtained through structural analysis.

3.7.2 Conversion of Earthquake Loading into Equivalent Static Load

- Calculate total vertical, horizontal and moment loads

$$H = H_g + H_e$$

$$V = V_g + V_e$$

$$M = M_g + M_e$$

- The combined effect of V and H can be replicated using an equivalent inclined load.
- The combined effect of M and V can be replicated using an equivalent eccentric load.
- The combined effect of V, H and M can be replicated using an equivalent eccentrically inclined load. (i.e. foundation is now an eccentrically obliquely loaded foundation).

3.7.3 Estimation of Bearing Capacity Under Earthquake Loading

(i) Find out effective dimensions of the foundation due to eccentric loading (L' and B'):

$$e_b = \frac{M_b}{V}$$

$$B' = B - 2e_b$$

$$e_l = \frac{M_l}{V}$$

$$L' = L - 2e_l$$

(ii) Calculate obliquity of loading 'i':

$$i = (H/V)$$

(iii) Use Generalised Bearing Capacity (GBC) equation with inclination factors to find q_{ult}.

$$q_{ult} = cN_c(s_c d_c i_c) + q' N_q(s_q d_q i_q) + 0.5\gamma B N_\gamma(s_\gamma d_\gamma i_\gamma) R'_w$$

where q_{ult}= Ultimate Bearing Capacity of soil,

N_c, N_q, N_γ are the bearing capacity factors,
s_c, s_q, s_γ are shape factors,
i_c, i_q, i_γ are inclination factors,
d_c, d_q, d_γ are depth factors.

Bearing capacity factors and other factors can be considered from any of the GBC theories or using the IS 6403 formula which is in fact based on Hansen, Vesic and other methods.

(i) From q_{ult}, calculate $q_{netsafe}$
(ii) Increase $q_{netsafe}$ value, as per Table 1 of IS 1893 Part 1.
(iii) Estimate, load carrying capacity, Q.
(iv) Compare Q with V:
    If $Q > V$, foundation is safe.
    Otherwise, redesign the foundation with new dimensions.
(v) Estimate settlements of underlying soils due to earthquake loading and then estimate total settlements. See whether this value is within the limits or not. If not, redesigning of foundation is necessary.

3.7.4 Accounting for Soil Strength Reduction

If soil is expected to liquefy or develop significant excess pore water pressures, it is required to carry out an effective stress analysis considering the increase in the pore water pressures during earthquake shaking. For this purpose, it is essential to estimate the liquefaction potential of the site by following the IS 1893 procedure, which has been adopted from [57]. For liquefied soil, use residual/steady-state strength of the soil in analysis.

3.7.5 Estimation of Sliding Failure

Calculate lateral resistance and sliding resistance offered by the soil and then compare it with the lateral force. Use earth pressure theories for calculation of lateral resistance of soils. This probably is not going to govern in the case of foundations resting on level ground. It is important only in the case of foundations resting on slopes where significant later sliding is expected.

3.7.6 Seismic Design Procedure as Per Indian Standards

The seismic coefficient $\frac{\left(\frac{Z}{2}\right)\times\left(\frac{Sa}{g}\right)}{\left(\frac{R}{I}\right)}$ represents inelastic design for DBE-level earthquake, whereas the seismic coefficient $\frac{Z\times\left(\frac{Sa}{g}\right)}{\left(\frac{R}{I}\right)}$ represents inelastic design for MCE-level earthquake.

Shallow Foundation Design as per IS codes (Geotechnical) is summarised below.

- Foundation Design is done using Working Stress Design.
- $q_{ns} = q_{nu}/Factor\ of\ Safety$ | Usually, $Factor\ of\ Safety = 3$.
- $q_a = \min(q_{ns}, q_{np})$.

where q_{nu} = Net Ultimate Bearing Capacity,
q_{ns} = Net Safe Bearing Capacity,
q_{np} = Net Soil Pressure for Specified Settlement,
q_a = Allowable Bearing Capacity.

The following discussion assumes that $q_a = q_{ns}$ in the design case under consideration. Under such assumption, seismic design of a shallow foundation for DBE is as follows:

- For Type A soil, $q_a{}^* = 1.5q_a = 1.5q_{ns} = 1.5q_{nu}/3 = q_{nu}/2$
- DL + LL → q_{ns}
- DL + LL + EQ (DBE) → $1.5q_{ns}$
- Therefore, EQ (DBE) ~ $0.5q_{ns}$ and EQ (MCE) ~ $1.0q_{ns}$
- Therefore, DL + LL + EQ (MCE) → $2.5q_{ns} < q_{nu}$,
- *where* $q_a{}^*$ = Allowable Bearing Capacity for earthquake load combination

However, the above procedure is strictly valid if and only if the applied loads are vertical only. An attempt is made here using a case study (see Sect. 3.7.7) to see if the foundation design as per the above procedure is valid when the foundation is loaded by horizontal forces and moments in addition to the vertical load.

Foundation design method used in practice is summarised below:

- Inclination factors are not used in the calculation of the bearing capacity. Horizontal loads are not considered. If considered, then it is ensured that Horizontal Load < Vertical Load × Coefficient of Friction

  For Axial + Moment,

  - When $e < B/6$, $P/A + M_x/Z_x + M_y/Z_y < q_a$ (Method 1a).
  - When $e > B/6$, $2P/(3L(0.5B\text{-}e)) < q_a$ (Method 1b).

Foundation design method as per IS 6403 is summarised below:

- Inclination factors are used in the calculation of the bearing capacity to account for the influence of horizontal loads on bearing capacity.

  For Axial + Moment,

– $P/(L' \times B') < q'_a$ (Method 2),
where $L' = L{-}2e_L$, $B' = B{-}2e_B$ and q'_{nu} is calculated using B'.

3.7.7 Case study

A building with the specifications mentioned below and the plan and the elevation as shown in Figs. 3.10 and 3.11 is modelled and analysed in ETABS. The analysis

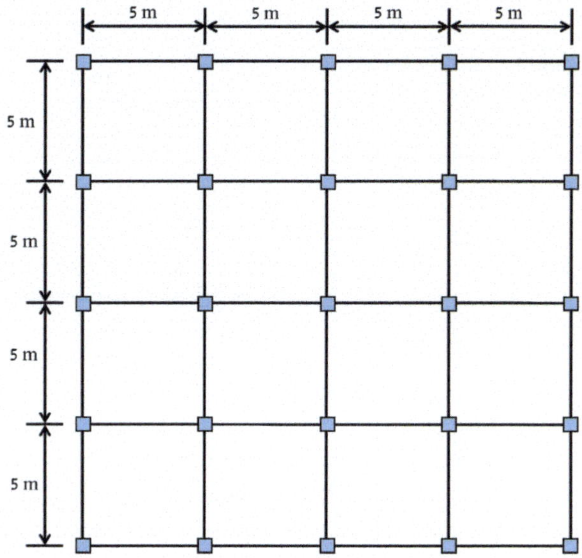

Fig. 3.10 Plan of the building

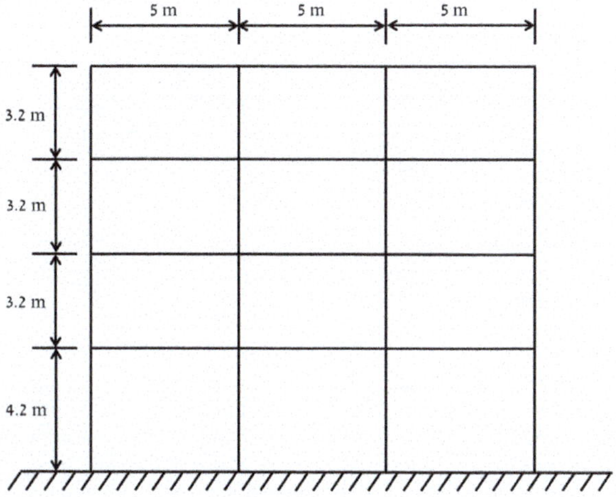

Fig. 3.11 Elevation of the building

Table 3.3 Results for central column from ETABS

Load combination	Vertical Load (in kN)	Shear (along Y) (in kN)	Moment (about X) (in kN·m)
DL + LL	−1521.1	0	0
DL + LL + EQX (DBE)	−1521.1	77.5	−167.7
DL + LL + EQX (MCE)	−1521.1	155	−335.4

$e_{BDL} = 0$ m, $e_{BDBE} = 0.110$ m, $e_{BMCE} = 0.221$ m
For a 2 m × 2 m foundation, e_{BDL} & e_{BDBE} & $e_{BMCE} < B/6$ (0.33 m)

Table 3.4 Results for edge column from ETABS

Load combination	Vertical Load (in kN)	Shear (along Y) (in kN)	Moment (about X) (in kN·m)
DL + LL	−690.1	7	−9.7
DL + LL + EQX (DBE)	−884.9	78.3	−168.8
DL + LL + EQX (MCE)	−1079.7	149.6	−327.9

$e_{BDL} = 0.014$ m, $e_{BDBE} = 0.191$ m, $e_{BMCE} = 0.304$ m
For a 1.8 m × 1.8 m foundation, e_{BDL} & $e_{BDBE} < B/6$ (0.30 m)
$e_{BMCE} > B/6$

results for the central column, an edge column and a corner column are reported in Tables 3.3, 3.4 and 3.5. Surface foundations for the central column, the edge column and the corner column are designed as per Methods 1 and 2 mentioned above and the same has been validated using OptumG3.

Table 3.5 Results for corner column from ETABS

Load combination	Vertical Load (in kN)	Shear (along Y) (in kN)	Moment (about X) (in kN•m)	Shear (along X) (in kN)	Moment (about Y) (in kN·m)
DL + LL	−690.1	7	−9.7	3.8	5.3
DL + LL + EQX (DBE)	−884.9	78.3	−168.8	4.3	5.9
DL + LL + EQX (MCE)	−1079.7	149.6	−327.9	4.7	6.5

$e_{BDL} = 0.017$ m, $e_{BDBE} = 0.327$ m, $e_{BMCE} = 0.472$ m
$e_{LDL} = 0.012$ m, $e_{LDBE} = 0.009$ m, $e_{LMCE} = 0.007$ m
For a 2 m × 2 m foundation, e_{BDL} & $e_{BDBE} < B/6$ (0.333 m), $e_{BMCE} > B/6$
For a 1.8 m × 1.8 m foundation, $e_{BDL} < B/6$ (0.3 m), e_{BDBE},$e_{BMCE} > B/6$
For a 1.6 m × 1.6 m foundation, $e_{BDL} < B/6$ (0.267 m), e_{BDBE},$e_{BMCE} > B/6$
For 2 × 2, 1.8 × 1.8, 1.6 × 1.6 foundations, e_{LDL} & e_{LDBE} & $e_{LMCE} < B/6$

The following data has been considered:
Seismic Zone V, Type A soil,
$Z = 0.36$ for DBE, $Z = 0.72$ for MCE, $R = 5$.
All Columns: 0.45 m × 0.45 m.
All Beams: 0.25 m × 0.5 m.
Dead Load: 12 kN/m^2, 10 kN/m^2 (on roof).
Live Load: 4 kN/m^2, 1.5 kN/m^2 (on roof).
For $\varphi = 38°$, $N > 30$ → Type A soil → 50% increase in $q_{na.}$
Hence, $c = 0$ and $\varphi = 38°$ are taken as soil parameters in OptumG3.

Three surface foundations (individual models) of size 2 m × 2 m, 1.8 m × 1.8 m and 1.6 m × 1.6 m with a vertical multiplier load at the centroid of the foundation are analysed using finite element limit analysis. The results from the analyses are presented below.

Vertical Load Multiplier for 2m × 2m: 4734.8 kN
Safe Load for 2m × 2m: 1578.3 kN
SBC for 2m × 2m: 394.6 kN/m^2
Vertical Load Multiplier for 1.8m × 1.8m: 3802.8 kN
Safe Load for 1.8m × 1.8m: 1267.6 kN
SBC for 1.8m × 1.8m: 391.2 kN/m^2
Vertical Load Multiplier for 1.6m × 1.6m: 2575.7 kN
Safe Load for 1.6m × 1.6m: 858.6 kN
SBC for 1.6m × 1.6m: 335.4 kN/m^2

3.7.8 Central Column

Method 1a:

$$\begin{aligned}
&\mathrm{DL + LL} \rightarrow 1522/4 = 380.5 < 394.6\,\mathrm{KN/m^2}(q_{ns})\\
&\mathrm{DL + LL + EQX(DBE)} \rightarrow P/A(1 + 6e_{BDBE}/B)\\
&= 380.5 \times (1 + 6 \times 0.110/2) = 506.1 < 391.2\,\mathrm{KN/m^2}(1.5q_{ns})\\
&\mathrm{DL + LL + EQX(MCE)} \rightarrow P/A(1 + 6e_{BMCE}/B)\\
&= 380.5 \times (1 + 6 \times 0.221/2) = 632.8 < 1183.8\,\mathrm{KN/m^2}(3q_{ns})
\end{aligned}$$

Method 2:

$$\begin{aligned}
&\mathrm{DL + LL} \rightarrow 1522/4 = 380.5 < 394.6\,\mathrm{KN/m^2}(q_{ns})\\
&\mathrm{DL + LL + EQX(DBE)} \rightarrow P/L(B - 2e_{BDBE}),\ B'_{DBE} = 1.78\,\mathrm{m}\\
&= 1522/2 \times (2 - 2 \times 0.110) = 427.5 < 526.8\,\mathrm{KN/m^2}\left(1.5q'_{ns}\right)\\
&\mathrm{DL + LL + EQX(MCE)} \rightarrow P/L(B - 2e_{BMCE}),\ B'_{MCE} = 1.558\,\mathrm{m}\\
&= 1522/2 \times (2 - 2 \times 0.221) = 448.5 < 922.2\,\mathrm{KN/m^2}\left(3q'_{ns}\right)
\end{aligned}$$

A 2 m × 2 m surface foundation is modelled in OptumG3 and checked for the MCE condition. The foundation is safe under the MCE loads.

3.7.9 Edge Column

Method 1a:

$$\begin{aligned}
&\text{DL} + \text{LL} \rightarrow P/A(1 + 6e_{BDE}/B) \\
&= 213 \times (1 + 6 \times 0.014/1.8) = 223 < 391.2\,\text{KN/m}^2(q_{ns}) \\
&\text{DL} + \text{LL} + \text{EQX(DBE)} \rightarrow P/A(1 + 6e_{BDBE}/B) \\
&= 213.1 \times (1 + 6 \times 0.191/1.8) = 446.8 < 586.8\,\text{KN/m}^2(1.5q_{ns}) \\
&\text{DL} + \text{LL} + \text{EQX(MCE)} \rightarrow P/A(1 + 6e_{BDBE}/B) \\
&= 333.2 \times (1 + 6 \times 0.304/1.8) = 670.6 < 1173.7\,\text{KN/m}^2(3q_{ns})
\end{aligned}$$

Method 2:

$$\begin{aligned}
&\text{DL} + \text{LL} \rightarrow P/L(B - 2e_{BDL}),\ B'_{DL} = 1.772\,\text{m} \\
&= 690/1.8 \times (1.8 - 2 \times 0.014) = 216 < 385.1\,\text{KN/m}^2\left(q'_{ns}\right) \\
&\text{DL} + \text{LL} + \text{EQX(DBE)} \rightarrow P/L(B - 2e_{BDBE}),\ B'_{DBE} = 1.418\,\text{m} \\
&= 885/1.8 \times (1.8 - 2 \times 0.191) = 346.7 < 462.3\,\text{KN/m}^2\left(1.5q'_{ns}\right) \\
&\text{DL} + \text{LL} + \text{EQX(MCE)} \rightarrow P/L(B - 2e_{BMCE}),\ B'_{MCE} = 1.192\,\text{m} \\
&= 1080/1.8 \times (1.8 - 2 \times 0.304) = 503.2 < 777.2\,\text{KN/m}^2\left(3q'_{ns}\right)
\end{aligned}$$

A 1.8 m × 1.8 m surface foundation is modelled in OptumG3 and checked for the MCE condition. The foundation is safe under the MCE loads.

3.7.10 Corner Column

Method 1a & 1b:

$$\begin{aligned}
&\text{DL} + \text{LL} \rightarrow P/A(1 + 6e_{BDL}/B) \\
&= 122.1 \times (1 + 6 \times 0.017/1.6) = 135.5 < 335.4\,\text{KN/m}^2(q_{ns}) \\
&\text{DL} + \text{LL} + \text{EQX(DBE)} \rightarrow 2P/3L(0.5B - e_{BDBE}) \\
&= 2 \times 493/(3 + 1.6(0.5 \times 1.6 - 0.327)) = 434.2 < 503\,\text{KN/m}^2(1.5q_{ns}) \\
&\text{DL} + \text{LL} + \text{EQX(MCE)} \rightarrow 2P/3L(0.5B - e_{BMCE}) \\
&= 2 \times 673/(3 + 1.6(0.5 \times 1.6 - 0.472)) = 855.1 < 1006.1\,\text{KN/m}^2(3q_{ns})
\end{aligned}$$

Method 2:

$$\begin{aligned}
&\text{DL} + \text{LL} \rightarrow P/L'_{DL}B'_{DL},\ L'_{DL} = 1.776\,\text{m}, \text{B}'_{DL} = 1.766\,\text{m} \\
&= 314/1.776 \times 1.766 = 99.7 < 383.8\,\text{KN/m}^2\left(q'_{ns}\right) \\
&\text{DL} + \text{LL} + \text{EQX(DBE)} \rightarrow P/L'_{DBE}B'_{DBE},\ L' = 1.782\,\text{m}, \text{B}' = 1.14\,\text{m} \\
&= 493/1.782 \times 1.146 = 241.3 < 373.6\,\text{KN/m}^2\left(1.5q'_{ns}\right) \\
&\text{DL} + \text{LL} + \text{EQX(DBE)} \rightarrow P/L'_{MCE}B'_{MCE},\ L' = 1.786\,\text{m}, \text{B}' = 0.856\,\text{m} \\
&= 673/1.786 \times 0.856 = 440.2 < 558.2\,\text{KN/m}^2\text{KN/m}^2\left(3q'_{ns}\right)
\end{aligned}$$

2 m × 2 m, 1.8 m × 1.8 m and 1.6 m × 1.6 m surface foundations are modelled in OptumG3 and checked for the MCE condition. Only the 2 m × 2 m foundation is safe under the MCE loads. The 1.6 m × 1.6 m and 1.8 m × 1.8 m foundations designed as per methods 1 and 2, as above can still survive the MCE loads, due to rocking. However, the methods do not provide any way to account for rocking explicitly.

3.7.11 *Important Points to Note*

The following are some important points to note:

1. Current practice of soil testing reports is not suitable as they do not suggest a reduction in bearing capacity due to horizontal loads and moments and they specify bearing capacity based on vertical load alone.
2. IS 1893 Part 1 does not explicitly specify which bearing capacity has to be increased. So, users are likely to increase the value of bearing capacity under vertical load alone. Bearing capacity estimated under the combined action of vertical, horizontal and moments is to be considered (see Sect. 7.3).
3. Methods 1 and 2 do not provide a way to account for rocking. Foundations designed using these methods are likely to undergo rocking under MCE loads, especially when the load eccentricity is high.

3.8 Calculation of Settlements Under Earthquake Loading

IS codes do not give any guidance on the calculation of settlements under earthquake loading. However, [54] mandates the calculation of settlements under earthquake loading in the case of cohesionless soils. Calculation of settlements under earthquake loading using methods typically used for static loads will be way off from the true settlement values. Hence, a well-established method of estimating settlements under earthquake loading is described below. Earthquakes induce settlements in both saturated and dry soils. The mechanism of earthquake-induced settlements is depicted in Fig. 3.12.

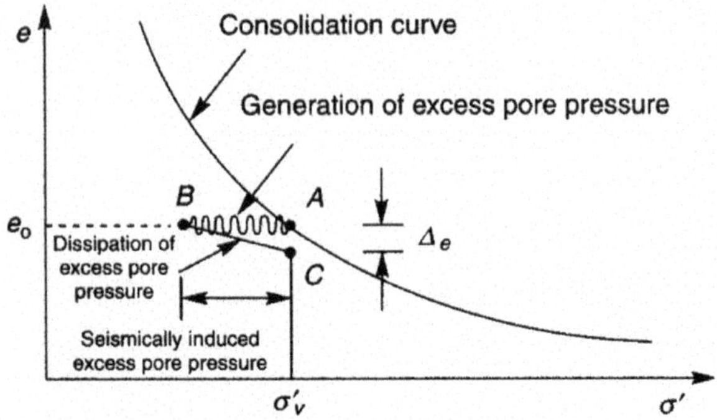

Fig. 3.12 Mechanism of earthquake-induced settlements [58]

3.8.1 Dry Sand Settlement

The current Indian code does not have any provision for the calculation of earthquake-induced soil settlements. In clays, it is not very important as clayey soils cannot experience significant settlements under vibrations induced by the earthquake loading. However, sands can experience significant earthquake-induced settlements. Dry sand settlements can be calculated by estimating cyclic shear strains induced on the soil during earthquake loading. Figure 3.13 presents a relation between the induced cyclic shear strain and expected volumetric strain of the soil for different densities of the soil and SPT-N value.

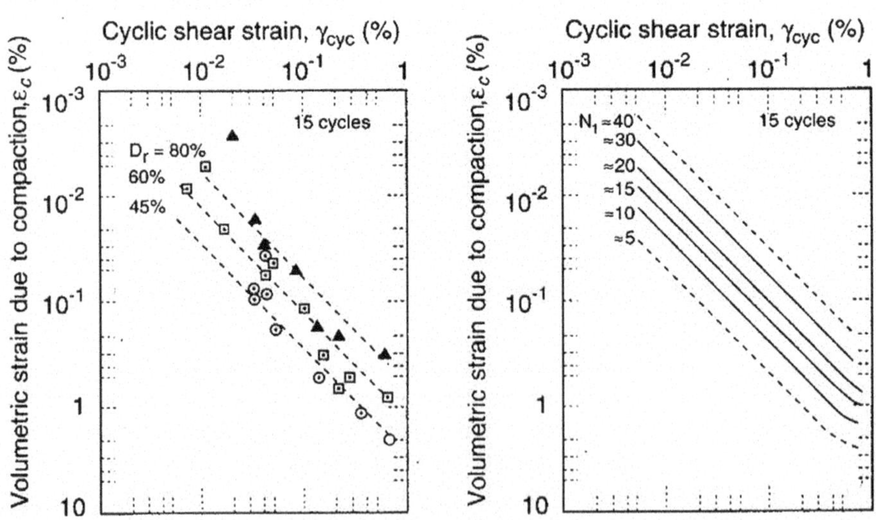

Fig. 3.13 Relationship between volumetric shear strain and cyclic shear strain in terms of **a** relative density and **b** standard penetration resistance [59]. (With permission from ASCE)

Table 3.6 Correction for volumetric settlement calculated [59]

Earthquake magnitude	$\frac{\varepsilon_{c,M}}{\varepsilon_{c,M=7.5}}$
5.25	0.4
6	0.6
6.75	0.85
7.5	1.0
8.5	1.25

Cyclic shear strain can either be estimated using detailed ground response analysis or a simplified procedure [59], using Fig. 3.13. Figure 3.13 provides volumetric strain for an earthquake magnitude of 7.5. Soil settlements under any other earthquake magnitude can be obtained using the correction given in Table 3.6.

$$\gamma_{cyc} = 0.65\frac{a_{max}}{g}\frac{\sigma_v r_d}{G(\gamma_{cyc})}$$

3.8.2 *Settlement of Saturated Sands*

Settlement of saturated sands can easily be obtained using the chart provided in Fig. 3.14. Figure 3.14 provides a relationship between liquefaction potential of the soil (Factor of safety of soil against liquefaction) and volumetric strain of the soil in terms of relative density and Standard Penetration Value.

3.9 Summary and Conclusions

Starts with a discussion on the role of seismic codes in seismic risk reduction and then goes on to explain the criteria for the selection of a suitable type of foundation. Further, it explains the various aspects of the design of a shallow foundation and the design philosophies used worldwide. The section ends with a flowchart depicting various stages in the seismic design and construction of a shallow foundation and a brief note on the various methods used worldwide to consider the combination of vertical, horizontal and moment loading simultaneously. Section 3.2 lists out the various components of earthquake loading, resisting mechanisms of the founding soil and possible shallow foundation failures when subjected to various components of earthquake loading. It then briefly mentions the various design methods employed for design against vertical loads, design against horizontal loads and design against moments. Section 3.3 discusses the various shallow foundation design alternatives. Section 3.4 discusses the underlying philosophy of Table 1 of IS 1893

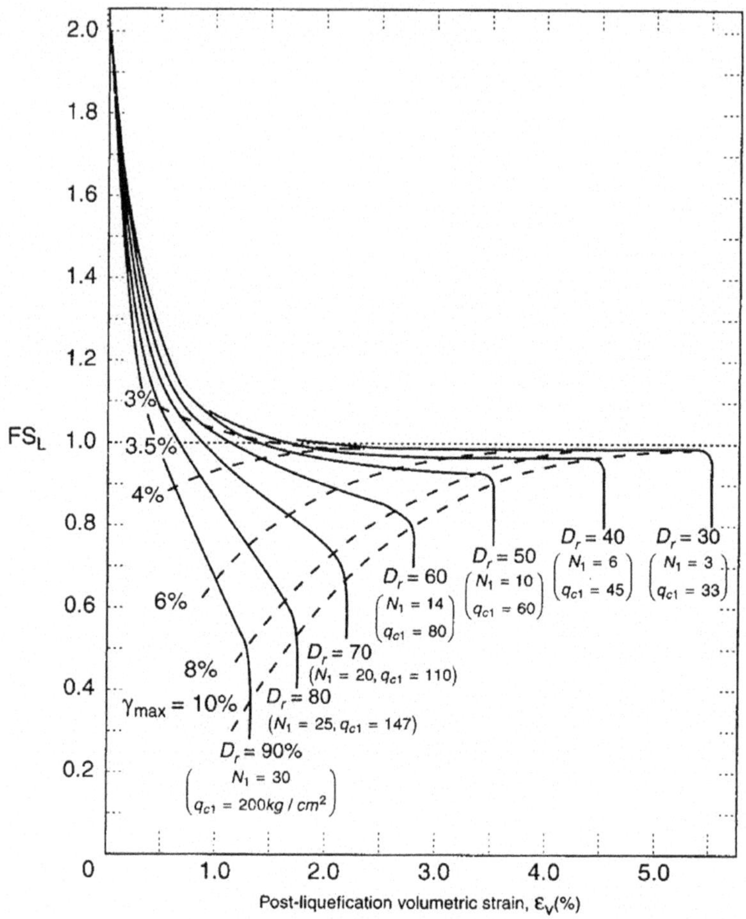

Fig. 3.14 Relationship between liquefaction potential of the soil (Factor of safety of soil against liquefaction) and volumetric strain [60]

Part 1. Section 3.5 discusses the seismic design philosophy of IS codes. Section 3.6 discusses the specific seismic effects to be considered in the computation of seismic load. Section 3.7 starts with a discussion on the Pseudostatic Method of Seismic Design as per IS 1893 Part 1 and ends with a case study illustrating the application of the IS code guidelines. Section 3.8 briefly discusses the procedures to be adopted for the computation of seismic settlements in dry sands as well as saturated sands.

3.9.1 Steps Involved in the Seismic Design of a Shallow Foundation

As mentioned in Sect. 3.5, shallow foundations shall be proportioned for the following load combinations:

1. Dead load + Live load and
2. Dead load + Live load + Seismic load.

For a symmetrical building, the minimum number of load combinations a foundation needs to be proportioned are as follows:

1. DL + LL.
2. DL + LL + EQ in positive X direction (DBE)*.
3. DL + LL + EQ in positive Y direction (DBE)*.

For an unsymmetrical building, the minimum number of load combinations* a foundation needs to be proportioned are as follows:

1. DL + LL.
2. DL + LL + EQ in positive X direction (DBE)*.
3. DL + LL + EQ in positive Y direction (DBE)*.
4. DL + LL + EQ in negative X direction (DBE)*.
5. DL + LL + EQ in negative Y direction (DBE)*.

* Accidental eccentricity must be considered, if significant.

The various steps involved in the seismic design of a shallow foundation are summarised below.

1. Select a suitable type of foundation ---> Strip/Individual/Combined/Raft.
2. Assume initial/trial dimensions of the selected foundation ----> Length (L), Width (B), Founding depth (D_f).
3. As outlined in Sect. 7.3, estimate $\boldsymbol{q_{ult}}$ (ultimate bearing capacity) using a bearing capacity theory: Terzaghi/Meyerhof/Hansen/Vesic, for each of the above relevant load combinations considering the influence of H_x,H_y,M_x,M_y, as appropriate.
4. Obtain $\boldsymbol{q_{netsafe}}$ and $\boldsymbol{Q_{netsafe}}$, for each of the above relevant load combinations.
5. Compare $\boldsymbol{Q_{netsafe}}$ with the vertical loading ($\boldsymbol{V}$) of the corresponding load combination.

   If $\boldsymbol{Q_{netsafe}} < \boldsymbol{V}$ ----------> Foundation is unsafe. Repeat steps 2 to 5, with new increased dimensions or increased depth (depending upon the type of soil).

   If $\boldsymbol{Q_{netsafe}} >> V$ ----------> Designed foundation is uneconomical. Repeat steps ii to v, with decreased dimensions or decreased depth (depending upon the type of soil).

   If $\boldsymbol{Q_{netsafe}} > \boldsymbol{V}$ ----------> Foundation is Safe. Proceed to the next step
6. For the permanent load combination, check whether $\boldsymbol{Q_{netsafe}}$ is greater than or equal to the safe bearing pressure value. If $\boldsymbol{Q_{netsafe}}$ is greater than or equal to

the safe bearing pressure value, then the foundation needs to be checked for its safety under seismic load combinations. Else, steps from 2 to 5 are to be repeated.

7. For safety against seismic load combinations, steps from 3 to 5 are to be repeated with an appropriate increase in the $q_{netsafe}$ value as per IS 1893 Part 1. The foundation dimensions need to be increased if needed.
8. Settlements in case of sands under seismic loading can be evaluated using the guidance available in Sect. 3.8 above. Evaluation of seismic settlements can help in finalising the actual foundation dimensions keeping in mind the consequences of settlement failures on the functionality of the structure in a post-earthquake scenario.

3.9.2 Concluding Remarks

Neither Peck's method (elastic stress method) nor Meyerhof's method (effective area method) provides a way to account for rocking. Foundations designed using either of these methods are likely to undergo rocking under MCE loads, especially when the load eccentricity is high.

IS 1904 [54] mandates the calculation of settlements under earthquake loading in case of cohesionless soils. However, IS codes do not give any guidance on the calculation of settlements under earthquake loading. Calculation of settlements under earthquake loading using methods typically used for static loads will be way off from the true settlement values. Hence, an appropriate method of estimating settlements under earthquake loading needs to be used for the computation of seismic settlements.

References

1. UNDRR (2015) Proposed updated terminology on disaster risk reduction: a technical review. Community, environment and disaster risk management, 7(August), pp 1–31. http://www.unisdr.org/we/inform/terminology. Accessed 24 November 2021
2. UNDRR (2017) A/RES/71/276 Report of the open-ended intergovernmental expert working group on indicators and terminology relating to disaster risk reduction | UNDRR. UN. https://www.undrr.org/publication/report-open-ended-intergovernmental-expert-working-group-indicators-and-terminology-0. Accessed 24 November 2021
3. UNISDR (2009) 2009 UNISDR terminology on disaster risk reduction, International Stratergy for Disaster Reduction (ISDR). www.unisdr.org/publications. Accessed 24 November 2021
4. Prevention Web (2021) Understanding disaster risk. https://doi.org/10.1016/c2018-0-02370-1
5. Becker DE (1996) Eighteenth Canadian geotechnical colloquium: limit states design for foundations. Part I. An overview of the foundation design process. Canad Geotech J 33(6):956–983. NRC Research Press Ottawa, Canada.https://doi.org/10.1139/t96-124
6. Meyerhof GG (1995) Development of geotechnical limit state design. Can Geotech J 32(1):128–136. https://doi.org/10.1139/t95-010
7. Poulos HG, Carter JP, Small JC (2002) Foundations and retaining structures—research and practice. In: Proceedings of the international conference on soil mechanics and geotechnical engineering. AA Balkema Publishers, pp 2527–2606

8. EN 1997-1 (2004) Eurocode 7: Geotechnical design—Part 1: General rules. European Committee for Standardization, Brussels
9. IS 6403 (1981) Code of practice for determination of bearing capacity of shallow foundations, Bureau of Indian Standards, New Delhi, India
10. ASCE 41 (2017) Seismic evaluation and retrofit of existing buildings, American society of civil engineers. Amer Soc Civil Eng.https://doi.org/10.1061/9780784414859
11. EN 1998-5 (2004) Eurocode 8: design of structures for earthquake resistance—Part 5: Foundations, retaining structures and geotechnical aspects, European Committee for Standardization, Brussels
12. ISO 19901–4 (2016) Petroleum and natural gas industries—Specific requirements for offshore structures—Part 4: geotechnical and foundation design considerations. Int Organi Standard
13. NCHRP 651 (2010) LRFD Design and construction of shallow foundations for highway bridge structures. National Academies Press. https://doi.org/10.17226/14381
14. Butterfield R, Gottardi G (1994) A complete three-dimensional failure envelope for shallow footings on sand. Geotechnique 44(1):181–184. https://doi.org/10.1680/geot.1994.44.1.181
15. Cocjin M, Kusakabe O (2013) Centrifuge observations on combined loading of a strip footing on dense sand. Geotechnique. Thomas Telford Ltd 63(5):427–433. https://doi.org/10.1680/geot.11.P.075
16. Gottardi G, Butterfield R (1993) On the bearing capacity of surface footings on sand under general planar loads. Soils Found. Elsevier 33(3):68–79. https://doi.org/10.3208/sandf1972.33.3_68
17. Gottardi G, Butterfield R (1995) Displacement of a model rigid surface footing on dense sand under general planar loading. Soils Found. Elsevier 35(3):71–82. https://doi.org/10.3208/sandf.35.71
18. Govoni L, Gourvenec S, Gottardi G (2010) Centrifuge modelling of circular shallow foundations on sand. Int J Phys Model Geotechn. Thomas Telford Ltd 10(2):35–46. https://doi.org/10.1680/ijpmg.2010.10.2.35
19. Montrasio L, Nova R (1997) Settlements of shallow foundations on sand: Geometrical effects. Geotechnique. Thomas Telford Ltd 47(1):49–60. https://doi.org/10.1680/geot.1997.47.1.49
20. Nova R, Montrasio L (1991) Settlements of shallow foundations on sand. Geotechnique. Thomas Telford Ltd 41(2):243–256. https://doi.org/10.1680/geot.1991.41.2.243
21. Chatzigogos CT, Pecker A, Salençon J (2011) A failure surface for circular footings on cohesive soils. Geotechnique. Thomas Telford Ltd 61(7):621–622. https://doi.org/10.1680/geot.10.D.005
22. Dormieux L, Pecker A (1995) Seismic bearing capacity of foundation on Cohesionless Soil.J Geotech Eng. Am Soc Civil Eng 121(3):300–303. https://doi.org/10.1061/(ASCE)0733-9410(1995)121:3(300)
23. Paolucci R, Pecker A (1997) Seismic bearing capacity of shallow strip foundations on dry soils. Soils Found 37(3):95–105. https://doi.org/10.3208/sandf.37.3_95
24. Pecker A (1998) Capacity design principles for shallow foundations in seismic areas. In: Proceedings 11th European conference on earthquake engineering, AA Balkema Publishing
25. Pecker A, Pender M (2000) Earthquake resistant design of foundations: new construction. ISRM International symposium, p ISRM-IS-2000–007
26. Salençon J, Chatzigogos CT, Pecker A (2009) Seismic bearing capacity of circular footings: a yield design approach. J Mech Mater Struct. Math Sci Publishers 4(2):427–440. https://doi.org/10.2140/jomms.2009.4.427
27. Salencon J, Pecker A (1995) Ultimate bearing capacity of shallow foundations under inclined and eccentric loads. I: ourely cohesive soil. Europ J Mech. A. Solids 14(3):349–375
28. Lesny K (2006) The role of Favourable and Unfavourable actions in the design of shallow foundations according to Eurocode 7. Amer Soc Civil Eng, 119–126. https://doi.org/10.1061/40865(197)15
29. Lesny K (2009) Safety of shallow foundations—Limit State Design according to Eurocode 7 vs. alternative design concepts. Georisk. Taylor & Francis Group 3(2):97–105. https://doi.org/10.1080/17499510802552877

30. Bransby MF, Yun GJ (2009) The undrained capacity of skirted strip foundations under combined loading. Geotechnique 59(2):115–125. https://doi.org/10.1680/geot.2007.00098
31. Feng X et al (2014) Design approach for rectangular mudmats under fully three-dimensional loading. Geotechnique 64(1):51–63. https://doi.org/10.1680/geot.13.P.051
32. Feng X et al (2019) Effectiveness of effective area method for assessing undrained capacity of shallow rectangular foundations. J Geotech Geoenviron Eng. Am Soc Civil Eng 145(2):06018013. https://doi.org/10.1061/(asce)gt.1943-5606.0002009
33. Gourvenec S (2007) Failure envelopes for offshore shallow foundations under general loading. Geotechnique. Thomas Telford Ltd 57(9):715–728. https://doi.org/10.1680/geot.2007.57.9.715
34. Gourvenec S (2007) Shape effects on the capacity of rectangular footings under general loading. Geotechnique 57(8):637–646. https://doi.org/10.1680/geot.2007.57.8.637
35. Gourvenec S (2008) Effect of embedment on the undrained capacity of shallow foundations under general loading. Geotechnique 58(3):177–185. https://doi.org/10.1680/geot.2008.58.3.177
36. Gourvenec S, Barnett S (2011) Undrained failure envelope for skirted foundations under general loading. Geotechnique 61(3):263–270. https://doi.org/10.1680/GEOT.9.T.027
37. Gourvenec S, Randolph M (2003) Effect of strength non-homogeneity on the shape of failure envelopes for combined loading of strip and circular foundations on clay. Geotechnique. ICE Publishing 53(6):575–586. https://doi.org/10.1680/geot.2003.53.6.575
38. Shen Z, Feng X, Gourvenec S (2016) Undrained capacity of surface foundations with zero-tension interface under planar V-H-M loading. Comput Geotechn. Elsevier 73:47–57. https://doi.org/10.1016/j.compgeo.2015.11.024
39. Taiebat HA, Carter JP (2000) Numerical studies of the bearing capacity of shallow foundations on cohesive soil subjected to combined loading. Geotechnique. Thomas Telford Services Ltd 50(4):409–418. https://doi.org/10.1680/geot.2000.50.4.409
40. Taiebat HA, Carter JP (2002) Bearing capacity of strip and circular foundations on undrained clay subjected to eccentric loads. Geotechnique 52(1):61–64. https://doi.org/10.1680/geot.2002.52.1.61
41. Taiebat HA, Carter JP (2010) A failure surface for circular footings on cohesive soils. Geotechnique 60(4):265–273. https://doi.org/10.1680/geot.7.00062
42. Ukritchon B, Whittle AJ, Sloan SW (1998) Undrained limit analyses for combined loading of strip footings on clay. J Geotechn Geoenviron Eng. Am Soc Civil Eng 124(3):265–276. https://doi.org/10.1061/(asce)1090-0241(1998)124:3(265)
43. Vulpe C, Gourvenec S, Power M (2014) A generalised failure envelope for undrained capacity of circular shallow foundations under general loading. Geotechnique Letters. Thomas Telford Services Ltd, 4(July–September), pp 187–196. https://doi.org/10.1680/geolett.14.00010
44. Yun G, Bransby MF (2007) The horizontal-moment capacity of embedded foundations in undrained soil. Canad Geotech J. NRC Res Press Ottawa Canada 44(4):409–424. https://doi.org/10.1139/T06-126
45. Loukidis D, Chakraborty T, Salgado R (2008) Bearing capacity of strip footings on purely frictional soil under eccentric and inclined loads. Can Geotech J 45(6):768–787. https://doi.org/10.1139/T08-015
46. Tang C, Phoon KK, Toh KC (2015) Effect of footing width on Nγ and failure envelope of eccentrically and obliquely loaded strip footings on sand. Canad Geotech J. Canadian Science Publishing 52(6):694–707. https://doi.org/10.1139/cgj-2013-0378
47. Jakka RS (2013) Earthquake resistant design of foundations: design principles. in Lecture notes, NDMA's NSSP Training Course, March 11-29, IIT Roorkee, India. https://doi.org/10.13140/RG.2.1.1492.6889
48. IBC (2018) 2018 International building code, international code council
49. AASHTO (2017) AASHTO LRFD bridge design specifications. American Association of State Highway and Transportation Officials
50. Rawat S, Mittal RK, Muthukumar G (2020) Isolated rectangular footings under biaxial bending: a critical appraisal and simplified analysis methodology. Pract Period Struct Des Constr 25(3):04020011. https://doi.org/10.1061/(asce)sc.1943-5576.0000471

51. Paulay T, Priestley MJN (1992) Seismic design of reinforced concrete and masonry buildings. Wiley, Seismic Design of Reinforced Concrete and Masonry Buildings. https://doi.org/10.1002/9780470172841
52. Fardis M et al (2015) Seismic design of concrete buildings to Eurocode 8, seismic design of concrete buildings to Eurocode 8. CRC Press
53. Moore WM, Darragh RD (1956) Some considerations in the design of foundations for earthquakes. World conference of earthquake engineering, pp 28.1–28.10. http://www.iitk.ac.in/nicee/wcee/first_conf_California/
54. IS 1904 (2021) Code of practice for design and construction of foundations in soils: general requirements, Bureau of Indian Standards, New Delhi, India
55. IS 1888 (1982) Method of load test on soils, Bureau of Indian Standards, New Delhi, India
56. IS 1893 Part 1 (2016) Criteria for earthquake resistant design of structures—Part 1: general provisions and buildings, Bureau of Indian Standards, New Delhi
57. Youd TL et al (2001) Liquefaction resistance of soils: summary report from the 1996 NCEER and 1998 NCEER/NSF workshops on evaluation of liquefaction resistance of soils. J Geotech Geoenviron Eng 127(10):817–833. https://doi.org/10.1061/(ASCE)1090-0241(2001)127:10(817)-
58. Kramer SL (1996) Geotechnical earthquake engineering. Pearson
59. Tokimatsu K, Seed HB (1987) Evaluation of settlements in sands due to earthquake shaking. J Geotechn Eng. Am Soc Civil Eng 113(8):861–878. https://doi.org/10.1061/(ASCE)0733-9410(1987)113:8(861)
60. Ishihara K, Yoshimine M (1992) Evaluation of settlements in sand deposits following liquefaction during earthquakes. Soils Found. Elsevier 32(1):178–188. https://doi.org/10.3208/sandf1972.32.173

Chapter 4
Seismic Induced Pounding of Structures and Its Mitigation

Aparna (Dey) Ghosh and Aviral Kumar

Abstract Limited availability of land resources and increasing population have led to closely spaced buildings in cities, in which the stipulated separation distance is most often not available between the buildings. When subjected to an external excitation, such as from wind or earthquake, the differences in the dynamic characteristics of these adjacent buildings cause phase differences in their responses, leading to chances of structural collision or pounding. Pounding between closely spaced buildings under earthquake excitation has been identified as a serious hazard, due to falling of building material, as well as a major cause of structural damage, that may range from minor, affecting non-structural components only, to heavy. There have been reports of significant pounding damage during several past earthquakes in not only buildings but between decks and abutments and at expansion joints of bridges as well. There is, thus, a necessity of mitigating the effects of pounding at the design stage, or in existing structures, through construction details or by the installation of vibration control devices. In this chapter, first, the various situations in which structural pounding can arise under seismic excitation are presented, followed by the types of pounding, such as one-sided pounding and two-sided pounding. A summary of the pounding damage that has been reported in past earthquakes is provided. The different pounding models developed by researchers are examined, and the effects of varying dynamic properties and separation distance on pounding are studied. Codal specifications on the minimum separation distance are highlighted and a discussion is made on the various mitigation strategies for seismic pounding.

A. (Dey) Ghosh (✉) · A. Kumar
Department of Civil Engineering, Indian Institute of Engineering Science and Technology (IIEST) Shibpur, 711103 Howrah, India
e-mail: aparna@civil.iiests.ac.in

T. G. Sitharam et al. (eds.), *Theory and Practice in Earthquake Engineering and Technology*, Springer Tracts in Civil Engineering,
https://doi.org/10.1007/978-981-19-2324-1_4

4.1 Introduction

Structural pounding refers to the lateral collision between adjacent buildings. In high-density metropolitan cities, the stipulated separation distance is often not available between buildings. This is chiefly because in the past, building codes did not specify any definite recommendations or guidelines to counteract pounding effects. This was taken to advantage and to maximize land usage as well as to avoid having large expansion joints, buildings that are adjacent to each other or within the same facility have been built very close to each other. In such closely spaced buildings, structural pounding under earthquake excitation has been identified as a serious hazard due to falling of building material as well as a major cause of structural damage that may range from minor, affecting non-structural components only, to heavy. It can lead to infill wall damage, plastic deformation, column shear failure, local crushing and possible collapse of the structure. In fact, pounding between adjacent buildings can sometimes be more dangerous than the effect of the earthquake on a single building. The damage from structural pounding can be categorized into four types, namely major structural damage, failure and falling of building scraps creating a life safety hazard, loss of building functions due to failure of mechanical, electrical or fire protection systems and architectural, non-structural and minor structural damage. An example of the effect of pounding between two closely spaced buildings is illustrated in Fig. 4.1.

When an earthquake occurs, due to the differences in the dynamic characteristics of buildings, adjacent buildings will vibrate out of phase and pounding will occur if sufficient separation distance is not provided between them (see Fig. 4.2). For neighbouring buildings with similar heights and structural properties, the effects of

Fig. 4.1 Pounding damage observed in Christchurch CBD, 2011 [1]

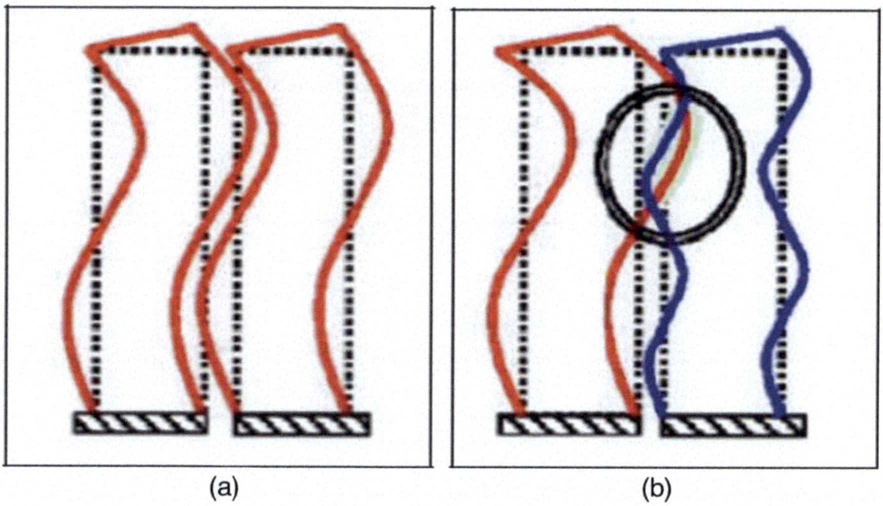

Fig. 4.2 **a** Similar seismic behavior. **b** Different seismic behaviour [2]

pounding will be limited to some local damage, mostly non-structural and light structural damage, and structural acceleration response will be higher in the form of short duration peaks. But the damage is severe when the colliding buildings have considerably different masses, periods and heights. Understandably, the pounding effect is more dangerous if the floor heights of the two adjacent buildings are different. If the floor heights of the two buildings are the same, then during the earthquake, the slab of one building hits the slab of the next building. But if the floor heights are different, then during the earthquake, the slab of one building acts like a battering ram into the column of the adjacent building and causes massive damage (Fig. 4.3). In this case, the column of one building may sustain repeated impacts along its height from the slab of the adjacent building, resulting in localized damage to the column and possibly partial collapse of the storey. Further, structures with different periods of vibration may sustain different levels of damage under identical pounding conditions during the same earthquake. Pounding may also occur because of structural irregularities that result in torsion.

There have been reports of significant pounding damage during several past earthquakes in not only buildings, but between decks and abutments, and at expansion joints of bridges as well. For the longer bridge structures, it is often the seismic wave propagation effect that is considered to be a dominant factor leading to the pounding of neighbouring superstructure segments (Fig. 4.4). This effect, due to time lag and spatial variation of seismic waves, results in different seismic inputs acting at the supports along the structure. Pedestrian bridges may also pose a potential pounding risk as they may collide against the structures to which they connect.

There is, thus, a necessity of mitigating the effects of pounding at the design stage, or in existing structures, through construction details or by the installation

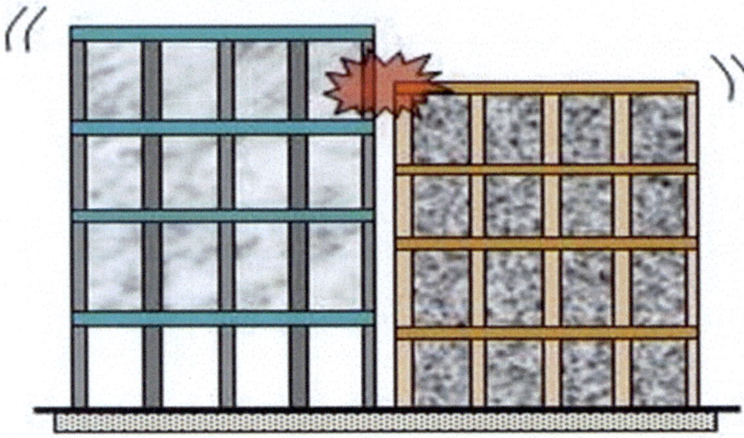

Fig. 4.3 Pounding effects: buildings with different floor heights [3]

of vibration control devices. In this chapter, first, the various situations in which structural pounding can arise under seismic excitation are presented, followed by the types of pounding, such as one-sided pounding and two-sided pounding. A concise summary of the pounding damage that has been reported in past earthquakes is included. Next, the phenomenon of structural pounding is investigated in detail and the treatment of the contact problem between two pounding structures made by researchers elaborated upon. The effects of varying structural dynamic properties and separation distance on the pounding forces generated are presented next. Lastly, various mitigation strategies for seismic pounding, along with codal specifications on the minimum seismic separation gap are highlighted.

4.2 Conditions and Types of Structural Pounding

It is clear from the foregoing discussion that structural pounding can arise under various conditions. These are enumerated below.

(a) Adjacent buildings with equal building height and equal floor height (Fig. 4.5a). This would be caused due to insufficient separation gap as well as differences in structural time periods and structural damping. This would mostly arise when the adjoining buildings are of different structural materials, e.g. one is a steel building and the other is a reinforced concrete building.

(b) Adjacent buildings with the same floor levels but different heights (Fig. 4.5b). Here, apart from the close proximity of the buildings, the marked difference in dynamic properties between the taller and more flexible building and the shorter and stiffer building would cause pronounced pounding.

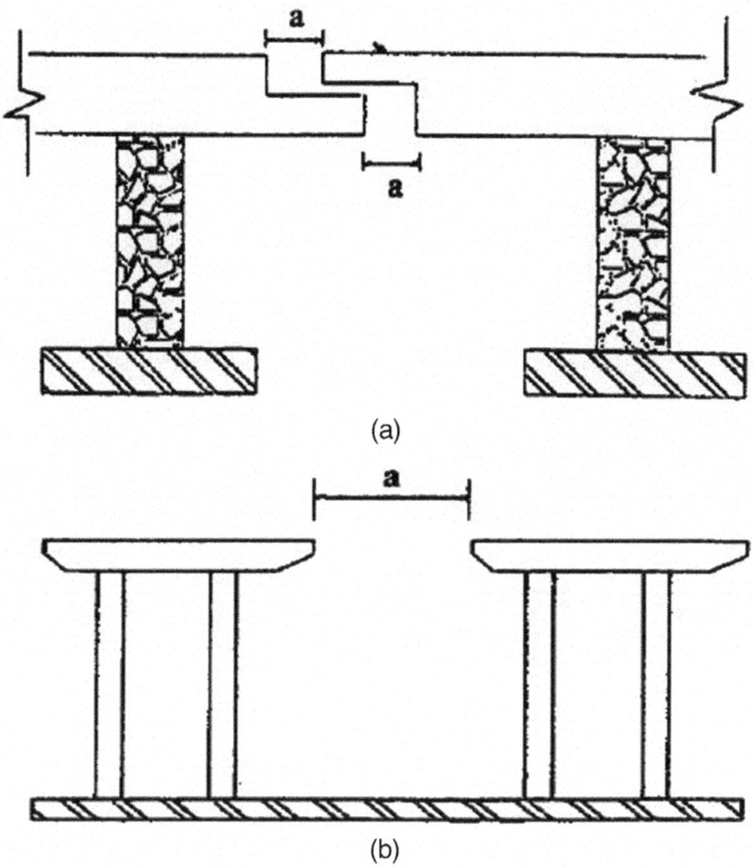

Fig. 4.4 **a** Horizontal pounding at expansion joints in bridges. **b** Lateral pounding at bridge piers [4]

(c) Adjacent structures with different total height and floor levels (Fig. 4.5c). The damage in this condition is expected to be severe due to the collision of the floor slab of one building with the column of the adjacent building at its mid-height.
(d) Structures are situated in a row (Fig. 4.5d). In this situation, two cases may arise, namely one-sided and two-sided pounding. This has been elaborated upon subsequently.
(e) Buildings having irregular lateral load resisting systems in 'plan' rotate during an 'earthquake' and due to the torsional rotations, pounding occurs near the building periphery against the adjacent buildings (Fig. 4.5e).
(f) Adjacent units of the same building, which are connected by one or more bridges, or through expansion joints, may undergo pounding if the gaps are not designed to account for the possibility of collision between the units.
(g) Possible settlement and rocking of structures located on soft soils may also lead to unanticipated pounding.

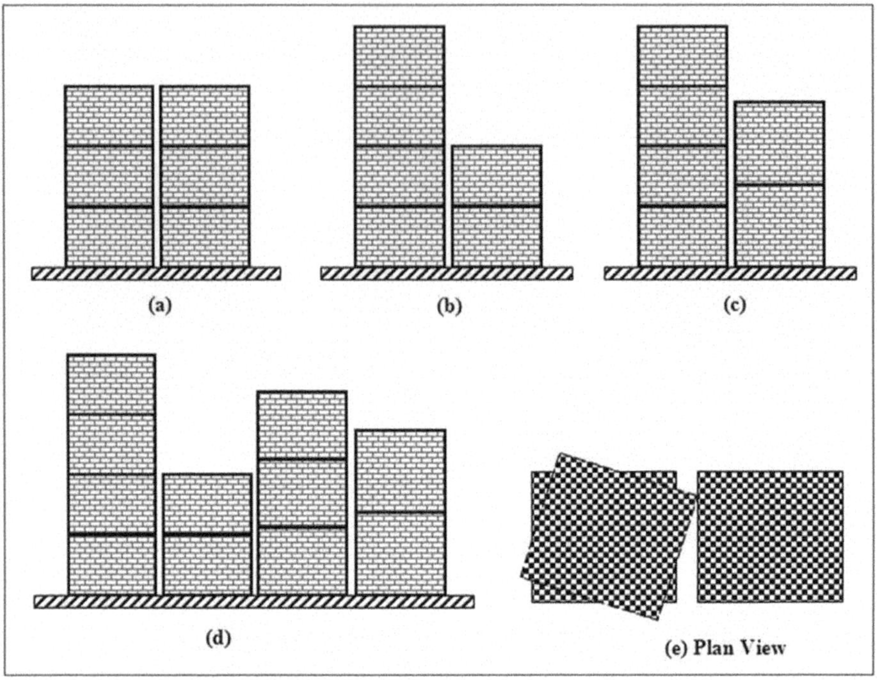

Fig. 4.5 Representation of different situations where pounding occurs [5]

One-sided and two-sided pounding

Anagnostopoulos [6] carried out a study on the seismic pounding of several adjacent buildings in a row and considered the effect of several parameters, such as building mass, time period and gap size, and reported that the end building structures experience substantial increment in response as compared to the interior one. This occurs because the building at the end of a row suffers one-sided pounding as it is free to move in the other direction.

On the other hand, if a building is situated between two other buildings, it will suffer pounding from both sides at the same time and it cannot move freely in either direction (see Fig. 4.6). For this reason, one-sided pounding is generally more hazardous than two-sided pounding.

4.3 Pounding Damage in Past Earthquakes

Some of the more notable worldwide observations on structural damage due to pounding are now discussed.

Fig. 4.6 Pounding between buildings in a row (two-sided pounding) [7]

In the Alaska earthquake of 1964, the tower of the Anchorage Westward Hotel was damaged by pounding with an adjoining three-storied ballroom portion of the hotel [8] (Fig. 4.7).

Pounding damage was also observed in Caracas during the Venezuela earthquake of 1967 (Fig. 4.8).

In the San Fernando earthquake of 1971, the second storey of the Olive View Hospital struck the outside stairway. In addition, the first floor of the hospital hit

Fig. 4.7 Anchorage Westward hotel damaged by pounding in the Alaska earthquake of 1964 [9]

Fig. 4.8 Pounding damage in Caracas during the Venezuela earthquake of 1967 [8]

against a neighbouring warehouse. The pounding of the main building against the stairway tower during the earthquake caused considerable damage at the contact points and caused permanent tilting of the tower [10] (Fig. 4.9). In subsequent years, in the 1972 Managua earthquake [11], the 1977 Romania earthquake [12], the 1977 Thessaloniki earthquake [13] and the 1981 Central Greece earthquake [14], many buildings were damaged because of structural pounding.

Mexico City, being a very congested city, recorded very significant pounding damages in the 1985 earthquake [16], so much so that structural pounding was assigned to be a leading cause of building collapse. More than 20% of the 114 affected structures were damaged because of pounding. To cite an instance, a 14-storey reinforced concrete residential building Nuevo León, located in the Nonoalco Tlatelolco apartment complex, consisted of three structurally independent units that were connected with non-structural expansion joints. After the earthquake, these joints were found to be destroyed, so it was supposed that pounding between the units had occurred and a significant impact load had been applied to the building structures.

In the Loma Prieta earthquake of 1989 [17], over 200 pounding occurrences involving more than 500 buildings took place at sites located over 90 km from the epicenter. Many old buildings with separation distances of barely 1.0–1.5 inches suffered. Structural pounding damage was observed in the form of large diagonal shear cracks in the masonry columns of a ten-storied building.

In the 1994 Northridge earthquake, the 4-inch wide seismic joint used to separate interstate 5 and state road 14 interchange of the Santa Clara River Bridge, located approximately 12 km from the epicenter, was insufficient to accommodate the relative displacements that were developed during the ground motion. This resulted in substantial pounding damage [8] (Fig. 4.10).

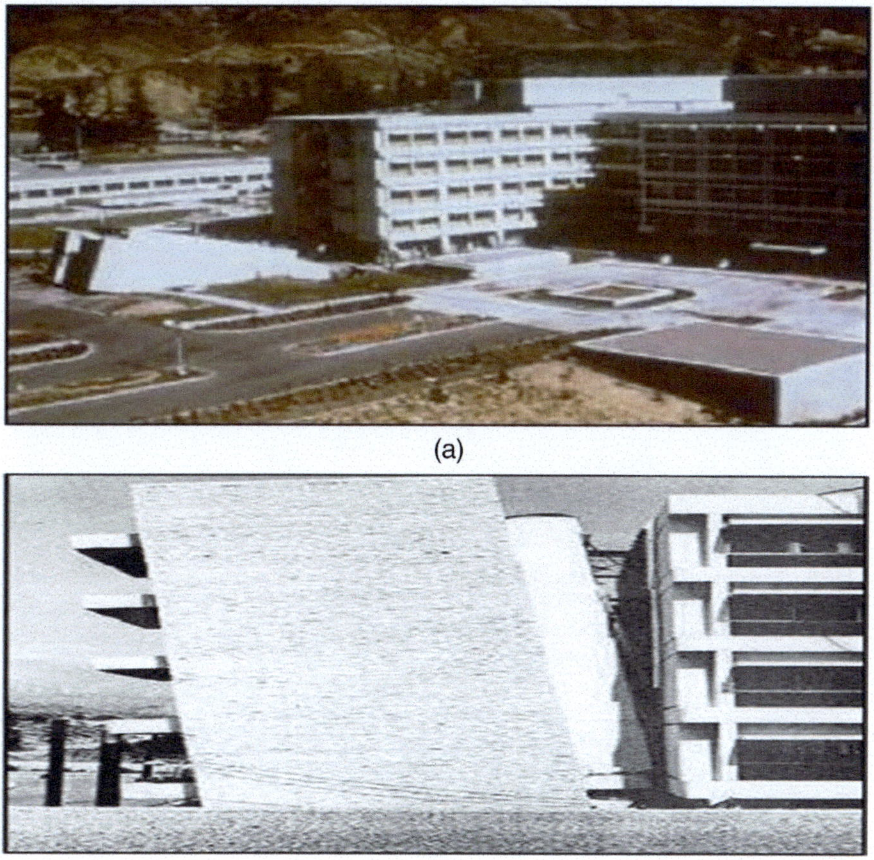

(a)

(b)

Fig. 4.9 Pounding damage of Olive View hospital. **a** View of Olive View hospital. **b** Permanent tilting of a stairway tower during the San Fernando earthquake, 1971 [15]

Poundings between adjacent decks or between a deck and an abutment occurred in the 1995 Hyogo-ken Nanbu earthquake [18], in which the expansion joints, contact faces of decks as well as the elastomeric bearings and columns were affected.

In 1999, in the Chi-Chi earthquake in central Taiwan, structural pounding was observed in several school buildings that had undergone expansion [20] (Fig. 4.11). These structures had older and newer portions with different fundamental vibration periods as well as insufficient gaps between them.

During the 2007 Niigata Chuetsu-Oki Japan earthquake too, pounding damage was observed in school buildings [20]. This type of damage occurred when the adjacent structures had floor slabs located at different elevations and insufficient separation distance between them (Fig. 4.12).

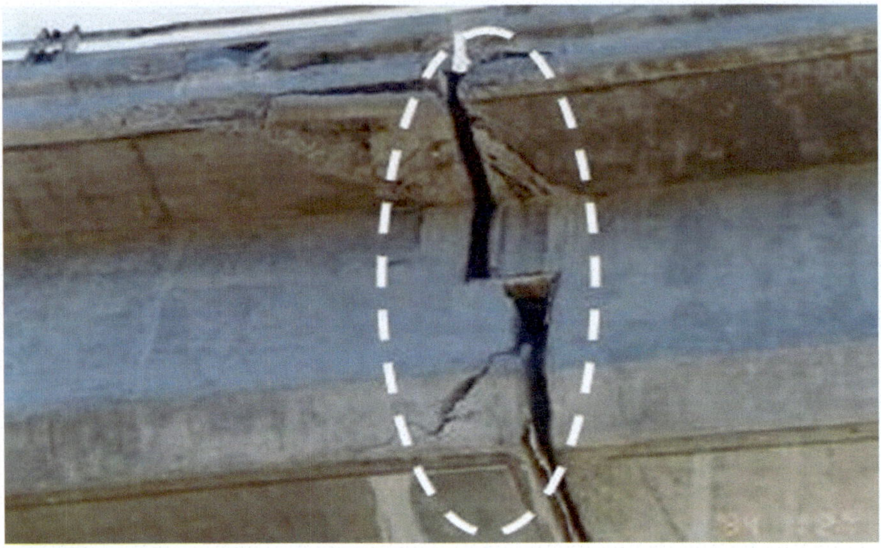

Fig. 4.10 Pounding damage in the Santa Clara River Bridge during the 1994 Northridge earthquake [19]

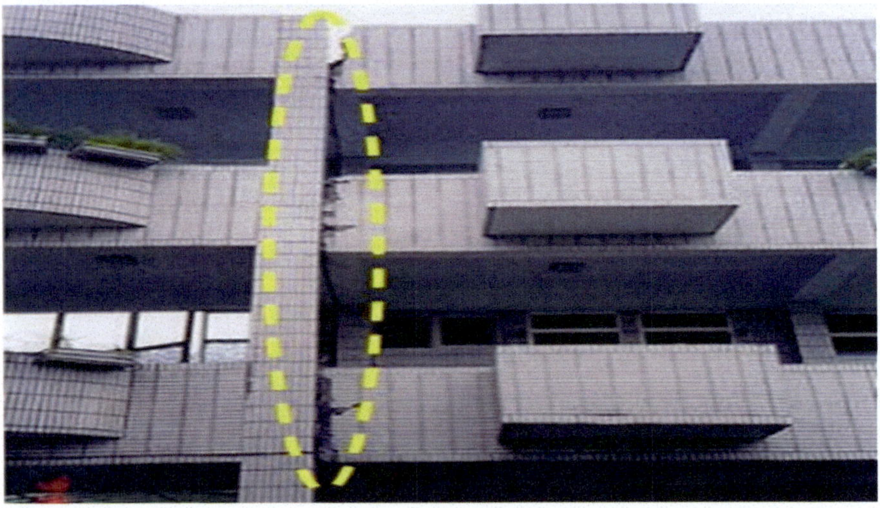

Fig. 4.11 Pounding occurrence between newer and older portions of a school building in the 1999 Chi-Chi earthquake, Taiwan [20]

Fig. 4.12 Pounding damage due to unequal slab levels during the 2007 Niigata Chuetsu-Oki Japan earthquake [21]

Other prominent observations of pounding in both building and bridge structures have been made in the 1988 Saguenay earthquake, 1992 Cairo earthquake, 1995 Kobe earthquake, 1999 Kocaeli earthquake, 2008 Wenchuan earthquake and 2011 Tohoku earthquake.

There have been several instances of structural pounding in Indian earthquakes as well. In the 2001 Bhuj earthquake [22], pounding of adjacent structures was evident in the Ayodhya apartment buildings in Ahmedabad with significant damages. In the following year, the Diglipur harbour sustained pounding damage at the intersection of the approach segment and the main berthing structure during the 2002 Diglipur earthquake (Fig. 4.13). During the Sumatra earthquake of 2004, pounding damage at junctions was noticed at the top ends of piles of the approach jetty [23]. In the 2006 Sikkim earthquake [24], pounding damages were observed between two long wings in the building and corridors connecting the wings of a nine-storey masonry infill reinforced concrete framed hostel building at the Sikkim Manipal Institute of Medical Sciences (SMIMS) Tadong, Gangtok, which caused severe damages in the walls and columns.

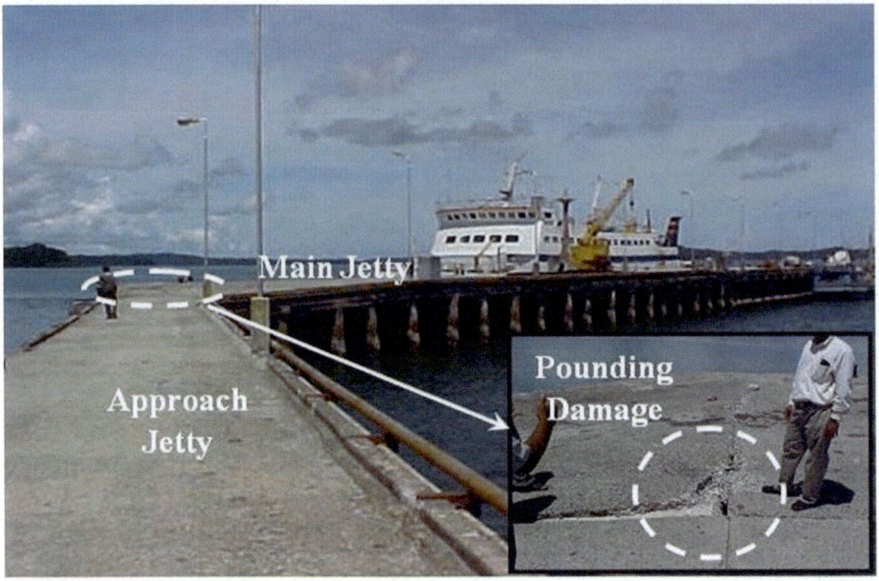

Fig. 4.13 Pounding damage at the intersection of approach to main jetty at Diglipur harbour during the 2002 Diglipur earthquake [25]

4.4 Pounding Models

Structural pounding is a complex phenomenon involving plastic deformations at contact points, local cracking or crushing, fracturing due to impact, friction, etc. The process of energy transfer during impact is highly complicated, which makes the mathematical analysis of this type of problem difficult. Forces created by collisions are applied and removed during a short interval of time, initiating stress waves which travel away from the region of contact. The collisions between adjacent buildings are simulated by means of contact elements that are activated when the bodies come in contact and deactivated if they are separated.

To facilitate an estimation of the structural responses under pounding, a simplified treatment of the contact problem between two pounding structures has been made by researchers. Anagnostopoulos [6] simulated the pounding of buildings through the use of linear, viscoelastic impact elements, and the same model was used by Jankowski et al. [26] while examining the pounding of superstructure segments in bridges. The linear, viscoelastic pounding model, though simple, assumes uniform energy dissipation during the entire pounding action, which is not a realistic one. Moreover, the contact problem is intrinsically non-linear, and this characteristic was considered by Davis [27] who modelled the interaction between two colliding structures by a Hertz contact force. The Hertzian impact model was also adopted by Pantelides and Ma [8] and Chau and Wei [28].

Though this model does not consider the energy dissipation at contact, it is a suitable model for cases where the coefficient of restitution is high, as may be assumed in colliding structures. The Hertzian impact stiffness parameter values are available for different structural materials such as concrete and steel, as well as for different contact surface geometries [29]. A more accurate non-linear viscoelastic impact element model that accounts for the energy dissipation that takes place during the approach period of the pounding was developed by Jankowski [30]. However, the impact stiffness and damping ratio parameters have to be evaluated iteratively, and their values for different material and surface geometry characteristics are needed for greater applicability of this model. An overview of the various modelling techniques is presented below.

The stereo-mechanical theory of impact is the classical formulation of the problem of impacting bodies. The stereo-mechanical approach assumes instantaneous impact and uses momentum balance and the coefficient of restitution to modify the velocities of the colliding bodies after impact. The original theory considered the impacting bodies as rigid; later, a correction factor to account for energy losses was introduced. The theory concentrates on determining the final velocities of two impacting bodies depending on their initial velocities and a coefficient of restitution to account for plasticity during impact. The final velocities of the bodies are determined from the following equations:

$$V_1' = V_1 - (1+e)\frac{m_2(V_1 - V_2)}{m_1 + m_2} \tag{4.1}$$

$$V_2' = V_2 + (1+e)\frac{m_1(V_1 - V_2)}{m_1 + m_2} \tag{4.2}$$

where V_1' and V_2' are final velocities, V_1 and V_1 are initial velocities of the colliding bodies, e is the coefficient of restitution and m_1 and m_2 denote the masses of the bodies.

Amongst the various contact force-based models, the simplest contact element consists of a linear elastic element (Fig. 4.14). The spring is assumed to have restoring force characteristics such that only when the relative distance between the masses becomes smaller than the initial distance, the spring contracts and generates forces

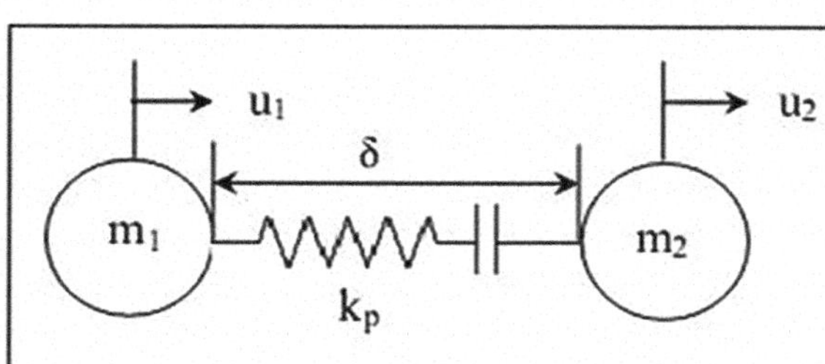

Fig. 4.14 Linear spring model

which enable us to consider the phenomenon of pounding. This collision spring is assumed to be the axial stiffness of the floors and the beams in each storey.

The force in the contact element can be expressed according to the equations given below.

$$\boldsymbol{F_c} = \boldsymbol{K_l}\left(\boldsymbol{u_1} - \boldsymbol{u_2} - \boldsymbol{g_p}\right); \boldsymbol{u_1} - \boldsymbol{u_2} - \boldsymbol{g_p} > 0 \tag{4.3}$$

$$\boldsymbol{F_c} = 0; \boldsymbol{u_1} - \boldsymbol{u_2} - \boldsymbol{g_p} \leq 0 \tag{4.4}$$

Here, u_1 and u_2 are the displacements of the impacting bodies, K_l is the spring constant of the element and $g_p(= \delta$ in Fig. 4.14) is the separation distance between the structures.

However, in this, the energy loss during impact cannot be modelled. Whenever two mechanical systems collide, there is an exchange of momentum and also energy is dissipated in the high stress region of contact. The energy loss is taken into account by the Kelvin–Voight element model.

The Kelvin–Voight element is represented by a linear spring in parallel with a damper. This impact model is thus capable of modelling the energy dissipation during impact and the impact force is represented by the equations given below.

$$\boldsymbol{F_c} = \boldsymbol{K_k}\left(\boldsymbol{u_1} - \boldsymbol{u_2} - \boldsymbol{g_p}\right) + \boldsymbol{C_k}(\boldsymbol{u_1} - \boldsymbol{u_2}); \boldsymbol{u_1} - \boldsymbol{u_2} - \boldsymbol{g_p} > 0 \tag{4.5}$$

$$\boldsymbol{F_c} = 0; \boldsymbol{u_1} - \boldsymbol{u_2} - \boldsymbol{g_p} \leq 0 \tag{4.6}$$

where u_1, u_2 and their derivatives are the displacements and velocities of the impacting bodies, K_k is the spring constant of the element and g_p ($= \delta$ in Fig. 4.15) is the separation distance between the structures. The damping coefficient C_k can be related to the coefficient of restitution (e), by equating the energy losses during impact.

$$\boldsymbol{C_K} = 2\zeta\sqrt{\boldsymbol{K_k}\frac{\boldsymbol{m_1 m_2}}{\boldsymbol{m_1} + \boldsymbol{m_2}}} \tag{4.7}$$

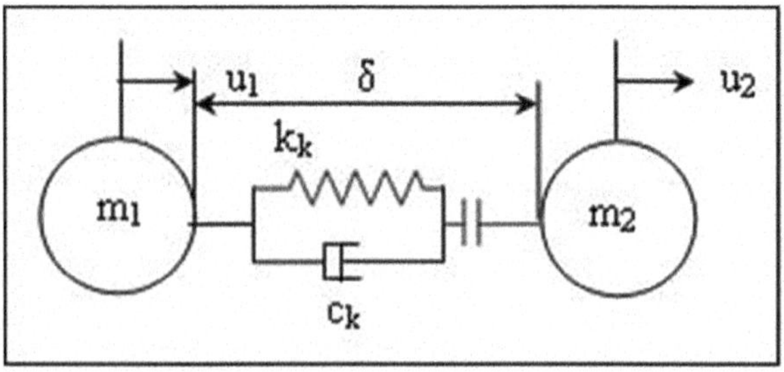

Fig. 4.15 Kelvin–Voight model

$$\zeta = \boldsymbol{lne}/\sqrt{\pi^2 + (\boldsymbol{lne})^2} \tag{4.8}$$

Here, m_1 and m_2 are the masses of the colliding bodies and ζ is the damping coefficient.

The disadvantage of the Kelvin model is that its viscous component is active with the same damping coefficient during the entire time of the collision. The damping forces cause negative impact forces that pull the colliding bodies together, during the unloading phase, instead of pushing them apart. Ye et al. [31] re-examined and modified the Kelvin model and the theoretical derivation has been verified with numerical experiment. The corrected damping ratio value can be expressed as

$$\zeta_k = 3\boldsymbol{K}_k \frac{(1+\boldsymbol{e})}{2\boldsymbol{e}(\mathbf{v}_1 - \mathbf{v}_2)} \tag{4.9}$$

where v_1 and v_2 are the initial velocities of colliding bodies.

One of the most popular models for representing pounding is the Hertz model, which uses a non-linear spring of stiffness (K_h). The impact force is represented by the following equations:

$$\boldsymbol{F}_c = \boldsymbol{K}_h(\boldsymbol{u}_1 - \boldsymbol{u}_2 - \boldsymbol{g}_p)^{\boldsymbol{n}};\ \boldsymbol{u}_1 - \boldsymbol{u}_2 - \boldsymbol{g}_p > 0 \tag{4.10}$$

$$\boldsymbol{F}_c = 0;\ \boldsymbol{u}_1 - \boldsymbol{u}_2 - \boldsymbol{g}_p \le 0 \tag{4.11}$$

The use of the Hertz contact law has a distinct advantage in modelling pounding, since one would expect the contact area between the colliding structures to increase as the contact force increases, leading to a non-linear stiffness described by the Hertz coefficient (n). The impact stiffness parameter, K_h, depends on the material properties of the colliding structures and the contact surface geometry. The Hertz coefficient, n, is typically taken as 1.5. Studies have shown that the system displacement response is relatively insensitive to the exponent, n, in the contact law. The contact force–displacement relationship for the various impact models is shown in Fig. 4.16.

The Hertz model suffers from the limitation that it cannot represent the energy dissipated during contact. Hence, an improved version of the Hertz model is considered, whereby a non-linear damper is used in conjunction with the Hertz spring. Similar models have been used in other areas such as robotics, and multi-body systems. However, its efficiency in structural engineering has not been considered.

In the Hertz damp model, the contact force can be expressed as

$$\boldsymbol{F}_c = \boldsymbol{K}_h(\boldsymbol{u}_1 - \boldsymbol{u}_2 - \boldsymbol{g}_p)^{\boldsymbol{n}} + \boldsymbol{C}_h(\boldsymbol{u}_1 - \boldsymbol{u}_2);\ \boldsymbol{u}_1 - \boldsymbol{u}_2 - \boldsymbol{g}_p > 0 \tag{4.12}$$

$$\boldsymbol{F}_c = 0;\ \boldsymbol{u}_1 - \boldsymbol{u}_2 - \boldsymbol{g}_p \le 0 \tag{4.13}$$

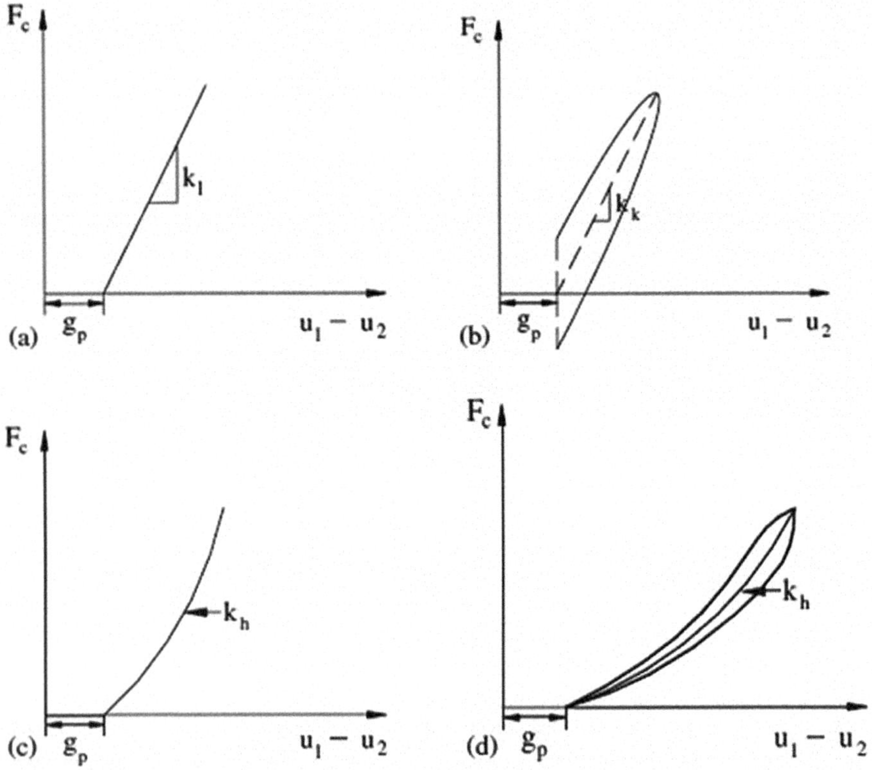

Fig. 4.16 Contact force–displacement relationship for various impact models: **a** Linear spring element; **b** Kelvin model; **c** Hertz non-linear spring; and (d) Hertz damp model [32]

where C_h is the damping coefficient, $(u_1 - u_2 - g_p)$ is the relative penetration and $(u_1 - u_2)$ is the penetration velocity. A non-linear damping coefficient is proposed so that the expected hysteresis loop during impact matches the one shown in Fig. 4.16d.

The damping coefficient is taken as follows:

$$C_h = \zeta \delta^n \tag{4.14}$$

where ζ is the damping constant, and δ is the relative penetration $(u_1 - u_2 - g_p)$.

Equating the energy loss during the stereo-mechanical impact to the energy dissipated by the damper, an expression for the damping constant () can be found in terms of the spring stiffness K_h, the coefficient of restitution (e) and the relative approaching velocity $(v_1 - v_2)$, as follows:

$$\delta = \frac{3K_h(1 - e^2)}{4(v_1 - v_2)} \tag{4.15}$$

Hence, the force during contact can be expressed as

$$F_c = K_h(u_1 - u_2 - g_p)^n \left[1 + \frac{3(1-e^2)}{4(\mathbf{v}_1 - \mathbf{v}_2)}(u_1 - u_2)\right]; u_1 - u_2 - g_p > 0 \tag{4.16}$$

$$F_c = 0; u_1 - u_2 - g_p \le 0 \tag{4.17}$$

4.5 Effect of Varying Structural Dynamic Properties and Separation Distance on Pounding

The pounding behaviour of structures is largely dependent on the predominant excitation frequencies, compared to the natural frequencies of the structure. It is seen that for structures with different natural periods, the same earthquake excitation can produce different magnitudes of pounding force and resulting structural responses. Again, pounding results in modification of the responses of the colliding structures. Papadrakakis et al. [33] performed shake table tests on pounding between two-storey reinforced concrete buildings, subject to sinusoidal excitation. The results indicated that pounding amplified the displacement responses of the stiffer structure and reduced the responses of the flexible structure. Anagnostopoulos et al. [34] extended their studies from a series of SDOF systems to MDOF. They idealized the buildings as lumped mass, shear beam type and found that if there are large differences in the masses of the colliding buildings, then pounding can cause more damage to the building with the smaller mass. Goltabar et al. [35] considered pounding between buildings of different heights and pointed out that the impact causes an increment in responses of taller buildings but the reverse in shorter ones. Adjacent multi-storied buildings subjected to seismic pounding were also considered by Abdel-Mooty et al. [36], and the results indicated that pounding force increases when the difference in dynamic properties of adjacent buildings increases. They also observed that the top colliding floors are subjected to maximum pounding force and that the number of pounding decreases as separation distance increases.

A systematic study was carried out by Pantelides and Ma [8] to examine the behaviour of a damped SDOF structural system with one-sided pounding during an earthquake. Their findings, which serve to illustrate the effect of the dynamic properties of a structure as well as that of the seismic separation distance on the pounding behaviour, are discussed in some detail now. The model proposed by Davis [27] was used. In Fig. 4.17, a simplified SDOF idealization of the damped elastic structural system undergoing one-sided structural pounding is made. The barrier in Fig. 4.17 is assumed to be rigid, and both the structure and the barrier are subjected to the same ground motion. The rigid barrier can be justified by considering the adjoining structural system to be a very stiff block of squat buildings.

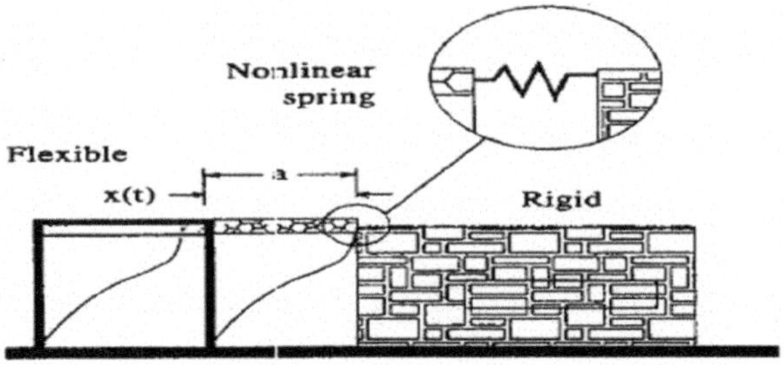

Fig. 4.17 Structural SDOF system model with adjacent barrier [8]

$x(t)$ is the lateral displacement of the flexible structure relative to ground, a is the separation distance between the flexible and rigid structure and xg is the earthquake ground acceleration. Here, the pounding force is denoted by $F(t)$ which is the impact force between the SDOF system and the neighbouring rigid structural model through a Hertz non-linear spring.

The impact force is expressed as (Pantelides et al. (1989)) [8]

$$F(t) = R[x(t) - a]^{3/2}, \quad x(t)a \tag{4.18}$$

$$F(t) = 0, \quad x(t) < a \tag{4.19}$$

where R is the impact stiffness parameter which depends on the material of the two structures that come in contact and also on the contact surface geometry. The value of the impact stiffness parameter is 80 kN mm$^{-3/2}$ for concrete structures.

The equation of motion for the mass of the flexible structure subjected to one-sided pounding is

$$m\ddot{x}(t) + c\dot{x}(t) + kx(t) + F(t) = -m\ddot{x}_g(t) \tag{4.20}$$

where m, k, c denote the mass, damping and elastic stiffness of the SDOF model of the structure.

Substituting Eqs. (4.18) and (4.19) in Eq. (4.20), the equation of motion considering pounding is

$$m\ddot{x} + c\dot{x}(t) + kx(t) + R[x(t) - a]^{\frac{3}{2}} = -m\ddot{x}_g \tag{4.21}$$

When pounding does not occur, the equation of motion is

$$m\ddot{x} + c\dot{x}(t) + kx(t) = -m\ddot{x}_g \tag{4.22}$$

Here, to examine the effect of the earthquake excitation's predominant frequency on the response of pounding of an elastic structural system, a sinusoidal base motion with a magnitude of 0.1 g and duration of 30 s is first considered, with two cases of excitation frequency, namely $f_1 = 2$ Hz and $f_2 = 0.67$ Hz. In addition, the effects of the separation gap and damping ratio of the elastic structure on the magnitude of the pounding force are investigated. The properties of the example SDOF elastic structure considered are mass $m = 350$ Mg, elastic stiffness $Ke = 10.5$ kN/mm and damping ratio $\zeta = 2\%$. The period of the structural system is 1.15 s (0.87 Hz). The separation distance is assumed to be $a = 25$ mm.

Figure 4.18a and b indicates the displacement and acceleration time history of the SDOF structure for the two excitations. Figure 4.18c shows the pounding force in the case of excitation f_2. The value of the peak pounding force is obtained as about 4.5 MN. As shown in Fig. 4.18a, when the excitation frequency is equal to f_1, the displacement never exceeds 25 mm and no pounding occurs. When the excitation has a frequency equal to f_2, both displacement and acceleration responses are significantly large and pounding occurs many times. This happens because in this case, the excitation frequency f_2 is closer to the natural frequency of the structural system than the frequency f_1. The maximum displacement for f_2 is 53 mm and the maximum acceleration is 1.29 g. The displacement and acceleration responses of the structure for no pounding case are given in Fig. 4.18d and e, respectively. The maximum displacement and peak acceleration for the no pounding case are 125.2 mm and 0.28 g, respectively. The maximum displacement of structure in the pounding case is much lower than that of the no pounding case, while the peak acceleration increases several times in the pounding case as compared to the no pounding case. This happens because, due to the rigid obstruction, the displacement response is reduced in the pounding case but the acceleration response is highly increased due to several collisions with the obstruction.

Next, in order to examine the effect of the separation gap on the maximum pounding force and number of impacts, a series of separation distances are studied from 12.5 mm to where pounding does not occur, at an interval of 12.5 mm. The results are given in Fig. 4.19a and b.

In Fig. 4.19a, at 12.5 mm separation distance, maximum pounding force occurs (7.5MN) and the number of pounding is also maximum (215 times). As the gap increases, the number of impacts decreases (Fig. 4.19b). But in Fig. 4.19a, as the separation gap increases up to 35 mm, the maximum pounding force decreases, but again increases giving a peak (6.8MN) when the gap is 75 mm and then it decreases gradually. The interesting observation from Fig. 4.19b is that there is a sharp decrease in the number of impacts with an initial increase in the separation gap, beyond which the number of poundings remains almost constant for a wide range of the value of the separation gap.

Next, pounding under a recorded earthquake is studied. Two different structural time periods, T = 1 s and 3 s, are subjected to the El Centro earthquake data of 18 May 1940. Figure 4.20a–c indicates the displacement response, the acceleration response and the pounding force generated, respectively.

Though pounding occurs in both cases, the peak displacement of the more flexible structure with period 3 s is approximately 2.46 times that of the structure with period 1 s, while the peak acceleration of the structure with period 1 s is approximately 1.34 times that of the structure with period 3 s. Further, the peak pounding force of

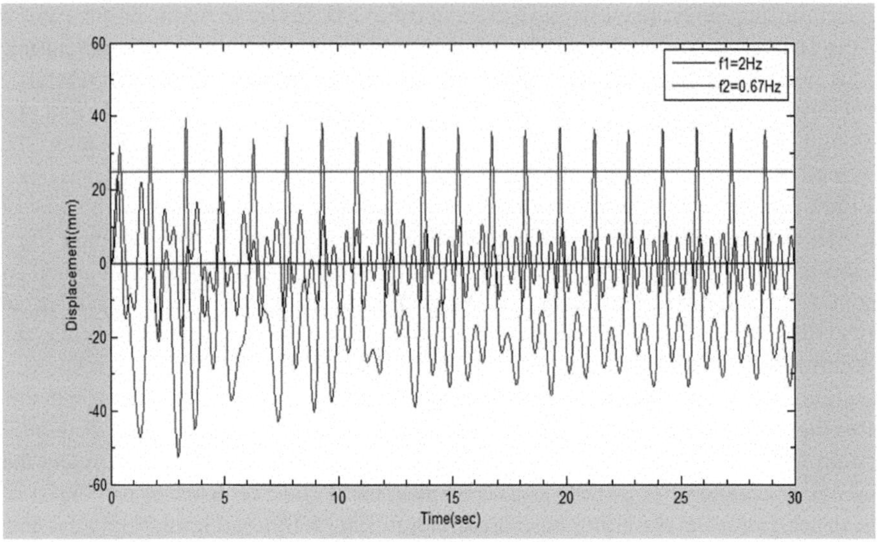

(a)

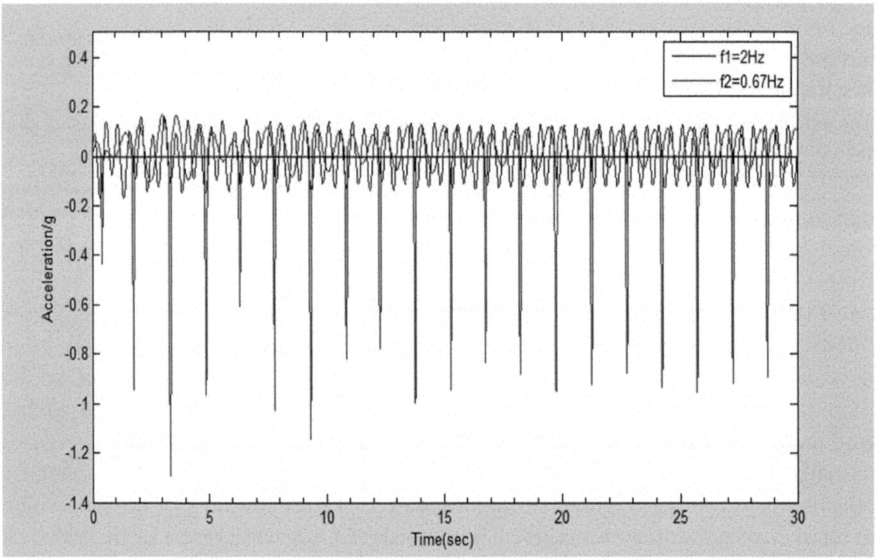

(b)

Fig. 4.18 Response of SDOF elastic structure during artificial sinusoidal earthquake when pounding occurs: **a** Displacement response, **b** Acceleration response, **c** Pounding force, **d** Displacement response when pounding does not occur **e** Acceleration Response when pounding does not occur

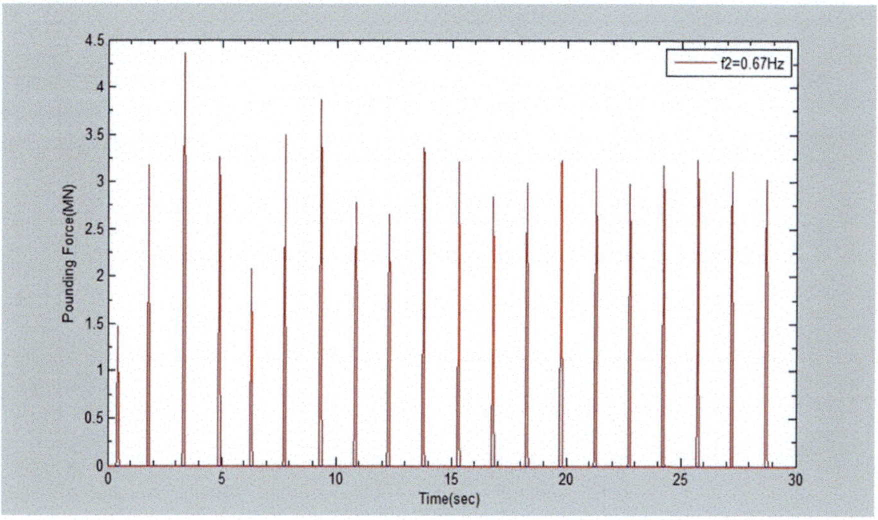

(c)

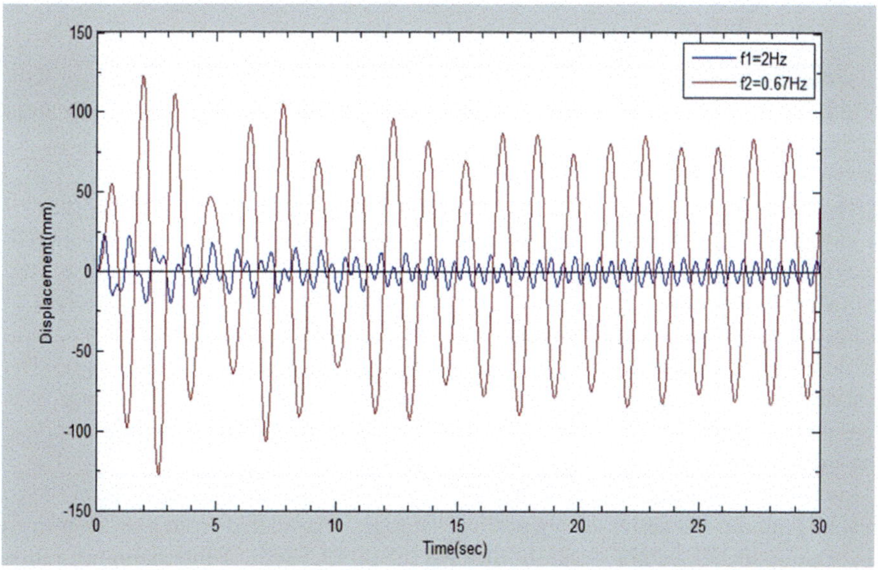

(d)

Fig. 4.18 (continued)

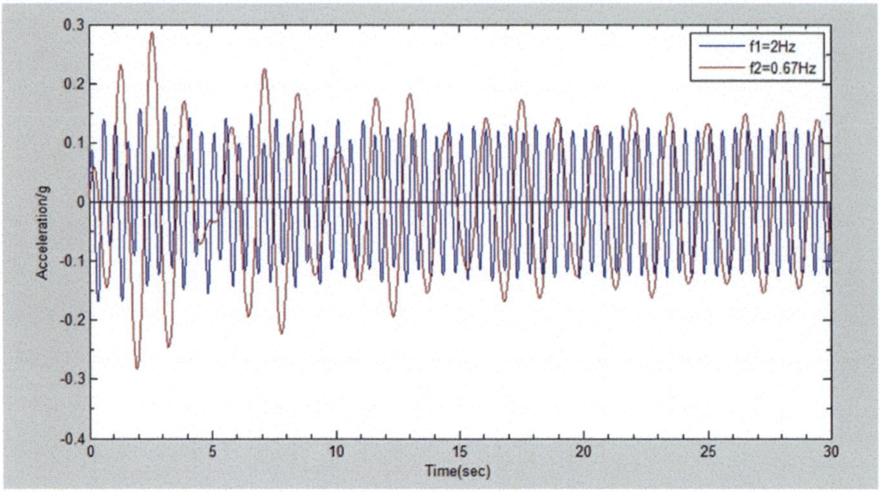

(e)

Fig. 4.18 (continued)

the structure with period 1 s is approximately 1.3 times that of the structure with period 3 s. These results indicate that structures with different periods of vibration may sustain different levels of damage under identical pounding conditions during the same earthquake.

The pounding response of the structure is also dependent on the damping ratio of the structure. The SDOF example structural system with period T = 1 s is considered. Two damping ratios are considered, namely 2 and 8%. Figure 4.21a–c indicates the displacement response, acceleration response and pounding force, respectively.

For the case of increased damping, peak displacement is reduced by 52.11%, peak acceleration is reduced by 65.12% and peak pounding force is reduced by 65.7%. In addition, the number of pounding occurrences is reduced by 65.3%. Hence, it is clear that supplemental damping in the structure can mitigate the pounding response very effectively.

In order to examine the effect of damping on the separation distance required to prevent pounding, a series of separation distances are studied from a = 25 mm to where pounding does not occur, at an interval of 12.5 mm. Three damping ratios are considered, namely 2, 8 and 20%. The results are shown in Fig. 4.22a and b. The input excitation is the 1940 El Centro earthquake. In Fig. 4.22a, the pounding force is not always reduced with an increase in the separation gap when the damping ratio is low. Even for structural damping as high as 8%, the pounding force remains constant over a large range of the separation gap. Only for very high damping, say 20%, the pounding force decreases linearly with an increase in separation gap. In Fig. 4.22b, it is observed that for lower structural damping, the number of poundings also does not decrease all the time with an increase in separation gap. Even for a high

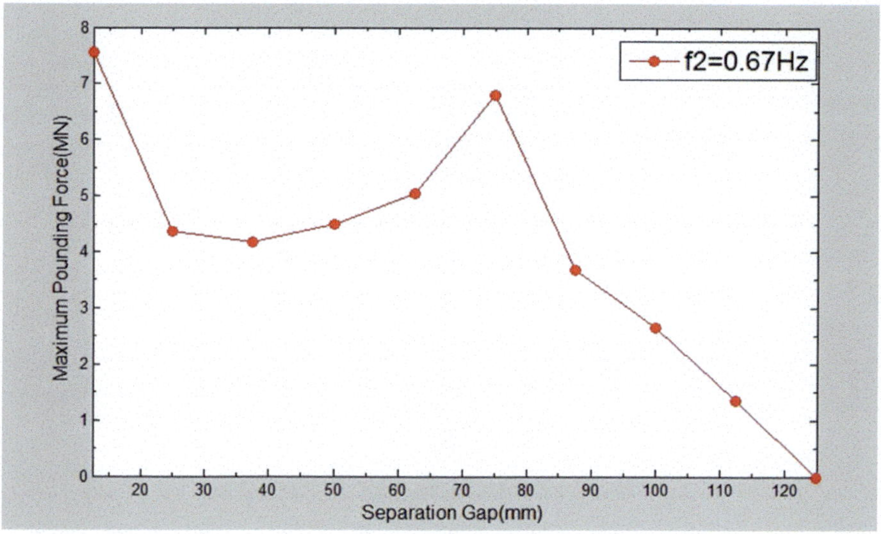

(a)

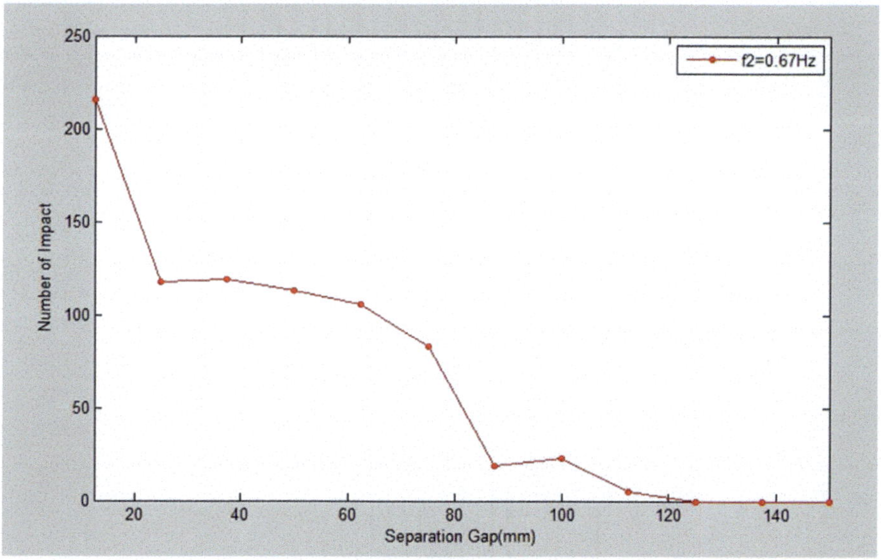

(b)

Fig. 4.19 Effect of seismic gap on maximum level of pounding force and number of impact

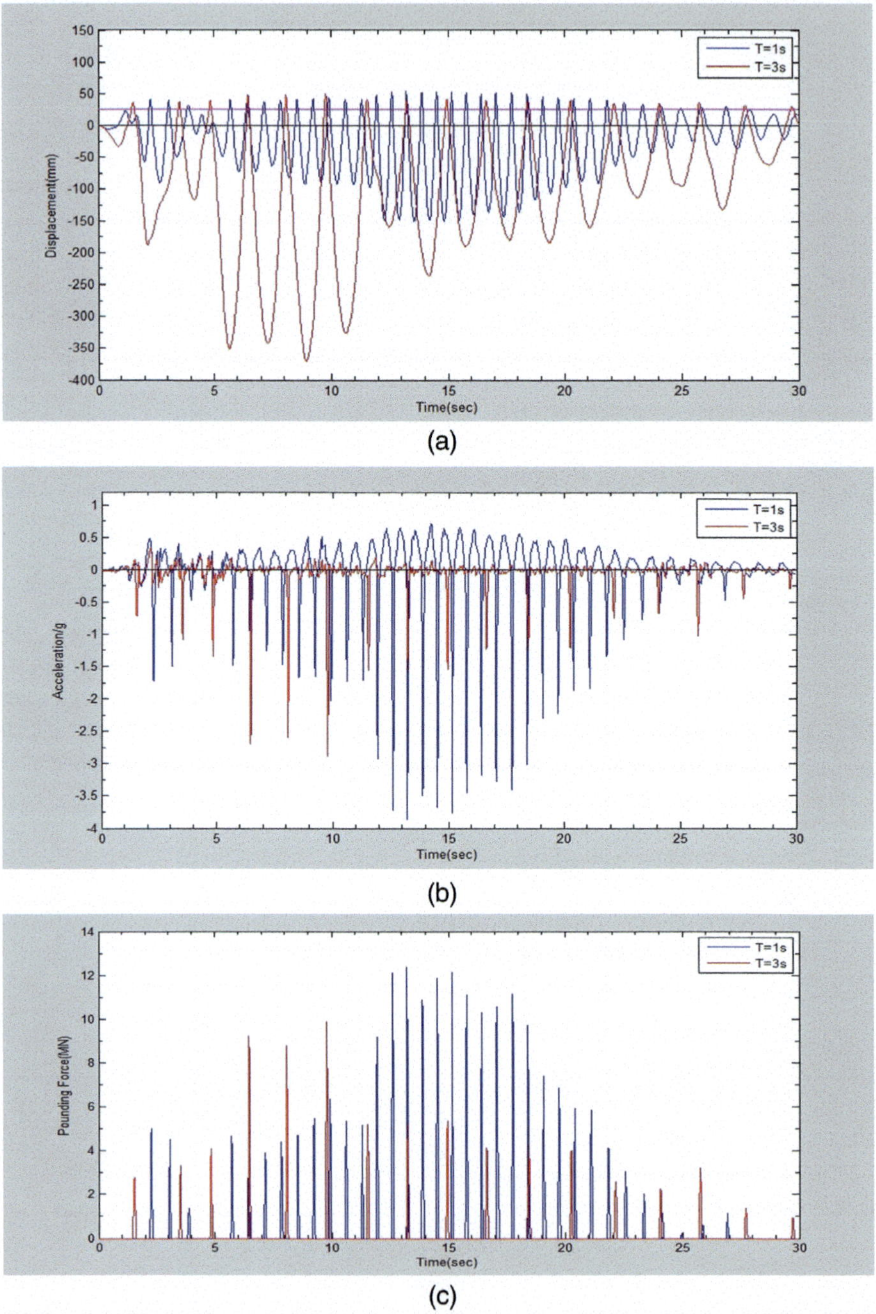

Fig. 4.20 Response of SDOF elastic structure during 1940 El Centro earthquake: **a** Displacement, **b** Acceleration and **c** Pounding force

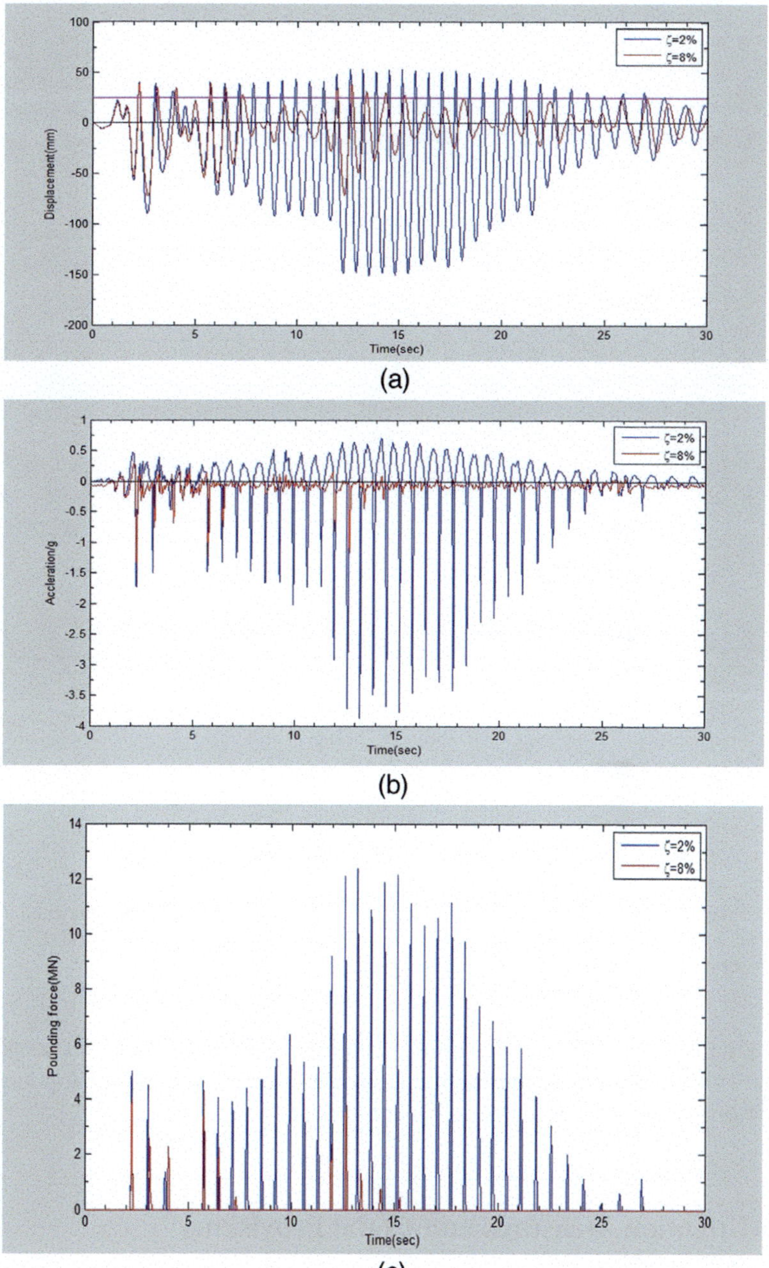

Fig. 4.21 Effect of damping on pounding of elastic structure: **a** Displacement, **b** Acceleration and **c** Pounding force

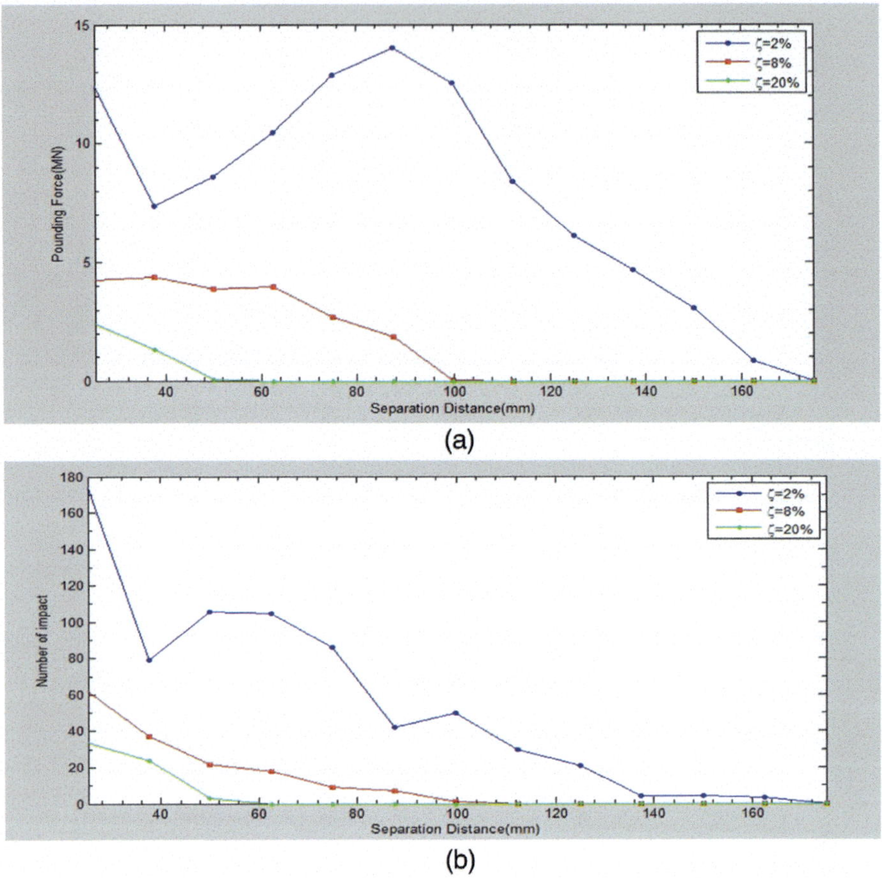

Fig. 4.22 Effect of damping on separation distance, maximum level of pounding force and number of impacts: **a** Effect of separation distance on maximum pounding force generated, **b** Effect of separation distance on number of pounding occurred

damping ratio, say 8%, it remains almost constant over a large range of separation gap. Only for a very high damping ratio, say 20%, does the number of poundings decrease linearly with an increase in separation gap.

4.6 Mitigation Measures and Codal Provisions for Pounding

It is clear from the foregoing sections that there is an utmost necessity of mitigating the effects of pounding both at the design stage as well as in existing structures. It is of course worth mentioning that buildings having simple regular geometry, uniformly distributed mass and stiffness in the plan as well as in elevation would perform better under pounding conditions than buildings with irregular configurations. Anagnostopoulos et al. [34] observed that the effects of pounding were reduced as the separation distance increases, and many building codes also suggest this method. Goltabar et al. [35] concluded that by maintaining proper distance between adjacent structures and by hardening buildings, the effect of impact could be reduced. The use of special ‘bumper’ shear walls has been suggested as a good alternative to the separation distance requirement, especially for situations where a new building is to be built next to an existing building with practically no gap in between the two. Bumper damper elements are link elements that are activated when the gap is closed. Such elements reduce energy transfer during pounding and high-frequency pulses. These bumper elements have already been incorporated in the Greek code and in Euro code-08 for earthquake-resistant design.

Different techniques to control pounding in existing structures have been proposed, ranging from simple retrofitting schemes to the provision of supplemental energy dissipation systems in the pounding structures. Westermo [37] recommended a simple reconstruction technique to avoid pounding by connecting building structures through a link and beam system which transmits the connection forces to the floors of the structures (see Fig. 4.23).

The mitigation of pounding response in bridges can be achieved by the use of restrainers, as in Fig. 4.24, and by variable dampers.

The use of crushable devices and shock transmission units to control pounding in elevated bridges has also been considered. Jankowski et al. [26] confirmed that the pounding in elevated bridges could be avoided by inserting hard rubber bumpers between segments and by linking them to one another. The results of their experiment indicate that for the workable application of such links, shock transmission units can be used. The bumper layer was also used by Cheng et al. [39] to mitigate the effect of pounding on concrete, liquid storage structures.

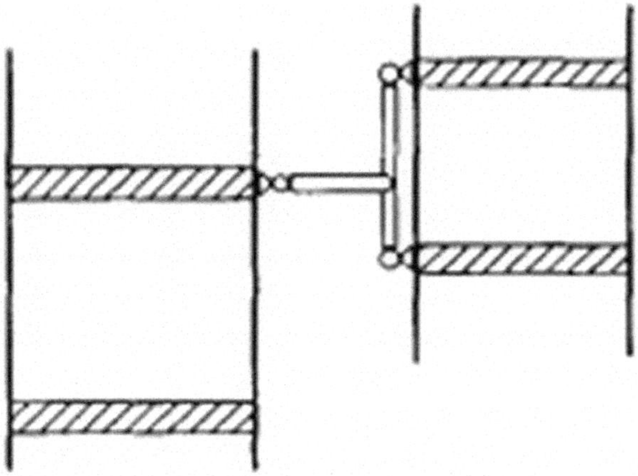

Fig. 4.23 Connecting buildings by a link and beam element [37]

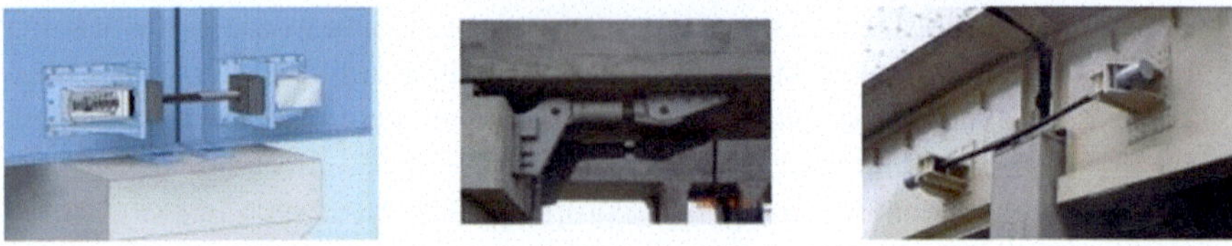

Fig. 4.24 Restrainers in bridges [38]

Structural pounding can be mitigated by the incorporation of some passive control device, such as friction/metallic/fluid/viscoelastic dampers or active control systems into the buildings. A theoretical study on the control of seismic pounding by connecting the adjacent structures with friction dampers was conducted by Dutta and Ghosh [40]. Friction dampers are popular due to their low capital and maintenance costs and ready commercial availability. The model in their study is shown in Fig. 4.25. They concluded that linking adjacent buildings with friction dampers could be a good option to control pounding.

In another study on the use of passive energy dissipation devices to control pounding, Das and Ghosh [41] studied tuned mass dampers (TMDs) and concluded that the TMD is equally effective in reducing both structural responses as well as pounding force. However, the TMD effectiveness is largely dependent on the mass ratio and the selection of the proper tuning ratio of the damper.

Codal provisions

The codes of Argentina, Australia, Canada, France, India, Indonesia, Mexico, Taiwan and the USA account for the pounding phenomenon by the specification of a minimum separation distance between the neighbouring buildings. It is to be noted that, apart from the structural displacement responses, the appropriate separation

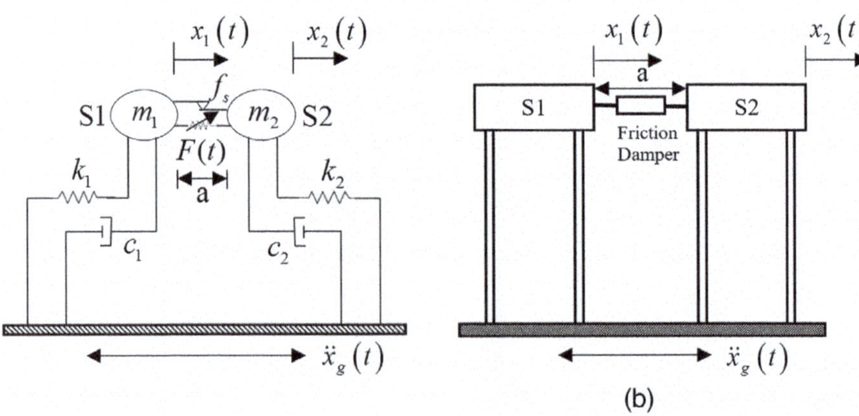

Fig. 4.25 **a** Two base-excited adjacent structures connected by friction damper and **b** its mechanical model [40]

distance is guided by several other factors, such as the importance factor and deflection amplification factor. The values of the separation distance depend on the type of soil and the seismic action as well.

Chenna and Ramancharla [42] have presented a review and comparison of the different worldwide codal provisions on the separation distance between adjacent buildings. It is observed that the procedure to determine the separation distance in the codes of these countries is not the same. For example, in Canada and Israel, the sum of the displacements of each building is specified as the separation distance to be maintained by the code. In France, it is a quadratic combination of the maximum displacements. In Taiwan, it depends on the building height and in Argentina, the minimum separation gap is 2.5 cm. According to UBC-1997 [43], the building separation distance for adjacent buildings located on the same property line is given by the square root of the sum of the squares of the maximum building displacements. FEMA 273-1997 [44] specifies the minimum separation distance as 4% of the building height to avoid pounding.

The New Zealand Loading Code NZS 4203:1992 [45] states the separation of adjacent structures as a distance equal to the sum of their maximum design lateral displacement for the ultimate limit state and 24 mm is the minimum separation gap requirement. According to IBC-2009 [46] and ASCE:7-2010 [47], the separation distance between two adjacent buildings is computed from

$$\delta_M = \frac{C_d \delta_{max}}{I} \tag{4.23}$$

where $\delta_{\max}$ is the maximum elastic displacement that occurs anywhere in a floor from the application of the design base shear to the structure.C_d is the deflection amplification factor and I is the importance factor for seismic loading.

Further, IBC-2009 stipulates that adjacent buildings on the same property shall be separated by a distance not less than δ_{MT}, determined by Eq. 4.24

$$\delta_{MT} = \sqrt{(\delta_{M1})^2 + (\delta_{M2})^2} \tag{4.24}$$

where δ_{M1}, δ_{M2} denote the maximum inelastic response displacements of the adjacent buildings.

The Indian code for seismic resistant design, IS-1893-Part1-2016 [48], stipulates that two adjacent buildings or two adjacent units of the same building with separation joint between them shall be separated by a distance equal to *R* times the sum of storey displacements of the two buildings or units, where *R* is the response reduction factor. However, when floors are at the same level, the separation is given by ($R_1\Delta_1$+$R_2\Delta_2$). Here, R_1 and R_2 are the response reduction factors and Δ_1 and Δ_2 are displacements corresponding to buildings 1 and 2, respectively.

Overall, it is gauged that the current separation distances are rather ad-hoc and at times found to be oversimplified or over-conservative. More in-depth studies are required to be dedicated to the evaluation of the separation distance between adjacent buildings to prevent the possibility of structural pounding at the design stage. Further research is also required to be directed to the mitigation of structural pounding, especially in existing closely spaced structures, through construction details and by the incorporation of structural vibration control systems.

References

1. Khatiawada et al (2011) Development of pounding model for adjacent structures in earthquakes. In: Proceedings of the ninth pacific conference on earthquake engineering building an earthquake–resilient society, Auckland, New Zealand
2. Kumar P, Kumar C (2015) Seismic pounding of the adjacent buildings with different heights. IJERST 2015
3. Asian Disaster Management News, vol 10, no 1, 2004
4. Pantelides CP, Ma X (1998) Linear and non-linear pounding of structural systems. Comput Struct 66:79–92
5. Karamadi AB, Tograsi R (2017) Analysis of seismic pounding between adjacent buildings. IRJET 2017
6. Anagnostopoulos SA (1987) Pounding of building in series during earth-quakes. Earthq Eng Struct Dyn 16:443–456
7. Huba et al (2012) Study on partial collapse of a five story reinforced concrete building during the 2010 Chile earthquake. 15 WCEE LISBOA 2012
8. Pantelides CP, Ma X (1998) Linear and nonlinear pounding of structural systems. Comput Struct 66:79–92
9. National Academy of Sciences, the Great Alaska Earthquake of 1964. Engineering, NAS Publication 1606, Washington, DC
10. Bertero VV, Collins RG (1973) Investigation of the failures of the Olive View stair towers during the San Fernando earthquake and their implications on seismic design
11. Earthquake Engineering Research Institute, Managua, Nicaragua Earthquake of December 23, 1972. Report EP-12, Oakland, CA, 1973
12. Tezcan SS, Yerlici V, Durgunoglu HT (1978) A reconnaissance report for the Romanian earthquake of 4 March 1977
13. Earthquake Engineering Research Institute, Thessaloniki, Greece earthquake of June 20, 1978. Reconnaissance Report, Report EP-32, Oakland, CA, 1978
14. Earthquake Engineering Research Institute, The Central Greece earthquakes of February-March 1981. Reconnaissance and Engineering Report, Report JP- 05, Oak-land, CA, 1982
15. Jankowski R (2009) Non-linear FEM analysis of earthquake-induced pounding between the main building and the stairway tower of the Olive View Hospital. Elsevier
16. Bertero VV (1987) Observations on structural pounding. In: Proceedings of international conference on Mexico earthquakes
17. Kasai K, Maison BF (1997) Building pounding damage during the 1989 Loma Prieta earthquake. Eng Struct 19(3):195–207
18. Kawashima K, Shoji G (2000) Effect of restrainers to mitigate pounding between adjacent decks subjected to a strong ground motion. In: Proceedings on 12th world conference on earthquake engineering
19. Jankowski R, Mahmoud S (2015) Earthquake-induced structural pounding
20. Chung LL, Jean WY, Yeh YK, Hwang SJ, Tsai KC (2007) Seismic upgrading of compulsory school buildings in Taiwan. In: Proceedings on 2nd international conference on urban disaster reduction
21. Global risk Miyamoto, Reconnaissance Report on 2007 Niigata Chuetsu-Oki Japan Earthquake

22. Jain SK et.al (2001) A field report on structural and geotechnical damages sustained during the 26 January 2001 M7.9 Bhuj Earthquake in Western India
23. Rai DC, Murty CVR (2005) Engineering lessons not learnt from 2002 Diglipur earthquake-a review after 2004 Sumatra earthquake. Curr Sci 89(10):1681–1689
24. Kaushik HB, Dasgupta K, Sahoo DR, Kharel G (2006) Performance of structures during the Sikkim earthquake of 14 February 06. Curr Sci 91(4):449–455
25. Rai DC, Murty CVR (2002) Reconnaissance report North Andaman (Diglipur) Earthquake of 14 September 2002
26. Jankowski R, Wilde K, Fujino Y (1998) Pounding of superstructure segments in isolated elevated bridge during earthquakes
27. Davis RO (1992) Pounding of buildings modeled by an impact oscillator. Earthq Eng Struct Dyn 21:253–274
28. Chau KT, Wei XX (2001) Pounding of structures modelled as non-linear impacts of two oscillators. Earthq Eng Struct Dyn 30:633–651
29. Chau KT, Wei XX, Guo X, Shen CY (2003) Experimental and theoretical simulations of seismic poundings between two adjacent structures. Earthq Eng Struct Dyn 32:537–554
30. Jankowski R (2005) Non-linear viscoelastic modelling of earthquake-induced structural pounding. Earthq Eng Struct Dyn 34:595–611
31. Ye K, Li L, Zhu H (2008) A note on the hertz contact model with non-linear damping for pounding simulation. Earthq Eng Struct Dyn
32. Muthukumar S, DesRoches R (2006) A hertz contact model with nonlinear damping for pounding simulation. Earthq Eng Struct Dyn 35:811–828
33. Papadrakakis M, Mouzakis HP (1994) Earthquake simulator testing of pounding between adjacent buildings. Earthq Eng Struct Dyn 24:811–834
34. Anagnostopoulos SA, Spiliopoulos KV (1992) An investigation of earthquake induced pounding between adjacent buildings. Earthq Eng Struct Dyn 21:289–302
35. Goltabar AM, Kami RS, Ebadi A (2008) Study of impact between adjacent structures during of earthquake and their effective parameters. Am J Eng Appl Sci 210–218
36. Abdel-Mooty M, Al-Atrpy H, Ghouneim M (2009) Modeling and analysis of factors affecting seismic pounding of adjacent multi-story buildings. In: Earthquake resistant engineering structures VII, p 127
37. Westermo BD (1989) The dynamics of inter-structural connection to prevent pounding. Earthq Eng Struct Dyn 18:687–699
38. Raheem A, Hayashikawa T (2013) Mitigation measures for expansion joint effects on seismic performance of bridge structures. In: Asia pacific conference
39. Cheng X, Jing W, Qi L, Gong L (2019) Pounding dynamic responses and mitigation measures of sliding base-isolated concrete rectangular liquid storage structures, KSCE J Civil Eng 23(7):3146–3161
40. Dutta NK, Ghosh AD (2021) Vibration control of seismically excited adjacent buildings prone to pounding by use of friction dampers. In: Advances in structural vibration, lecture notes in mechanical engineering, select proceedings of ICOVP 2017, Springer
41. Das M, Ghosh (Dey) A (2016) Mitigation of structural pounding by the tuned mass damper. In: Proceedings SEC–2016 (Structural Engineering Convention) CSIR-SERC Chennai, India
42. Chenna R, Ramancharla PK (2012) Study of impact between adjacent buildings: comparison of CODAL provision. 15 WCCE, LISBOA
43. Uniform Building Code (1997)
44. NEHRP Guidelines for the Seismic Rehabilitation of Buildings FEMA:273–1997
45. NZS 4203:1992 New Zealand standard code of practice for general structural design and design loadings for buildings
46. International Building Code (2009)
47. ASCE:7–2010 Minimum Design Loads and Associated Criteria for Buildings and Other Structures
48. IS-1893-Part 1- 2016 Criteria for Earthquake Resistant Design of Structures, General Provisions and Buildings

Chapter 5
Influence of Soil-Structure Interaction on Yielding of Pile Embedded in Stratified Soil

Arup Bhattacharjee and Bidisha Borthakur

Abstract The safety and serviceability of an engineering structure rely on the prediction of response and failure mechanism of the structure. The accurate prediction of the response of the structures subjected to earthquake loading is a difficult and great concern of the researchers. The prediction of response of pile foundations which are surrounded by different soil types throughout their length becomes difficult due to the complex soil-structure interaction mechanism during earthquake. This has led to the failure of many structures due to earthquake. This further emphasizes the importance of the non-linear soil-structure interaction for designing pile foundations. The pushover analysis is a static non-linear seismic analysis which can be used to predict the seismic response of the pile foundation with the incorporation of soil-structure interaction. In this research work the modeling of the pile-soil system is done in OpenSees PL, a user-friendly interface of OpenSees. The piles embedded in layers of cohesionless and cohesive soil having different relative densities and layer thickness are pushed to a pile head displacement of 2% of pile length. From the response of pile obtained from pushover analysis, the yield moments are determined. The yielding moment generated in pile is influenced by the stratification of soil. The density of the soil and soil type influences the yield moment generated in pile.

Keywords Pushover analysis · Pile foundation · Stratified soil · Yield moment

5.1 Introduction

Past earthquakes are evidence to the severity of damage that can be caused to important structures like bridges. Any amount of damage can question the serviceability of bridges which can bring a halt to the normal lives of people. Since the foundation is the most important part of a structure, proper studies must be conducted while designing the foundation especially in earthquake prone areas. The post damage inspection and

---

A. Bhattacharjee (✉) · B. Borthakur
Department of Civil Engineering, Jorhat Engineering College, Jorhat, Assam, India
e-mail: bhatta_arup@yaoo.com

T. G. Sitharam et al. (eds.), *Theory and Practice in Earthquake Engineering and Technology*, Springer Tracts in Civil Engineering,
https://doi.org/10.1007/978-981-19-2324-1_5

repair of pile foundations are difficult and require proper designing with the consideration of inelastic behavior of structure. It is very important to study the failure mechanism of the pile foundation based on location of yielding of pile foundation. Development of such a design is possible only with the consideration of actual site conditions which includes the effect of soil-structure interaction along with inelastic behavior of structure. The non-linear pushover analysis is one such method which is an effective tool to be used to evaluate the expected non-linear behavior and failure pattern of the structure. The static pushover analysis is a partial and relatively simple intermediate solution to the complex problem of predicting force and deformation demands imposed on structures and their elements by severe ground motions [5]. It is an efficient method to analyze any component of a structure or the structure as a whole, which, unlike dynamic analysis, takes into account the non-linear behavior of the structure arising from deformation caused by lower magnitude of loading. The ability of static pushover analysis to estimate the non-linear behavior of the pile in terms of displacement and bending moment when subjected to monotonically increasing lateral load helps to determine the response of the pile embedded in soil in a bridge foundation. Sun and Zhang [8] after conducting pushover analysis on pier-pile-soil system concluded that a reasonable estimation of the maximum response of the structure as well as the status of destruction of the structure can be obtained from pushover analysis. Mukhopadhyay et al. [7] conducted force-based pushover analysis to study the response of pile foundation when embedded in stratified soil with different positions of liquefiable soil layer using OpenSees PL. The response of a bridge subjected to seismic loading is affected by inelastic deformation of the soil surrounding the foundation as a result of lateral spreading due to liquefaction of soil [10]. The mode of pile deflection relative to the surrounding soil is dependent on the deformed shape of the soil profile along with pile foundation stiffness and the load from the non-liquefied crust [3]. Berill et al. [2], Hamada [4] concluded that permanent displacement of non-liquefied soil lying above the liquefied soil is a governing factor of pile damage. This shows that the surrounding soil conditions play a major role in the response of pile foundation along with the load from the superstructure and thus consideration of soil-structure interaction effects becomes mandatory for designing. In this research work, non-linear force-based pushover analysis is conducted to study the response of the pile surrounded by different soil combinations. The effect of surrounding stratified soil conditions on yielding of pile is studied.

5.2 Numerical Modeling of Soil-Pile System

The element formulation, material relations, and solution strategies for earthquake engineering simulations can be solved using finite element programs. The finite element program OpenSees program uses finite element code for non-linear static as well as dynamic analyses. The user-friendly Finite Element graphical interface of OpenSees is OpenSees PL. The generation of mesh and associated boundary conditions, material parameters and execution of static and dynamic computation can be performed using OpenSees PL [9]. OpenSees PL allows conducting pushover analysis as well as seismic simulations.

5.2.1 Numerical Modeling of Pile

A 10 m long circular pile of 0.8 m diameter is considered to be fully embedded in a stratified soil domain. The soil domain consists of a single type of soil with water table up to the top most surface of soil. The pile is fixed at the pile head as well as at the base with rigid rock lying at the base of the pile. A dead load of 10 tons is acting on the pile head. The pile is modeled using linear beam-column elements. Rigid beam-column elements are used for representing the cross-sectional diameter and the interface with the soil elements surrounding the pile which were 10^4 times stiffer than pile elements axially and flexurally. The mass density of the pile is taken as 2400 kg/m^3. Young's modulus is 3×10^7 kPa and the shear modulus is 1.154×10^7 kPa. The moment of inertia of the pile and torsion constant is 0.020106 m^4 and 0.0402123 m^4, respectively.

5.2.2 Numerical Modeling of Soil

The soil domain consisted of stratified soil layers. The following four cases of different soil types surrounding the pile are considered to study the influence of each soil type on the pile.

1. Saturated cohesionless very loose sand,
2. Saturated cohesionless medium sand,
3. Saturated cohesionless dense sand, and
4. Cohesive stiff soil.

In order to capture seismic events, the soil domain is modeled with 8-node brick elements using *MultiYield* material. The cyclic behavior of soil can be captured well by MultiYield model as compared to other models. Silt and sand is modeled using *PressureDependMultiYield* soil model. It is an elastio-plastic material model for simulating response characteristics such as dilatancy and liquefaction exhibited during monotonic and cyclic loading. Clay can be modeled using *PressureIndepend-MultiYield* model which is implemented to simulate monotonic or cyclic response of materials whose shear behavior is insensitive to confinement change. The details about the soil elastic properties, soil non-linear properties, fluid properties, dilatancy properties and liquefaction properties for the saturated cohesionless soil and cohesive soil are given in Table 5.1. The water table is considered up to the pile head so as to consider liquefaction analysis while conducting dynamic or static analyses.

Table 5.1 Soil properties

	Cohesion-less very loose sand	Cohesion-less Dense sand	Cohesive medium soil	Cohesive stiff soil
Soil elastic properties				
Saturated mass density (Mg/m^3)	1.7	2.1	1.5	1.8
Reference pressure (kPa)	80	80	100	100
Reference Shear modulus (kPa)	55,000	130,000	60,000	150,000
Reference Bulk modulus (kPa)	150,000	390,000	300,000	750,000
Soil nonlinear properties				
Friction (deg)	29	40	0	0
Cohesion (kPa)	0.2	0.3	37	75
Fluid Properties				
Fluid mass density (Mg/m3)	1	1	1	1
Horizontal permeability (m/s)	6.6E-05	6.6E-05	1.00E-0.9	1.00E-0.9
Vertical permeability (m/s)	6.6E-05	6.6E-05	1.00E-0.9	1.00E-0.9
Dilatancy/liquefaction properties				
Phase transformation angle (deg)	29	27	–	–
Contraction parameter	0.21	0.03	–	–
Dilation parameter 1	0	0.8	–	–
Dilation parameter 2	0	5	–	–
Liquefaction parameter 1	10	0	–	–
Liquefaction parameter 2	0.02	0	–	–
Liquefaction parameter 3	1	0	–	–

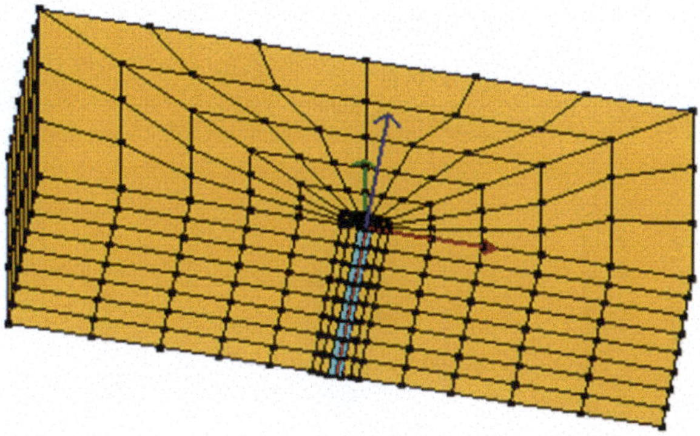

Fig. 5.1 3D View of half meshed pile-soil model in OpenSees PL

5.2.3 Boundary Conditions

The boundary condition is rigid box type and fixed at the bottom in all directions. The plane of symmetry for half mesh configuration is fixed in Y direction while keeping it free in Z and X direction to model 3D full mesh scenario. The half meshed pile-soil model embedded in soil as modeled in OpenSees PL is shown in Fig. 5.1.

5.2.4 Validation of Numerical Model

The seismic analysis of pile-soil configuration of single pile embedded in soil modeled in OpenSees PL is compared with results of Lu et al. [6] and analytical results obtained by Abedzedah et al. [1]. A circular free-head pile of 10.15 m length and radius 203.20 mm, fully embedded in a 20.12 m soil domain of submerged unit weight 9.87 kN/m^3 as considered by Lu et al. [6] is modeled for validation. Figure 5.2 shows the pile deflection and the bending moment experienced by the pile throughout its length due to pushover loading. The results obtained from the present analysis, results of numerical modeling in OpenSees PL by Lu et al. [6] and analytical results of pushover analysis obtained by Abedzedah et al. [1] are in good accuracy.

5.3 Pushover Analysis of Pile Embedded in Stratified Soil

Single fixed head concrete piles of length 10 m and 0.8 m diameter is considered to be embedded fully in stratified soil. Static pushover analysis is conducted by applying monotonically increasing horizontal load of 10kN at the pile head incrementally. The pile is pushed until a pile head displacement of 2% of pile length is obtained. The soil domain of 10 m depth consists of layers of cohesionless and cohesive soils

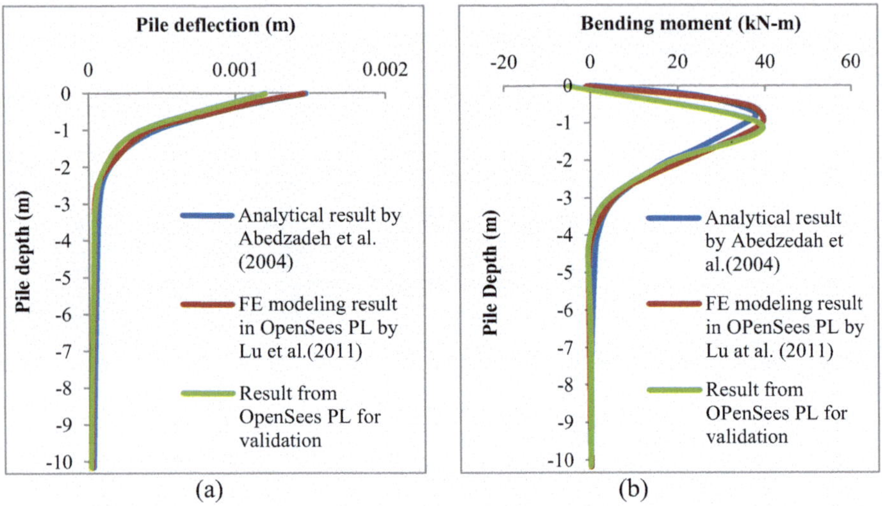

Fig. 5.2 Comparison of analytical results and FE modeling results in terms of **a** Pile deflection and **b** bending moment of pile for seismic analysis of pile-soil system for validation

Fig. 5.3 Pile embedded in different soil combinations consisting of **a** double-layered soil and **b** triple-layered soil

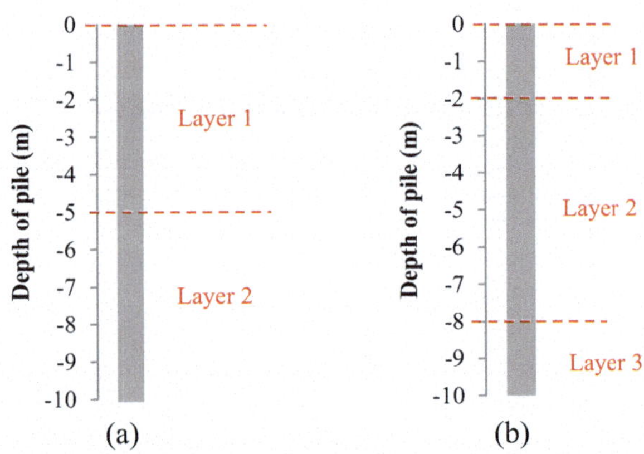

in different two-layered or three-layered combinations as shown in Fig. 5.3 and Table 5.2. The water table is considered up to the pile head such that liquefaction analysis is considered while conducting pushover analyses.

5.3.1 Pile Response Due to Pushover Analysis

The piles surrounded by different soil conditions require a different number of pushover loading steps to attain a pile head displacement of 2% of its pile length.

Table 5.2 Soil combinations of double- and triple-layered soil profile

Combination	Layer	Soil	Depth (m)
Double layered soil			
1	Layer 1	Cohesionless dense sand	5
	Layer 2	Cohesive medium soil	5
2	Layer 1	Cohesive medium soil	5
	Layer 2	Cohesionless dense sand	5
3	Layer 1	Cohesionless very loose sand	5
	Layer 2	Cohesive medium soil	5
4	Layer 1	Cohesive medium soil	5
	Layer 2	Cohesionless very loose sand	5
Triple-layered soil			
5	Layer 1	Cohesive medium soil	2
	Layer 2	Cohesionless dense sand	6
	Layer 3	Cohesive medium soil	2
6	Layer 1	Cohesive stiff soil	2
	Layer 2	Cohesionless dense sand	6
	Layer 3	Cohesive stiff soil	2
7	Layer 1	Cohesive medium soil	2
	Layer 2	Cohesionless very loose sand	6
	Layer 3	Cohesive medium soil	2
8	Layer 1	Cohesive stiff soil	2
	Layer 2	Cohesionless very loose sand	6
	Layer 3	Cohesive stiff soil	2

As per the bending moment profiles shown in Fig. 5.4 for pile embedded in different double-layered and triple-layered soil conditions, it is seen that pile response is different for all the different conditions. The bending moment is seen to vary not only in magnitude, but also in occurrence location throughout the pile length. It is seen that presence of cohesionless very loose sand in any layer results in lower magnitude of bending moment witnessed by pile. Similar is the case with cohesive medium soil present at any layer of surrounding soil. The pile response is thus seen to depend not only on the pushover loading but also on the surrounding soil condition.

5.3.2 *Effect of Soil Type on Yield Moment of Pile*

The response of a pile is dependent on the surrounding soil conditions. The response of the pile is quantified in terms of yield moment of the pile. The yield moment of pile embedded in soil is determined by the moment–curvature graphs drawn for all depths

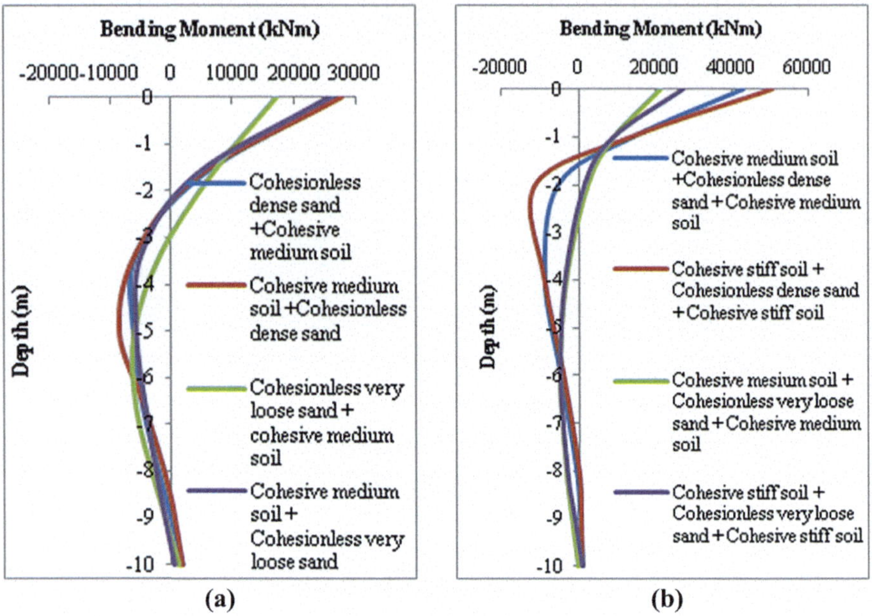

Fig. 5.4 Bending moment profile of 0.8 m diameter pile embedded in **a** double-layered and **b** triple-layered soil

of pile. The double tangent is drawn on the moment–curvature graphs to obtain the yield moment. Figure 5.5 shows the maximum yield moments generated in the pile embedded in double-layered soil with different layer combinations. The magnitude of yield moment for soil containing cohesionless dense sand layer placed above cohesive medium soil is 1750 kN-m and occurs within a depth zone of 3.5–6.5 m. The moment–curvature relation above this zone is linear and almost insignificant below it. The moment–curvature within the yielding zone is non-linear and yield moment is determined by drawing double tangent to it. The magnitude of yield moment for soil containing cohesive medium soil layer placed above cohesionless dense sand is 2000 kN-m and occurs within depth zone of 3.5–6.5 m. Similarly magnitude of yield moment for soil containing cohesionless very loose sand + cohesive medium soil and cohesive medium soil + cohesionless very loose sand is 1500 kN-m and 2000 kN-m, respectively. The yield moment of cohesionless very loose sand + cohesive medium soil is least among the four different combinations considered. However, for soil containing cohesive medium soil layer at the top results in the same yield moment irrespective of the density of sand present at the bottom layer.

Figure 5.6 shows the maximum yield moments generated in the pile embedded in triple-layered soil with different layer combinations. The magnitudes of yield moment for soil containing cohesive medium soil + cohesionless dense sand + cohesive medium soil and cohesive stiff soil + cohesionless very loose sand + cohesive stiff soil is 1500 kN-m and occurs within depth zone of 2.0–6.0 m. Similarly magnitude of yield moment for soil containing cohesive stiff soil + cohesionless dense sand

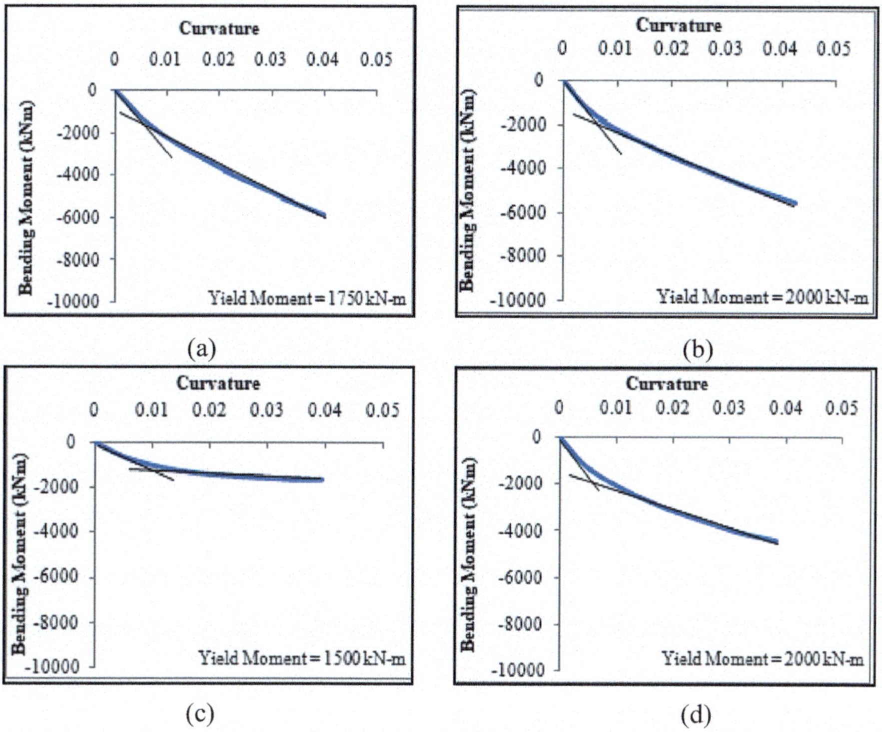

Fig. 5.5 Moment–curvature graphs of 0.8 m diameter pile embedded in double-layered soil consisting of **a** Cohesionless dense sand + Cohesive medium soil **b** Cohesive medium soil + Cohesionless dense sand **c** Cohesionless very loose sand + Cohesive medium soil **d** Cohesive medium soil + Cohesionless very loose sand

+ cohesive stiff soil and cohesive medium soil + cohesionless very loose sand + cohesive medium soil is 2000 kN-m and 1000 kN-m, respectively. It is observed from the graph that the yield moment of pile surrounded by cohesionless very loose sand sandwiched between cohesive medium soil is the least among the different combinations of triple-layered soil considered. The soil containing cohesionless dense sand sandwiched between cohesive stiff soil is the highest yield moment.

From the yield moments obtained for piles surrounded by different soil conditions, it is seen that the magnitude of yield moment of pile depends on the soil type and the position of soil layers. Presence of cohesionless dense sand in any layer of double or triple-layered soil results in higher magnitude of yield moment. However, presence of cohesionless very loose sand results in lower yield moment of pile. It is also seen that cohesive medium soil, when present at the top, results in lower yield moment when soil contains layer of cohesionless very loose sand. However, cohesive stiff soil increases the yield moment of pile. Thus it can be said that the relative density of soil influences the yielding of pile greatly.

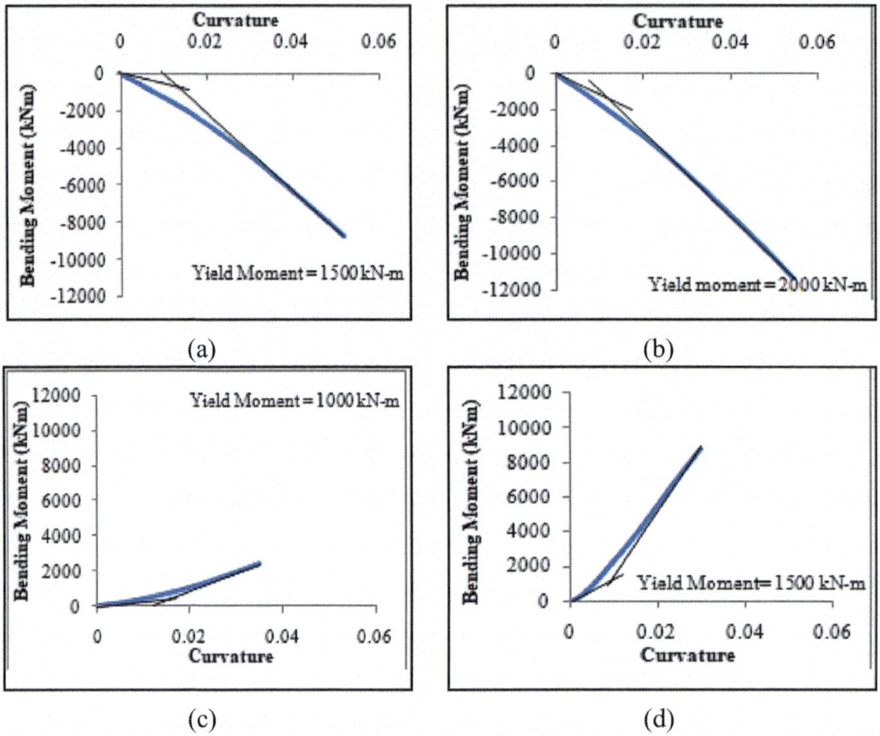

Fig. 5.6 Moment–curvature graphs of 0.8 m diameter pile embedded in triple-layered soil consisting of **a** Cohesive medium soil + Cohesionless dense sand + Cohesive medium soil **b** Cohesive stiff soil + Cohesionless dense sand + Cohesive stiff soil **c** Cohesive medium soil + Cohesionless very loose sand + Cohesive medium soil **d** Cohesive stiff soil + Cohesionless very loose sand + Cohesive stiff soil

5.4 Conclusion

From the pushover analysis response of pile embedded in stratified soil with different surrounding soil conditions, it can be concluded that:

- Static pushover analysis can be used to estimate the pile behavior and yielding of pile due to seismic loading for pile embedded in stratified soil.
- The yield moment of pile depends on the surrounding soil combination along with positioning of soil layers in stratified soil.
- Relative density of cohesionless and stiffness cohesive soil is seen to influence the yield moment of piles surrounded by different multilayered soil conditions. Piles surrounded by soil layers having high-density results in higher yield moment of pile. Also, combining low-density soil layer with high-density soil increases the yield moment of pile.

References

1. Abedzadeh F, Pak RYS (2004) Continuum mechanics of lateral soil-pile interaction. J Eng Mech 130(11):1309–1318
2. Berrill JB, Christensen SA, Keenan RP, Okada W, Pettinga JR (2001) Case studies of lateral spreading forces on a piled foundation. Geotechnique 51(6):501–517
3. Brandenberg SJ, Singh P, Boulanger RW, Kutter BL (2001) Behavior of piles in laterally spreading ground during earthquakes. Proceedings 6th caltrans seismic research workshop, CA, Paper 02-106
4. Hamada M (2000) Performances of foundations against liquefaction-induced permanent ground displacements. Proceedings of the 12th world conference on earthquake engineering. Auckland, New Zealand, paper no 1754
5. Krawinkler H, Seneviratna GK (1998) Pros and cons of a pushover analysis for seismic performance evaluation. engineering structures. Eng Struct 20(4–6):452–464
6. Lu J, Elgamal A, Yang Z (2011) OpenSees PL: 3D lateral pile-ground interaction. University of California, San Diego, User Manual
7. Mukhopadhyay M, Choudhury D, Phanikanth VS, Reddy GR (2008) Pushover analysis of piles in stratified soil. Proceeding of 14Th world conference on earthquake engineering
8. Sun L, Zhang C (2004) Improvement of pushover analysis taking account of pier-pile-soil interaction. Proceedings. of 13Th world conference on earthquake engineering, Canada. Paper No. 659
9. Wang N (2015) Three-dimensional modeling of ground-pile systems and bridge foundations. Phd Thesis, University of California, San Diego
10. Zhang SY, Conte JP, Yang Z, Elgamal A, Bielak J, Acero G (2008) Two dimensional nonlinear Earthquake response analysis of a bridge-foundation-ground. Earthq Spectra 24(2):343–386, Earthquake Engineering Research Institute. https://doi.org/10.1193/1.2923925.

Chapter 6
Development of Liquefaction Susceptibility Maps for Vishakhapatnam (India)

Neelima Satyam and Swathi Priyadarsini

Abstract Historic earthquakes followed by liquefaction failures are familiar and are being studied since the past years [1, 2, 3]. Development of liquefaction susceptibility maps is considered to be prominent as most of the damages during past earthquakes are confirmed to be due to liquefaction phenomenon. In the present study, Liquefaction Severity Index (LSI) and Liquefaction Potential Index (LPI) are used for developing susceptibility maps. LSI and LPI are used to examine the performance of the soil under seismic excitation. In this paper, an attempt has been made to construct response spectrum for the site from the results of earthquake hazard assessment provided by Putti and Satyam (in response analysis and liquefaction hazard assessment for Vishakhapatnamcity. Innov Infrastruct Solut 3:12, [4]). Based on the maximum credible earthquake intensity from the site-specific response spectra Bhuj earthquake strong motion data has been scaled to 0.1 g and is used as an input ground motion for response analysis and liquefaction assessment. The outcomes of the present study suggest that some of the sites in the study area are prone to medium liquefaction probability where silty sands and marine clays with high silt content are found to be predominant. The peak ground accelerations (PGA) are observed to be varying from as low as 0.6–0.14 g. As the study area Vishakhapatnam (India) is the financial capital of Andhra Pradesh state as well as the rapidly growing industrial area, the PGA, LPI and LSI hazard maps are further helpful for infrastructure development and solving future engineering problems in the study area.

Keywords Liquefaction index · Severity · Standard penetration number—N · Hazard

---

N. Satyam
Department of Civil Engineering, IIT Indore, Simrol, India

S. Priyadarsini (✉)
Department of Civil Engineering, Lords Institute of Engineering and technology, Hyderabad, India
e-mail: Pspdarshini@gmail.com

T. G. Sitharam et al. (eds.), *Theory and Practice in Earthquake Engineering and Technology*, Springer Tracts in Civil Engineering,
https://doi.org/10.1007/978-981-19-2324-1_6

6.1 Introduction

During the planning and construction of the structures at the locations that are susceptible to settlement, estimation of liquefaction capacity/potential of the soil is the major task in that has to be taken care in geotechnical earthquake engineering. Numerous studies have been carried out on ground failures due to liquefaction and liquefaction severity in the site by number of investigators after the historic earthquakes. Slope failures, bridge and foundation failures have been observed to be the major ground failures due to liquefaction. Therefore, liquefaction assessment and quantification have become prominent in the recent times. Seed and Idriss [1] have given a methodology for liquefaction assessment which has been improved by many other researchers in the recent times. Energy methods and stress-based methods are the prominent methodologies applied for liquefaction potential and severity index quantification.

In this study, seismic and geotechnical data have been combined to quantify the hazard in terms of LPI and LSI for the study area. Liquefaction of the soil during seismic excitation has been estimated using LPI-based approach for the city of Vishakhapatnam (India). The data used includes geotechnical characteristics from logs of 210 boreholes such as density of the soil, fine contents, porosity, standard penetration number—N values and the depth of groundwater table of soil strata at every borehole. The strong motion data of Bhuj earthquake (2001) scaled to 0.1 g has been used as input according to the response spectrum developed in this research and PGA values attained from seismic hazard assessment. The recent methodology proposed by Boulanger and Idriss [4] has been employed to estimate the liquefaction potential of the soil strata at all the borehole points. Further LPI has been evaluated for the soil columns at all the locations using the method given by Iwasaki et al. [2]. Liquefaction severity index and potential index were obtained to assess the liquefaction probability using the empirical formula given by Papathanassiou [5]. Liquefaction hazard assessment results have been represented in terms of hazard maps with zonation with respect to liquefaction susceptibility and probability while subjected to strong ground motion. It has been observed that few locations in the urban Vishakhapatnam are susceptible to liquefaction due to the presence of silt beds and marine clay sediments. Most of the locations in central and southern Vishakhapatnam are identified to be liquefiable. Vishakhapatnam is the rapidly developing coastal city with one major and one minor port and is highly industrialized because of its trading feasibility. Therefore, the outcomes from the research are useful in design and analysis of prominent structures along with retrofitting, further solving the practical problems in structural and geotechnical engineering.

6.2 Seismotectonic Details of the Study Region

Visakhapatnam City falls under zone-II as per Indian seismic code [6] with 0.10 g zone factor and is located in intra-plate region. The seismotectonic data within the range of 300 km radius making the study area as centre have been shown in Fig. 6.1. Andhra Pradesh (India) has come through the major earthquake events of magnitude 5.5, 5.4 and 5.7 at Vijayanagaram in 1917, Ongole in 1967 and Bhadrachalam in 1969, respectively. Many faults and folds along with fractures characterize the seismicity of the eastern coastal belt region, making it susceptible to major and minor seismic activities. Seismic activities are observed to be increasing in the Eastern Ghats Region along the coast line because of subduction of Burman plate into Bay of Bengal. This phenomenon has led to activation of older inactive faults as well as creation of new faults [7]. Major faults of Nagavali, Bobbili-Polavaram, Kannada Kumali, etc. are found to be the potential sources of seismic activities around the study. The constant stress built up in the northwest and northeast trending faults is making them highly vulnerable to reactivation. Along with this increasing seismicity, 1040 km long coastal belt of Andhra is further exposed to tsunami risk. Vishakhapatnam has experienced the tidal waves of height up to 4 m during the tsunami which was caused by the Great Indian Ocean earthquake of magnitude 9.1 (M_W) during 2004. The coastal belts of Ongole, Vizag and Vijaynagaram are identified as weak seismic zones due to the tectonic activities that are observed in the recent times [8]. 1.4 m high tide gauge has been recorded at the port of Vishakapatnam during the tsunami and it took 2 h 20 min for the first propagated seismic wave to reach the Indian Coast from the source.

Therefore, though it is evident that no devastating earthquakes of higher magnitude are not experienced by the study area, it is advised to conduct seismic hazard studies and liquefaction response studies of the soil by considering the recent hike in the tectonic activities in Bay of Bengal.

6.3 Development of Response Spectra for Vishakhapatnam

A design response spectrum is developed and adopted as an input for design of earthquake-resistant structures is, in general, of smooth shape than other spectra such as UHRS, and spectrum which is constructed specifically for a site. It helps in evaluating and quantifying the possible lateral loads and dynamic loads that are imposed on to a structure during along its life time due to seismic excitation. The intensity of input ground motion is selected according to its functional use and importance of the structure manner according to the owner's choice. ATC40, IS 1893 [9] and many others have recommended hazard analysis for soils of D type, for Design-based earthquake and maximum credible earthquake intensities. Development of response spectra requires mapped spectral accelerations that are site specific for a short period of 0.2 s and a long period of 1 s for 2475 years of return time period at damping of 5%

(a)

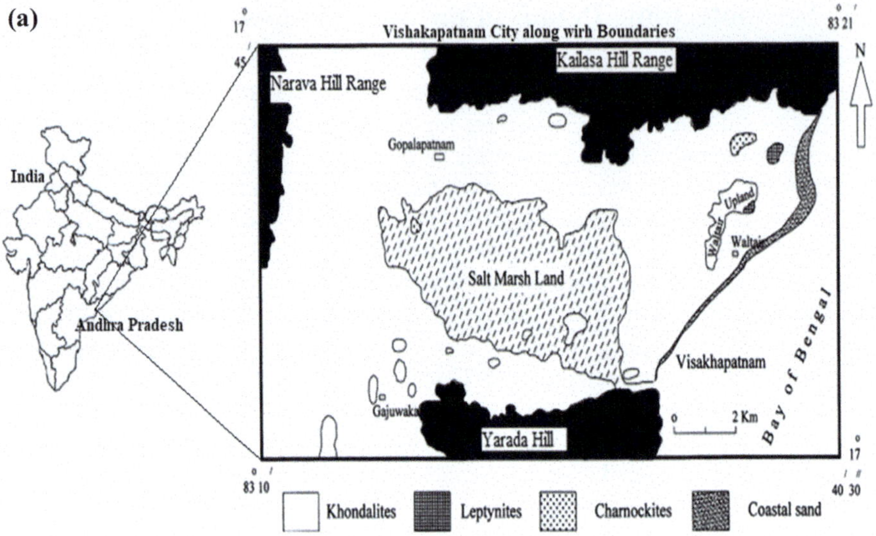

(b)

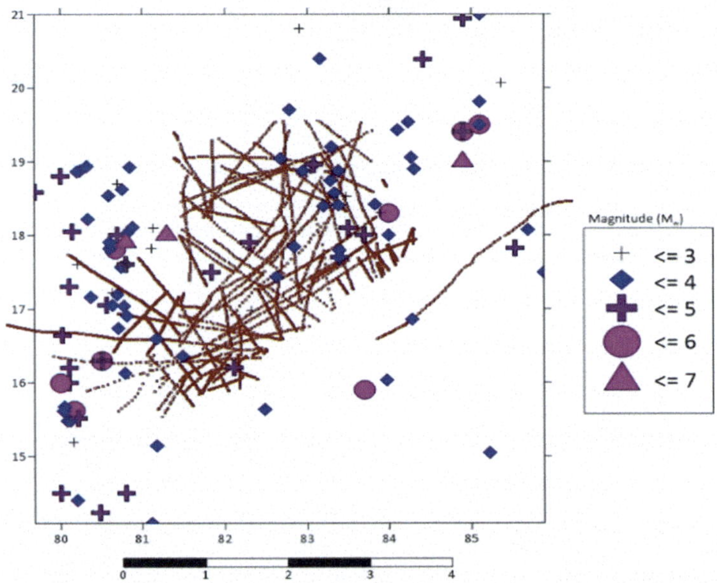

Fig. 6.1 **a** Location map of Vishakhapatnam along with geological features. **b** Distribution of seismotectonic features in 300 km radius around Vishakhapatnam

at bedrock level, which is identified to be 0.176 g and 0.0485 g, respectively, as shown in Table 6.2. The above values are taken from the results of hazard assessment carried out for the region of Vishakhapatnam by Putti and Satyam [10]. The procedure for development of response spectra has been followed as per the Sect. 11.4 of ASCE 7-05. Figure 6.2 shows the site-specific spectra obtained for Level-2 (design-based earthquake (DBE)) and Level-3(maximum credible earthquake (MCE)) earthquakes.

Table 6.1 Major seismic events around the study area [11]

S. no.	Location	Year	M_w
1	Ongole, Andhra Pradesh	1800	5.1
2	Nellore, Andhra Pradesh	1820	4.8
3	Vizag, Andhra Pradesh	1827	4.8
4	Vijayanagaram, Andhra Pradesh	1858	4.8
5	Guntur, Andhra Pradesh	1859	5.1
6	Guntur, Andhra Pradesh	1859	4.8
7	Guntur, Andhra Pradesh	1859	5.0
8	Chittoor, Andhra Pradesh	1860	4.9
9	Parchuru, Andhra Pradesh	1867	4.9
10	Bay of Bengal	1869	5.1
11	Rajahmundry, Andhra Pradesh	1869	4.8
12	Kakinada, Andhra Pradesh	1869	4.7
13	Vizag, Andhra Pradesh	1870	5.0
14	Sironcha, Andhra Pradesh	1872	5.1
15	Secunderabad-Hyderabad, Andhra Pradesh	1876	5.1
16	Bay of Bengal	1917	6.1
17	Bay of Bengal	1918	5.1
18	Bhadrachalam, Andhra Pradesh	1969	6.1
19	Bay of Bengal	1982	4.9
20	Bay of Bengal	1985	5.1

Table 6.2 Spectral accelerations at rock level for 5% damping ratio

Time period (Sec)	Peak spectral acceleration (g)	
	Level III (L3)	Level II (L2)
0	0.106	0.07
0.05	0.11	0.123
0.1	0.125	0.176
0.2	0.264	0.096
0.5	0.144	0.069
1	0.072	0.048

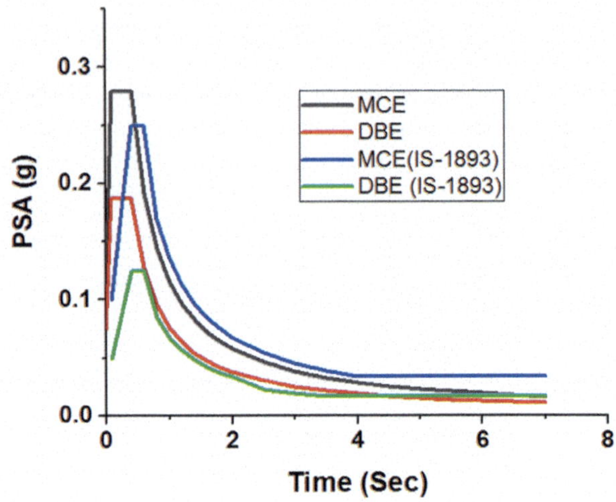

Fig. 6.2 Comparison obtained as per site and IS-1893:2016 [6]

The peak accelerations attained at DBE and MCE intensities are 0.19 g and 0.279 g, respectively. The PGA according to the response spectra given by Indian standard code IS-1893 [6] for DBE and MCE intensities are 0.05 and 0.1 g. There are no values provided by the Indian standard code for service earthquake intensity.

There is a variation of 52% for design earthquake and 12% for maximum credible earthquake in spectral acceleration values between the developed response spectra and default response spectra given by Indian code indicating that IS code is underestimating the ground motions in the site. Thus, it is noted that development of response spectrum for structures that are prominent is essential and a revised version can be made in the Indian codal provisions.

6.4 Peak Ground Acceleration Hazard Maps

The most common and prominent issue in earthquake geotechnics is the quantification of site response. Response analysis at a particular site helps in identifying natural site periods, assessment of ground motion amplification, provides details of ground motion intensities for construction of response spectra, quantify liquefaction severity and to estimate the forces induced by the seismic excitations which can lead to instability of earth slopes. Strong motion intensity along with local site effects influences the characteristics for the damage caused to a structure. Soil behaviour or response under seismic excitation can be interpreted better in terms of ground accelerations and site effects for a location. An exact or appropriate soil model which represents the dynamic characteristics of the soil is required to quantify the site response of the soil strata. Field test data from seismic cross hole, down hole and SPT-N values and other parameters from cyclic triaxial tests and geotechnical laboratory tests are required for response analysis. Using the field and laboratory test outcomes a soil model can be prepared for response analysis. Here, in this study, 1D response analysis has been employed for different locations to quantify the local site conditions which modify ground response using equivalent linear approach. PGA at bedrock level and at surface have been obtained from the analysis at all the proposed locations. The ground acceleration at surface has been identified to be in the range of 0.08–0.14 g. Site amplification factor is identified between 1.0 and 1.4. Figure 6.3 shows the variation of PGA across the study area.

These acceleration values identified at different levels were further used to calculate cyclic stress ratio (CSR) in liquefaction assessment as the site-specific value gives reliable results.

6.5 Liquefaction Hazard Maps

Past ground failures and the damages due to liquefaction in saturated soil deposits were identified to be occurring post liquefaction due to sand boiling and further causing ground subsidence and uneven tilts, cracks and even collapse of a structure. Major earthquakes since past have resulted in extensive ground damage due to soil liquefaction and further illustrated the need for assessment and mitigation of liquefaction through engineering procedures [12]. Experimental and numerical studies have been rigorously carried out by many researchers since 1963 for better understanding of this phenomenon.

6.5.1 Liquefaction Potential Index Using Stress-Based Approach

Iwasaki et al. [2] have proposed using an index which is directly related to the thickness/depth of the liquefiable and non-liquefiable soil strata (cap) and the calculated factor of safety value against liquefaction (FS_L). According to the scale given by him, liquefaction failure potential has been considered as high if the LPI is between 5 and 15 and is considered to be low at sites where LPI ranges from 0 to 5. The liquefaction potential considered to be extremely low where LPI is equal to 0 and is considered to be extremely high at sites where LPI is higher than 15.LPI scale has been modified and updated by numerous researchers for more applicability and better interpretation. The basic advantage of estimation of potential index is that it quantifies the likelihood of liquefaction for a site with a single value for the entire soil column instead of different values for each layer similar to factor of safety. Moreover, LPI calculated at different sites can be further used for development of hazard maps, which is considered to be preliminary zonation tool of the liquefaction potential and can be used by decision-makers at different levels for planning, construction and analysis. Several liquefaction hazard maps have been developed for

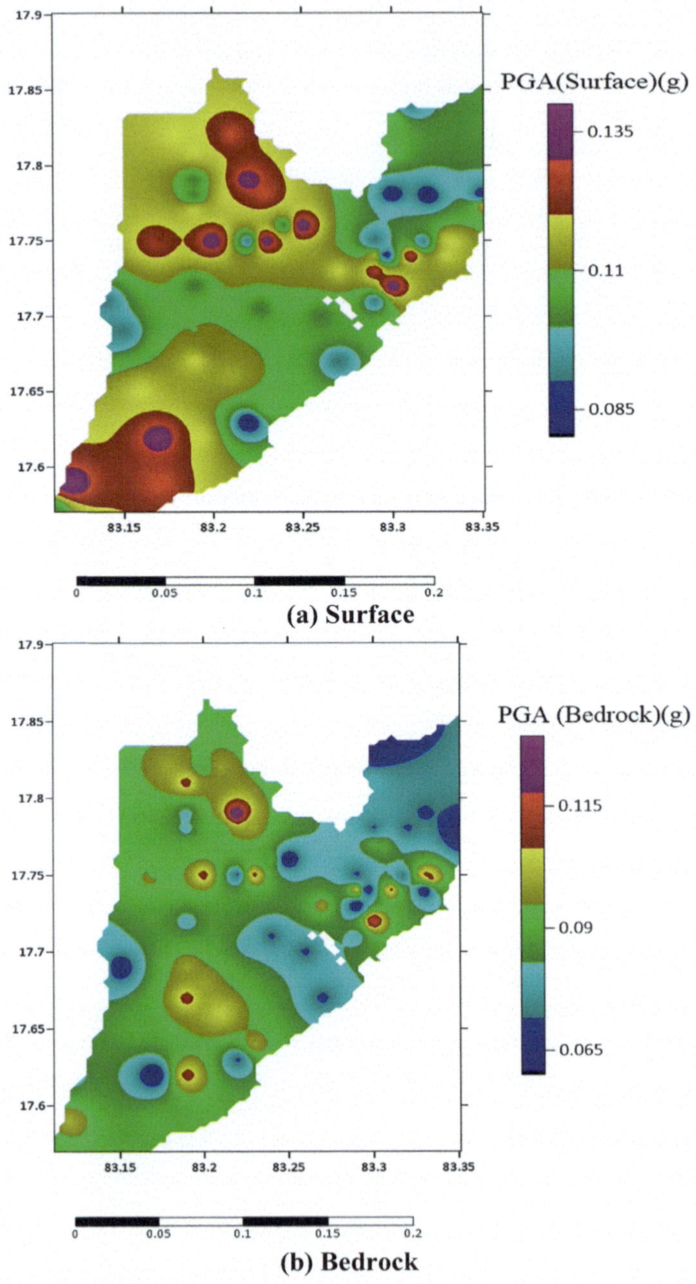

Fig. 6.3 Peak ground acceleration (g) maps of Vishakhapatnam

different urban cities and locations around the world such as for Dhaka, Bangladesh [13], California, USA [14], Turkey [15, 16], Turkey [17] and cities in Greece [18]. Deterministic approach towards LPI-based evaluation was initially proposed by Seed and Idriss [1] and has been upgraded by various other researchers since then [19, 20]. The procedure adopted is given as follows.

6.5.2 Estimation of Cyclic Resistance Ratio (CRR)

The CRR, according to Boulanger and Idriss [4], is estimated using the following equation:

$$\mathrm{CRR} = \mathrm{CRR}_{M=7.5} \,.\, \mathrm{MSF} \tag{6.1}$$

where

$\mathrm{CRR}_{\mathrm{M=7.5}}$ is the resistance ratio calculated for the earthquake of magnitude 7.5 and is estimated by

$$\mathrm{CRR}_{M=7.5} = \frac{1}{34-(N1)60} + \frac{(N1)60}{135} + \frac{50}{[10.(N1)60+45]^2} - \frac{1}{200} \tag{6.2}$$

MSF stands for magnitude scaling factor and is calculated by the equation given by Seed and Idriss [21]:

$$\mathrm{MSF} = \frac{10^{2.24}}{Mw^{2.56}} \tag{6.3}$$

$(\mathrm{N1})_{60}$ is the normalized blow count for surcharge load of 100 kPa for a hammer efficiency of 60% and is estimated by using the following formula:

$$(N_1)_{60} = N_m \,.\, C_N \,.\, C_E \,.\, C_B \,.\, C_R \,.\, C_S \tag{6.4}$$

where N_m is the measured SPT at the site; C_N, C_E, C_B, C_R and C_S are the correction factors.

6.5.3 Estimation of Cyclic Stress Ratio (CSR)

Boulanger and Idriss [4] formulated an equation for estimation of the cyclic stress ratio:

$$\text{CSR} = \frac{\tau_{cyc}}{\sigma' v} = 0.65\left(\frac{a_{\max}}{g}\right)\left(\frac{\sigma v}{\sigma' v}\right) r_d \tag{6.5}$$

where a_{max} is the maximum acceleration at the ground level generated by the earthquake and rd is the stress reduction factor. σ_{v0}, σ'_{v0} are quantified using the equations given by Kayen et al. [22]:

$$\sigma_{VO} = \gamma_d Z_w + \gamma_{\text{sat}}(Z - Z_w) \tag{6a}$$

$$\sigma'_{VO} = \gamma_d Z_w + \gamma_b(Z - Z_w) \tag{6b}$$

where γ_d —dry density of the soil; γ_{sat}—saturated density of the soil; γ_b—floating unit weight of soil; z—calculation depth of σ_{v0} and σ'_{v0} in m and Z_m—water depth, in m. The coefficient r_d accounts for soil strata flexibility and is determined using the equation given by Liao and Whitman [23]:

$$r_d = 1 - 0.00765z \text{ for } z \leq 9.15\,\text{m} \tag{7a}$$

$$r_d = 1.17 - 0.0267z \text{ for } 9.15\,\text{m} < z \leq 9.15\,\text{m} \tag{7b}$$

6.5.4 Factor of Safety for Each Sediment Layer

Factor of safety is estimated for each soil strata as follows:

$$\text{FS} = \frac{\text{CRR}}{\text{CSR}} \tag{6.8}$$

Using the collected SPT-N data collected from 74 locations extended throughout the study area, a detailed liquefaction hazard assessment is carried out using the recent methodology proposed by Seed et al. [24] which was developed after the simple assessment method given by Putti and Satyam [10]. Factor of safety has been evaluated for all layers of the soil strata until a depth of 20 m at all the proposed locations. The Ground Acceleration Values (PGA) obtained from the ground response analysis are used for the calculation of cyclic stress ratio at all the sites. Figure 6.4 shows the factor of safety against liquefaction (F_L) distribution across the city. Calculations for two sites have been presented in Table 6.3 and the spectral acceleration graphs and soil profiles for the same locations are presented in Fig. 6.5.

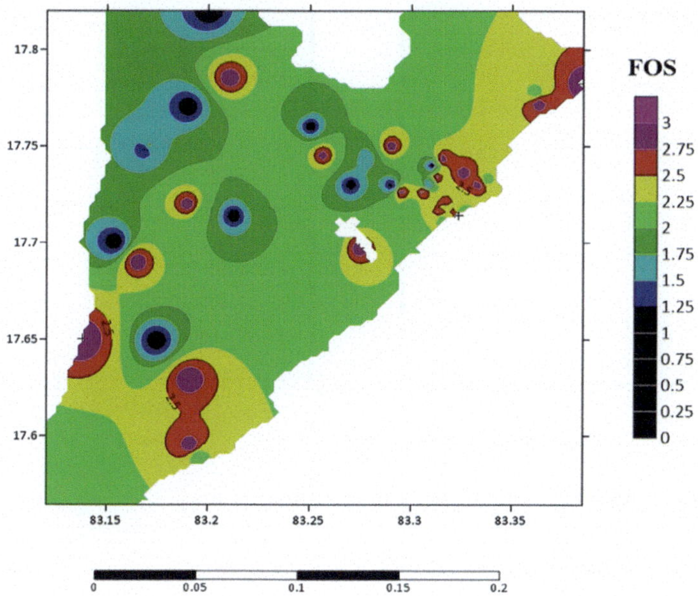

Fig. 6.4 Liquefaction hazard map in terms of factor of safety from stress-based approach

Table 6.3 Liquefaction susceptibility calculations for fishing harbour and Chinna Waltair locations

Location: near the port (fishing harbour)				Location: Chinna Waltair			
Depth (m)	CSR	CRR	F_L	Depth (m)	CSR	CRR	F_L
0	0.067	0.026	0.387	3	0.070659	0.246541	3.489168
4.05	0.083	0.094	1.131	5	0.086548	0.70334	8.126622
7	0.089	0.152	1.712	8	0.074264	0.70334	9.470775
10	0.091	0.026	0.288	8.34	0.06717	0.70334	10.47109
15	0.09	0.583	6.462	10	0.06717	0.70334	10.47109

6.5.5 *Liquefaction Potential Index (LPI)*

The LPI calculated for each bore-log is evaluated using Eq. (6.2):

$$\text{LPI} = \int_0^z F(z)w(z)dz \tag{6.9}$$

where z is the depth of the soil layer below the ground level and is estimated as

$$w(z) = 10 - 0.5z$$

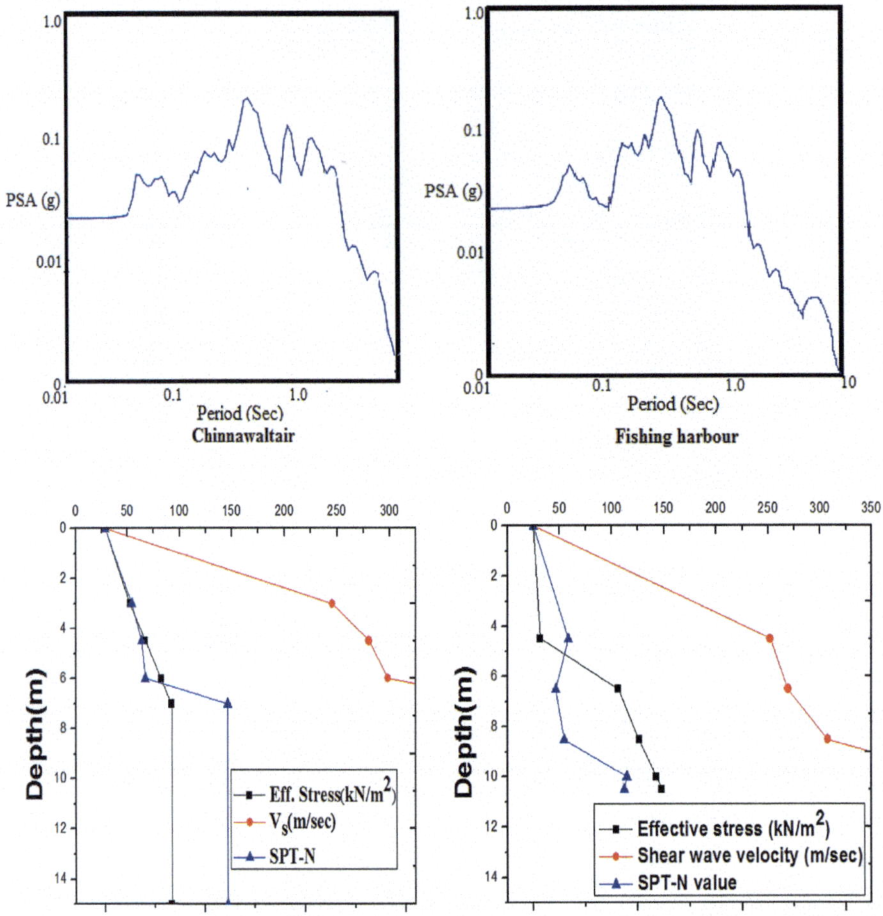

Fig. 6.5 Spectral acceleration plots and soil profiles for fishing harbour and Chinna Waltair locations

where $F(z) = 1 - \text{FS}; \text{FS} \leq 1$

$$F(z) = 0; \text{FS} > 1$$

The empirical relationship between liquefaction probability and LPI proposed by Papathanassiou [5] has been employed for hazard mapping for the study area:

$$Pg = \frac{1}{1 + e^{3.092 - 0.218\,\text{LPI}}} \tag{6.10}$$

Liquefaction severity index can be determined using the below equation:

$$L_s = \int_0^{20} P_L(z).W(z)dz \tag{6.11}$$

$$P_L(z) = \frac{1}{1 + (F_L/0.96)} \text{ for } F_L \leq 1.411 \tag{6.12}$$

$$P_L(z) = 0 \text{ for } F_L > 1.411.$$

The city of Vishakhapatnam (population 2,226,000 and area 681.96 km^2) is located along the Bay of Bengal coast in the Eastern Ghats Region [25]. Once highly active fault to the east of north Andhra Pradesh with its maximum activity focussed in the vicinity of offshore Vishakhapatnam [26]. Though rock outcrops were observed at locations scattered all over the city, presence of silt beds and marine clay deposits along the coast and in the southern part of the city makes the city vulnerable to liquefaction. The geotechnical data required for the liquefaction assessment is obtained from extensive field and laboratory investigation reports collected from different private and government organizations. The factor of safety required for determination of liquefaction severity index and liquefaction potential index calculated using the stress-based methodology are presented in the hazard maps above. Geotechnical data between 0 and 20 m has been considered for further LSI and LPI calculation as the depth below 20 m is considered to be non-liquefiable because of the over burden pressure.

The severity classes based on LSI and LPI for all boreholes are given in Tables 6.4 and 6.5, respectively. The liquefaction maps are developed using the factor of safety, LSI (Fig. 6.6) and LPI (Fig. 6.7) for the study area by using the technique of linear interpolation and triangulation method considering the obtained index values and the

Table 6.4 Liquefaction severity classification [27, 28]

Liquefaction severity index	Liquefaction intensity
85–100	Very high
65–84	High
35–64	Moderate
15–34	Low
0–15	Very low

Table 6.5 Classification based on liquefaction potential index [27]

LPI (P_G)	Liquefaction intensity/potential
0.35	Low
0.35–0.75	Medium
0.75–1.0	Very high

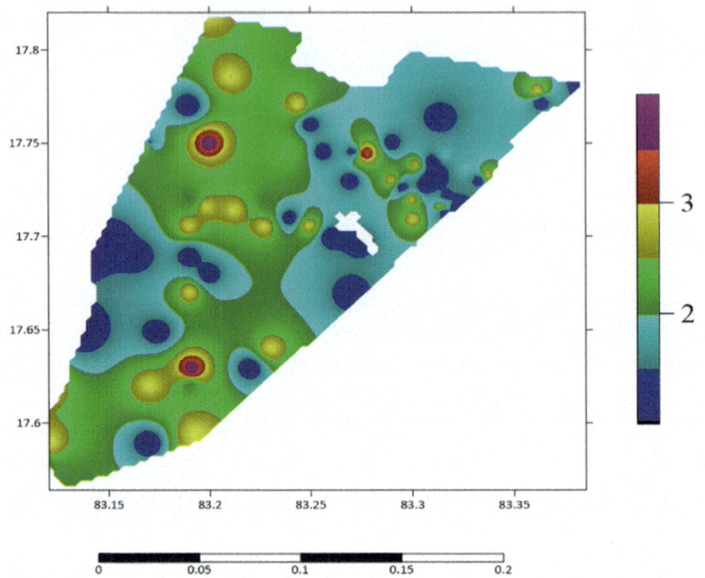

Fig. 6.6 Liquefaction hazard map in terms of Liquefaction Severity Index (LSI)

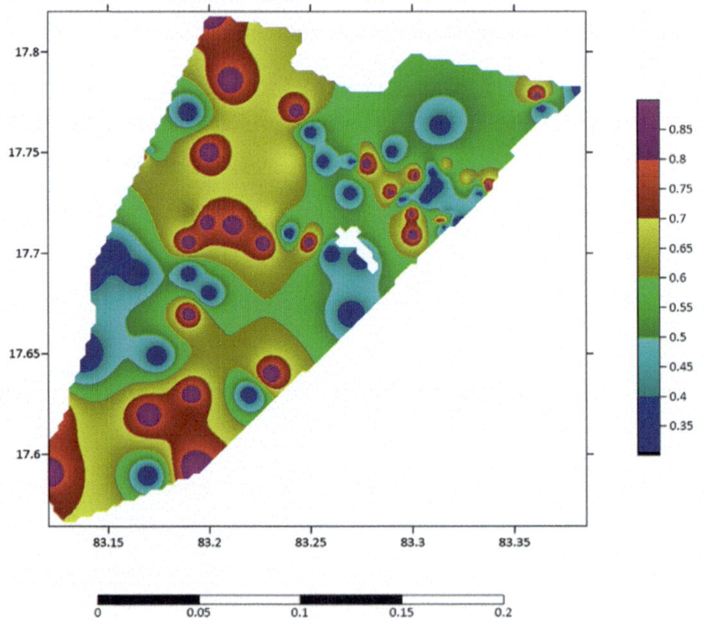

Fig. 6.7 Liquefaction Potential Index (LPI) hazard map

GPS coordinates of the test locations. Liquefaction was observed to be of moderate-to-high severity in few locations along the coast. Although the southern locations have recorded moderate liquefaction severity and the soil at the ground level has not shown any response.

6.6 Results and Conclusions

From the present liquefaction study, the following are the results and conclusions:

- To overcome the limitations in the existing liquefaction susceptibility parameters, liquefaction potential and severity have been used in the present study for classification.
- Site-specific response spectrum from the study shows that IS code has underestimated values of ground accelerations which leads to unreliable results.
- Site response analysis at all the locations has shown PGA values varying from 0.6 to 0.14 g throughout the city and amplification values ranging between 0.9 and 1.5 showing that the ground motion is amplified because of presence of soft sediments. Such locations are highly susceptible to liquefaction under the presence of groundwater table.
- To estimate the liquefaction susceptibility in the study area, where silt beds and marine clay sediments are observed along the coast line, LSI and LPI have been calculated and hazard maps have been developed accordingly. From the liquefaction assessment, it is identified that MVP colony, Thungalam, Simhadri, Parawada, NTPC, Akkayapalem and few other areas in the port area are certainly liquefiable.
- From the hazard maps, it is evident that the central and northeastern part of the city have shown no signs of liquefiable soil strata with factor of safety against liquefaction (F.L) > 2.0. For such locations with no traces of soil liquefaction, the LPI has been considered to be 1.0, whereas liquefaction is certain (F.L < 1.0) in few locations of southern city. Deepanjali Nagar, Seethammadhara, R.K beach, Sheelanagar, Aganampudi, Old Municipal office and few locations in the port have shown the probability of liquefaction (F.L) to be in the range of 1.0–2.0.

References

1. Seed HB, Idriss IM (1971) Simplified procedure for evaluating soil liquefaction potential. ASCE J Soil Mech Found Division 97:1249–1273
2. Iwasaki T, Tokida K, Tatsuoka F, Watanabe S, Yasuda S, Sato H (1982) Microzonation for soil liquefaction potential using simplified methods. In: Proceedings 3rd international conference on microzonation, Seattle, USA, pp 1319–1330
3. Juang CH, Jiang T, Andrus RD (2002) Assessing probability-based methods for liquefaction potential evaluation. J Geotech Geoenviron

4. Boulanger RW, Idriss IM (2004) Evaluating the potential for liquefaction or cyclic failure of silts and clays. Center for Geotechnical Modeling, p 131
5. Papathanassiou G (2008) LPI-based approach for calibrating the severity of liquefaction-induced failures and for assessing the probability of liquefaction surface evidence. Eng Geol 96(1–2):94–104
6. Rao KS, Neelima Satyam D (2006) Seismic microzonation of Delhi NCR. In: National conference on earthquake diaster technology and management, MNNIT Allahabad, 11–12 February 2006
7. Priyadarsini PS, Satyam N (2021) Pushover analysis of pile-supported wharf structure in Visakhapatnam (India). In: Patel S, Solanki CH, Reddy KR, Shukla SK (eds) Proceedings of the Indian geotechnical conference 2019. Lecture notes in civil engineering, vol 138. Springer, Singapore. https://doi.org/10.1007/978-981-33-6564-3_58
8. Putti SP, Satyam N (2020) Evaluation of site effects using HVSR microtremor measurements in Vishakhapatnam (India). Earth Syst Environ 4(2): 439–454
9. IS 1893 (Part 1): 2016 Indian standard criteria for earthquake resistant design of structures Part 1 general provisions and buildings
10. Putti SP, Satyam N (2018) Ground response analysis and liquefaction hazard assessment for Vishakhapatnam city. Innov Infrastruct Solut 3:12. https://doi.org/10.1007/s41062-017-0113-4
11. Swathi Priyadarsini P, Neelima S (2021) Design of pile foundation system for wharf structure in liquefiable soils. In: Latha Gali M, Raghuveer Rao P (eds) Geohazards. Lecture notes in civil engineering, vol 86. Springer, Singapore. https://doi.org/10.1007/978-981-15-6233-4_30
12. Ansary MA, Rashid MA (2000 July) Generation of liquefaction potential map for Dhaka, Bangladesh. In: Proceedings of the 8th specialty conference on probabilistic mechanics and structural reliability
13. Toprak S, Holzer TL (2003) Liquefaction potential index: field assessment. J Geotech Geoenviron Eng 129(4):315–322
14. Sonmez H (2003) Modification of the liquefaction potential index and liquefaction susceptibility mapping for a liquefaction-prone area (Inegol, Turkey). Environ Geol 44(7):862–871
15. Sonmez H, Gokceoglu C (2005) A liquefaction severity index suggested for engineering practice. Environ Geol 48(1):81–91
16. Koyuncu NP, Ulusay R (2004) Geo-engineering evaluation with prime consideration to liquefaction potential for Eskisehir city (Turkey). In: Engineering geology for infrastructure planning in Europe. Springer, Berlin, Heidelberg, pp 125–132
17. Papathanassiou G, Pavlides S, Ganas A (2005) The 2003 Lefkada earthquake: Field observations and preliminary microzonation map based on liquefaction potential index for the town of Lefkada. Eng Geol 82(1):12–31
18. Bolton Seed H, Tokimatsu K, Harder LF, Chung RM (1985) Influence of SPT procedures in soil liquefaction resistance evaluations. J Geotech Eng 111(12):1425–1445
19. Youd TL, Idriss IM (2001) Liquefaction resistance of soils: summary report from the 1996 NCEER and 1998 NCEER/NSF workshops on evaluation of liquefaction resistance of soils. J Geotech Geo Environ Eng 127(4):297–313
20. Idriss IM, Boulanger RW (2010) SPT-based liquefaction triggering procedures. Rep. UCD/CGM-10, 2, pp 4–13
21. Seed HB (1982) Ground motions and soil liquefaction during earthquakes. Earthquake Engineering Research Insititue

22. Kayen RE, Mitchell JK, Seed RB, Lodge A, Nishio SY, Coutinho R (1992, May) Evaluation of SPT-, CPT-, and shear wave-based methods for liquefaction potential assessment using Loma Prieta data. In: Hamada M, O'Rourke TD (eds) Proceedings of the 4th Japan-US workshop on earthquake resistant design of lifeline facilities and countermeasures for soil liquefaction
23. Liao SS, Whitman RV (1986) Overburden correction factors for SPT in sand. J Geotechn Eng 112(3):373–377
24. Seed HB, Whitman RV, Dezfulian H, Dobry R, Idriss IM (1972) Soil conditions and building damage in 1967 Caracas earthquake. J Soil Mech Found Division 98(8):787–806
25. Putti SP, Devarakonda NS, Towhata I (2019) Estimation of ground response and local site effects for Vishakhapatnam, India. Nat Hazards 97:555–578. https://doi.org/10.1007/s11069-019-03658-5
26. Priyadarsini PS (2021) Probabilistic seismic hazard assessment and dynamic behavior of port structures (Doctoral dissertation, International Institute of Information Technology, Hyderabad)
27. Manne A, Satyam N (2015) A review on the discrete element modeling of dynamic laboratory tests for liquefaction assessment. Electron J Geotech Eng 20(1):21–46
28. Manne A, Satyam ND (2011) Geotechnical site characterization for Vijayawada urban. In: Proceedings of 3rd Indian young geotechnical engineers conference (3IYGEC), New Delhi. Indian Geotechnical Society, New Delhi, pp 191–196

Chapter 7
Effectiveness of Base Isolation Systems for Seismic Response Control of Masonry Dome

Pushkar G. Kakade, Hema K. Munot, and Suhasini N. Madhekar

Abstract Domes are constructed historically over the last many centuries. They are doubly curved structures, without angles and corners. The most important advantage of dome structures is that they enclose an enormous amount of column-free interior space, in addition to providing decent aesthetic sight. Historically, domes were built of masonry material. Masonry structures have very low ductility, and hence they are weak in resisting the lateral loads. Most of the domes are designed for gravity loads using simple geometrical rules, considering the dome as an arch of identical section. Due to the absence of reinforcement in the masonry domes, their thickness must be kept high to resist the tensile stresses. Because of their large thickness, masonry domes attract a large magnitude of seismic forces due to higher mass thus, making them vulnerable to earthquake excitations. Due to earthquake forces, the masonry domes are subjected to tensile forces at the bottom rings and as a result, cracks are developed in the bottom parts of domes. The conventionally designed and constructed masonry domes are vulnerable to severe damage or total collapse under strong seismic excitations. To preserve these ancient structures of historic importance from being damaged due to seismic excitations, base isolation can prove to be a very effective technique. In the present research, seismic response of the case study masonry dome of span 25 m, located in Maharashtra, India, is investigated analytically. The specific objectives of the study are (i) to analyse the seismic performance of the fixed base masonry dome structure under real earthquake ground motions, (ii) to analyse the seismic performance of the masonry dome installed with base isolation systems, viz. lead rubber bearings (LRB) and friction pendulum systems (FPS) and (iii) to compare the seismic performance of the fixed base masonry dome with that, installed with LRB and FPS. The response of the base-isolated dome is obtained using SAP2000 by performing nonlinear time history analysis and is compared with the corresponding response of the conventional dome without base isolators. The nonlinear time history analysis is performed considering real earthquake ground

P. G. Kakade · H. K. Munot · S. N. Madhekar (✉)
College of Engineering Pune, Pune 411 005, India
e-mail: suhasinimadhekar@gmail.com

P. G. Kakade
e-mail: kakadepg18.civil@coep.ac.in

T. G. Sitharam et al. (eds.), *Theory and Practice in Earthquake Engineering and Technology*, Springer Tracts in Civil Engineering,
https://doi.org/10.1007/978-981-19-2324-1_7

motions of PGA ranging between 0.1 g and 0.35 g. The effectiveness of base isolation technique in improving the response of the dome is explored. The major evaluation criteria considered are tensile stresses, base shear and displacement at the apex point of the dome. It is observed that the seismic response of the base-isolated dome diminishes significantly in comparison with the conventionally constructed dome, depicting the effectiveness of the base isolation strategy. Both, the elastomeric and sliding systems, are found to be very effective in decreasing the response quantities, substantially. The force–displacement loops for both the isolators show considerable energy dissipation. The original uniqueness and aesthetic value of the historical monumental dome are maintained unaltered, even after employing base isolators at the foundation level of the dome.

Keywords Masonry domes · Base isolation · Lead rubber bearing · Friction pendulum system · Nonlinear time history analysis · SAP2000

7.1 Introduction

Domes have been labelled as the 'kings' of all roofs, as they cover some of the most important structures. Domes enclose an enormous amount of space, thus providing a large column-free area. Despite their slimness, they are some of the strongest and stiffest structures in existence. Historically, domes were built of masonry, and they were designed only for gravity loads. The masonry dome is built without any supporting shuttering with small mud bricks laid in a mud mortar. Masonry domes are very weak in resisting lateral loads. The analysis of a shell is concerned with two stresses, (i) the stress that acts in the meridional direction and (ii) the stress that acts in the parallel direction. Meridional forces (like the meridians, or lines of longitude, on a globe) are compressive and increase towards the base, while hoop forces (like the lines of latitude on a globe) are compressive at the top and tensile at the base. The hoop compressive stresses are maximum at the top of the dome and go on reducing towards the bottom. At a roll-down angle of 51.8°, the hoop compressive forces become zero and then the hoop tensile forces start increasing. The hoop tensile forces are thus maximum at the bottom ring of the dome, and as a result, the dome is subjected to maximum tensile stresses at its bottom ring. Thus, over a period, severe tensile cracks are seen at its bottom portion.

Croci [1] studied the theory of the design of masonry dome and its failure modes. The seismic behaviour of masonry domes, viz. Vaults of Hagia Sophia in Istanbul and St. Francis in Assisi, are discussed. The study was performed on a dome of rise 54 m and diameter 43 m. Seismic elastic analysis showed that the behaviour of the dome was symmetrical in the two major directions. Matsagar and Jangid [2] analytically investigated the seismic responses of structures retrofitted using base isolation devices. The retrofitting of various important structures using seismic isolation technique, by incorporating the layers of isolators at suitable locations is studied.

Historical buildings are selected to investigate the effectiveness of the base isolation in seismic retrofitting, using isolation devices, such as elastomeric bearings and sliding systems. It is observed that the seismic response of the retrofitted structures reduces significantly in comparison with the conventional structures depicting the effectiveness of the retrofitting done through the base isolation technique.

Narayanan and Sirajuddin [3] described the properties of the masonry elements which are to be used in the software to perform the nonlinear analysis. Brick masonry exhibits distinct directional properties due to mortar joints, which act as planes of weakness, resulting in brick masonry structures showing complex and nonlinear mechanical behaviour. For the experimental study, the authors considered three varieties of brick and three mixed proportions of mortar. Compressive strength, water absorption, modulus of elasticity and Poisson's ratio of bricks; and compressive strength, modulus of elasticity, Poisson's ratio and density of different mortars were determined. The appropriate values of parameters for nonlinear FE analysis of masonry structures were recommended. Michiels et al. [4] performed a parametric study of the masonry roof shells. Singly curved (cylindrical shell) and doubly curved (spherical shell) were analysed. Real earthquake ground motions were applied, and the deformations and maximum principal stresses were computed. It was found that the key parameters which influenced the seismic response of the structure are the rise, span and thickness of the dome. Lupasteanu et al. [5] studied the behaviour of byzantine churches under seismic response and a shake table study. Preservation of historic structures was their main goal and hence several retrofitting techniques were not allowed, as they would change the architectural view of the heritage structure. The structural system of these churches consists of columns, walls, vaults, and sustained compressive gravity load stresses. Consequently, because the domes located at the top of these churches are prone to earthquake damage, the seismic protection techniques were extremely important. The authors compared the seismic response of the fixed base byzantine churches, strengthened byzantine churches and the base-isolated churches, in which it was found that the performance of the base-isolated byzantine churches under seismic loading was superior.

Masonry domes have represented the monumental structures thus increasing their grandeur. For the preservation of these massive, monumental structures against the strong seismic excitations, meticulous analysis of the masonry domes becomes a necessity. The specific objectives of the study are (i) to analyse the seismic performance of the fixed base masonry dome structure, (ii) to analyse the seismic performance of the masonry dome structure with lead rubber bearings, (iii) to analyse the seismic performance of the masonry dome structure with friction pendulum system and (iv) to compare the seismic performance of the fixed base masonry dome structure, LRB base-isolated structure and FPS base-isolated structure.

7.2 Base Isolation for Masonry Dome

Base isolation is one of the most powerful tools of passive structural vibration control technologies. The isolators decouple the superstructure from its substructure, thus protecting it from the damaging effects of an earthquake. The isolation can be achieved by using elastomeric and friction bearings. As masonry domes are unreinforced, their thickness is kept high to resist the tensile stresses. Due to their heavy mass, masonry domes attract large seismic forces, making them vulnerable to seismic excitations. Due to earthquake-induced forces, masonry domes are subjected to tensile stresses at the bottom rings, leading to the development of cracks at the bottom. The seismic performance of masonry domes can be significantly improved by implementing base isolators. The design basis report, architectural drawings and the structural drawings of the case study Sabhamandap dome located at Aurangabad, Maharashtra, India, are used for this research. The case study dome is isolated by the Lead Rubber Bearing (LRB) system and Friction Pendulum System (FPS).

7.3 Analysis Method

The masonry dome is analysed using nonlinear time history method. The method employed in SAP2000 is an extension of the Fast Nonlinear Analysis (FNA) developed by Wilson. The method is extremely efficient and is designed to be used for structural systems which are primarily linear elastic but have a limited number of pre-defined nonlinear elements. In the FNA method, all nonlinearity is confined to the link/support elements. Table 7.1 presents the particulars of the masonry dome considered for the study.

Figure 7.1 presents the elevation of the masonry dome and Fig. 7.2 presents the plan view.

Table 7.1 Particulars of the masonry dome

Parameter	Value
Rise (m)	5.93
Span/radius of dome (m)	24.78
Radius of curvature (m)	15.88
Roll-down angle	52°
Height of the wall/cylinder height (m)	6.61
Thickness of the wall (m)	0.47

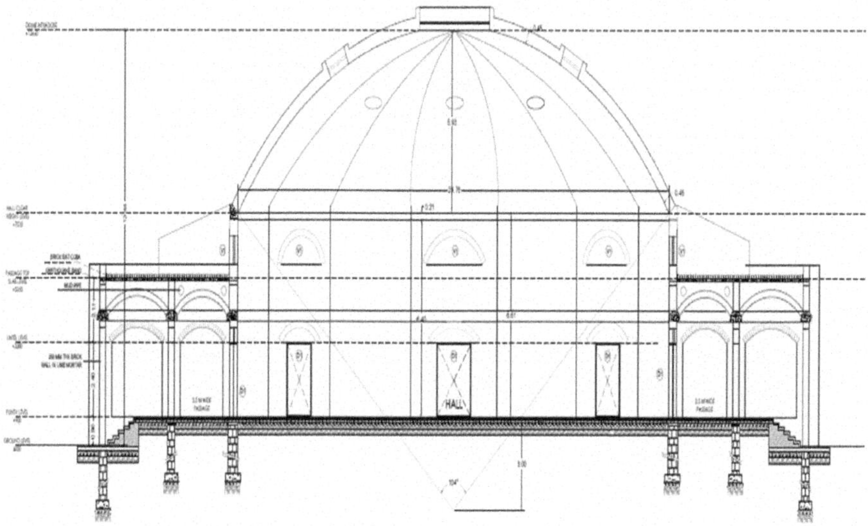

Fig. 7.1 Elevation of masonry dome. (Courtesy: Nandadeep Designers and Valuers Pvt Ltd, Aurangabad)

7.4 Simulation of Masonry Dome in SAP2000

The masonry dome structure is modelled using the shell element in SAP2000, as this element resists in-plane as well as out-of-plane bending moments. The radius of curvature of the dome, the roll-down angle (the angle with respect to vertical, up to which the dome extends) and the shell thickness are specified in SAP2000. The masonry material properties considered for simulation are modulus of elasticity = 2000 MPa, unit weight = 20 kN/m^3 and Poisson's ratio = 0.1. The masonry wall is simulated using a cylindrical shell element, which resists in-plane as well as out-of-plane bending moments. The geometry of the whole structure is created and gravity loads are applied. Further, nonlinear time history analysis is performed, considering three real earthquake ground motions, viz. El Centro (1940, PGA 0.33 g), Chamoli (1999, PGA 0.35 g) and Bhuj (PGA 0.1 g).

7.5 Masonry Dome Installed with Lead Rubber Bearings

Tables 7.2 and 7.3 respectively present the properties and dimensions of LRBs.

Parameters affecting the seismic performance of masonry dome, isolated by LRB are shear modulus of rubber and stiffness and damping of LRB, which are provided by the lead core. Table 7.4 presents the engineering properties of LRB used for nonlinear analysis.

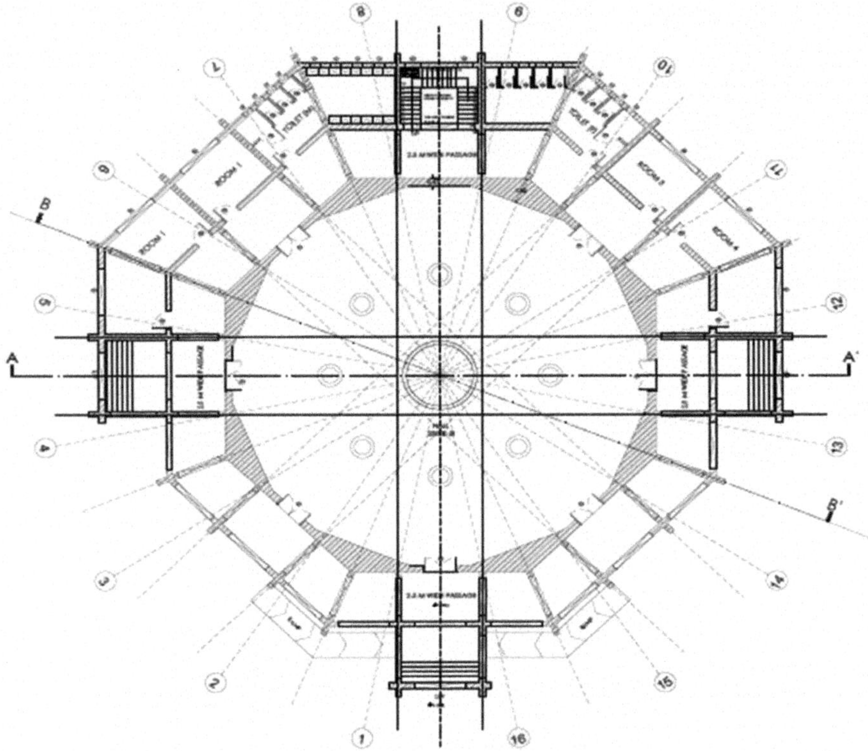

Fig. 7.2 Plan of masonry dome. (Courtesy: Nandadeep Designers and Valuers Pvt Ltd, Aurangabad)

Table 7.2 Properties of lead rubber bearings

Parameter	Value	Parameter	Value
Shear modulus (G) (MPa)	0.4	Bulk modulus (K) (MPa)	1500
Ultimate elongation (ε)	6.5	Elasticity modulus (E) (MPa)	1.35
Material constant (k)	0.87	Lead yield strength (σ_y) (MPa)	8

Table 7.3 Dimensions of lead rubber bearings

Property	Value	Property	Value
Plan dimensions (mm^2)	550 × 550	Side cover (mm)	10
Rubber layer thickness (mm)	12	Steel shim thickness (mm)	3
Number of rubber layers	20	Load plate thickness (mm)	25.5
Lead core diameter (mm)	120	Total height (mm)	348

Table 7.4 Engineering properties of LRB for nonlinear analysis

Parameter	Value
Initial horizontal stiffness (kN/m)	4663.19
Yield force (kN)	99.10
Post-yield stiffness ratio	0.08704

7.6 Masonry Dome Installed with FPS

Parameters affecting seismic analysis of masonry dome structure, isolated by FPS are the coefficient of friction (μ), radius of sphere (R), and the restoring force provided by the system. The post-sliding stiffness is determined by the geometry and supported weight as (W/R). The total force resisted by a spherical slider bearing is directly proportional to the supported weight. The engineering properties of FPS used for nonlinear analysis are presented in Table 7.5.

Figure 7.3 shows the elevation of the masonry dome modelled in SAP2000.

Table 7.5 Engineering properties of FPS for nonlinear analysis

Parameter	Value
Radius of pendulum (m)	1.924
Coefficient of friction (fast)	0.05
Coefficient of friction (slow)	0.11
Rate parameter	1.5
Maximum vertical load acting on each FPS (kN)	962
Effective horizontal stiffness (kN/m)	1000

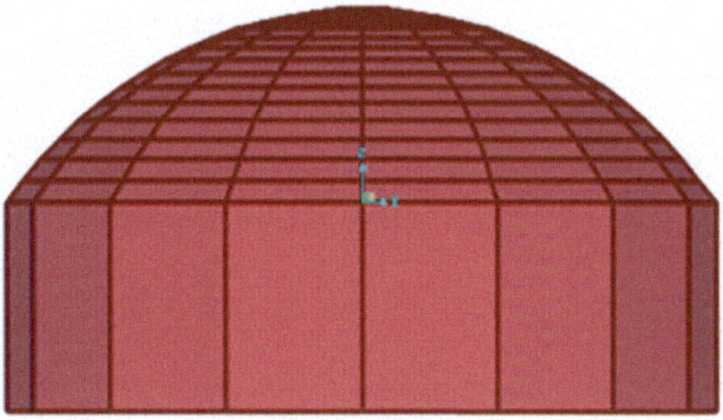

Fig. 7.3 Elevation of the dome structure (SAP2000)

7.7 Nonlinear Time History Analysis

In the nonlinear time history analysis in SAP2000, the isolators are modelled as link elements, and all nonlinearity is restricted to the link (support) elements. The results of the nonlinear time history analysis, viz. base shear, bending moments, tensile stresses, isolator forces and displacements are obtained for three cases, viz. (i) the fixed base dome, (ii) dome installed with LRB and (iii) dome installed with FPS. The analysis is performed considering three earthquake ground motions; however, the time history results are presented for El Centro (1940) earthquake.

(i) **Seismic Performance of Fixed Base Masonry Dome**

From modal analysis, the fundamental time period of the fixed base masonry dome is found to be 0.1838s, which reduces to 0.1097 s the in the 12th mode. Figure 7.4 and Table 7.6 present the response of fixed-base masonry dome.

(ii) **Seismic Performance of Masonry Dome isolated with Lead Rubber Bearings**

From modal analysis, the fundamental time period of the LRB-isolated masonry dome is found to be 1.8702 s, which reduces to 0.1063 s in the 12th mode.

From Fig. 7.5c, it is observed that up to a yield force of 99.10 kN, the initial stiffness of the isolator is high, i.e. 4663.19 kN/m. Hence, the deflection is very less up to the yield force. However, as the lead yields, it undergoes plastic deformation and thus deflection goes on increasing (see Table 7.7).

(iii) **Seismic Performance of Masonry Dome isolated with Friction pendulum System**

From modal analysis, the fundamental time period of the FPS-isolated masonry dome is found to be 1.7427 s, which reduces to 0.0701 s in the 12th mode. The time period is same in both directions, as the dome structure is symmetric in two directions. For the first two modes itself, the mass participation is 100%. As a result, the dynamic response is concentrated in these two modes itself (see Fig. 7.6 and Table 7.8).

7.8 Comparison of Fixed Base, LRB-Isolated and FPS-Isolated Masonry Dome

(a) Time period

From Fig. 7.7, the highest natural fundamental time period is found to be that of the LRB-isolated structure followed by FPS. However, it is observed that the first two modes have same time period as the structure is symmetric in nature. After first three modes, it can be observed that Fixed base, LRB, FPS have almost the same time periods for the next subsequent modes.

Fig. 7.4 Response of fixed base dome: El Centro earthquake (1940)

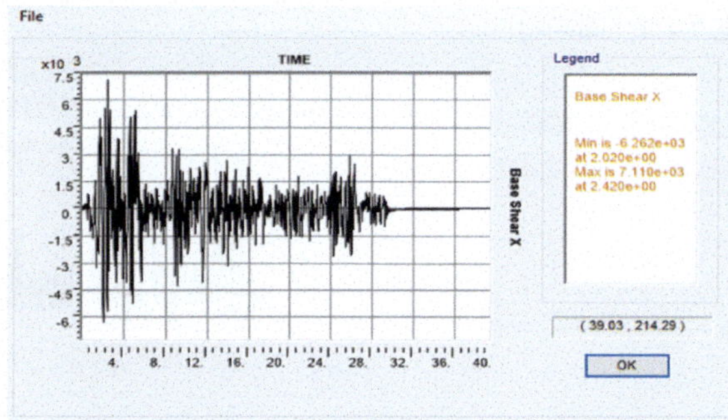

(a) Base shear response of dome

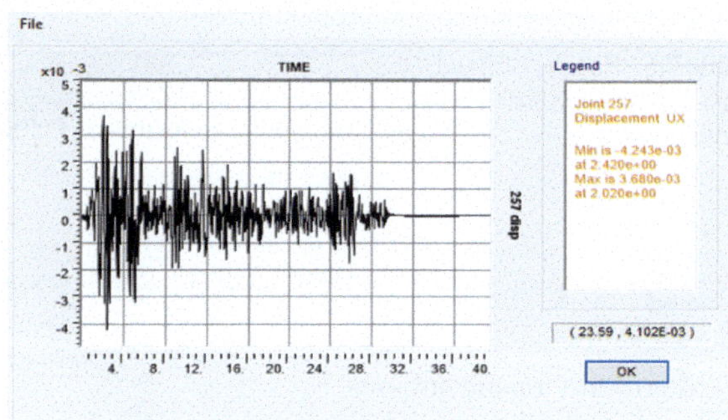

(b) Displacement at top joint of the dome

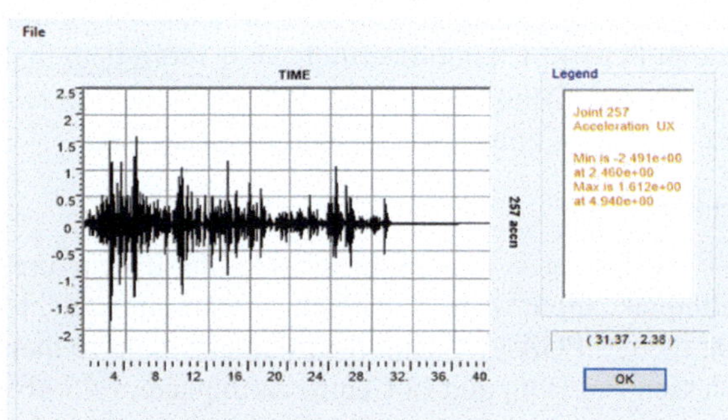

(c) Acceleration at top joint of the dome

Fig. 7.4 (continued)

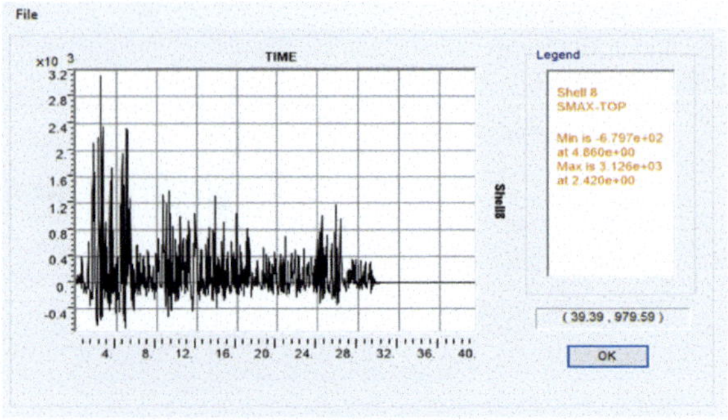

(d) Shell stresses at the top of the dome

Table 7.6 Results of fixed base dome under different earthquakes

Parameter	El Centro	Chamoli	Bhuj
Maximum shell stresses (MPa)	3.12	0.33	1.152
Base shear (kN)	7110	6754	3311
Top displacement (mm)	4.24	3.97	1.964
Top acceleration (m/s^2)	2.491	0.84	1.08

(b) Maximum tensile stresses

The maximum tensile stresses are found at the bottom ring of the masonry dome structure. The stresses are maximum for fixed base structure when subjected to El Centro earthquake input with 3.12 MPa the maximum value. It can be observed that though PGA of Chamoli earthquake is more than that of El Centro, tensile stresses are lesser. For Bhuj earthquake, FPS gives the best results, while LRB gives the best results for El Centro and Chamoli (see Fig. 7.8).

(c) Base shear

The base shear is maximum for fixed base structure when subjected to El Centro earthquake input with 7110 kN the maximum value. It can be observed from Fig. 7.9 that though PGA of Chamoli earthquake is more than that of El Centro, base shear is lesser. For Bhuj and El Centro earthquake, FPS gives the best results, while LRB gives a maximum reduction in base shear for Chamoli earthquake ground motion.

(d) Displacement at the top of the dome

The displacement at the top of the dome is maximum for fixed base structure when subjected to El Centro earthquake input with 4.24 mm as the maximum value. It can be observed that though PGA of Chamoli earthquake is more than that of El Centro, displacement at the top is lesser. For Bhuj earthquake, FPS gives the best results, while LRB gives the best results for Chamoli and El Centro (see Fig. 7.10).

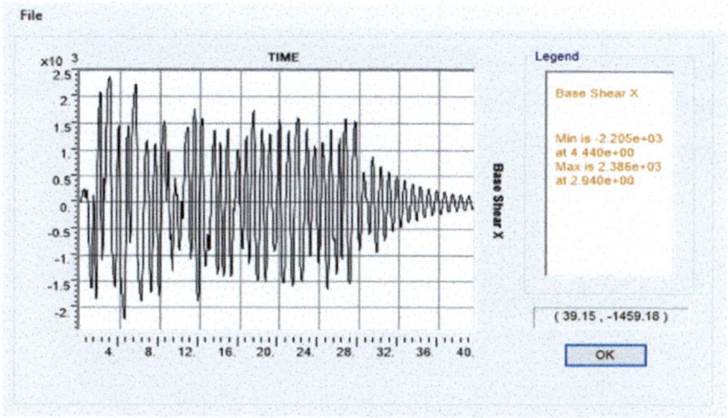

(a) Base shear response of LRB-isolated dome

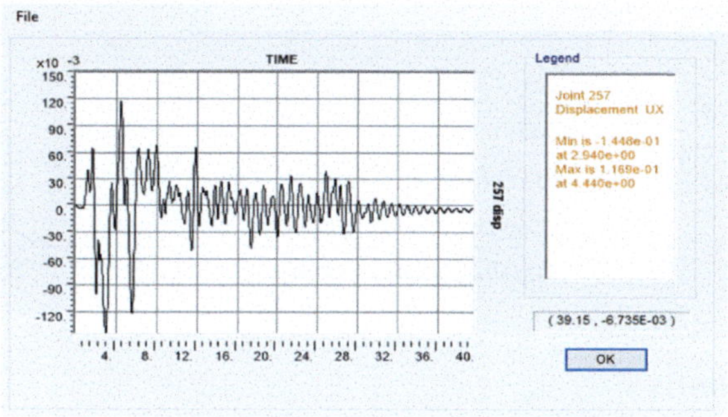

(b) Displacement at top joint of the LRB-isolated dome

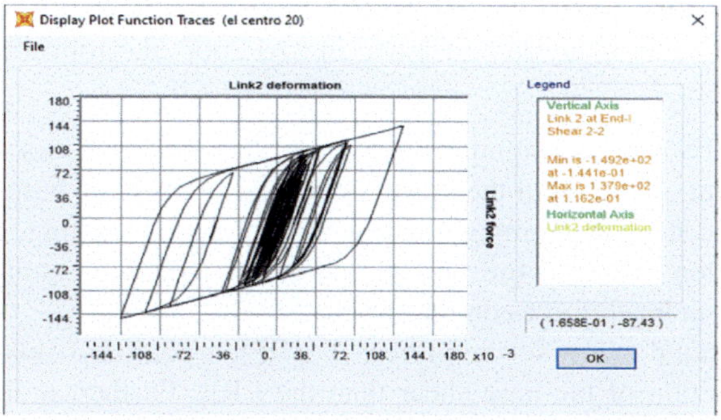

(c) Hysteresis loop of the LRB isolator

Fig. 7.5 Response of LRB-isolated dome: El Centro earthquake (1940)

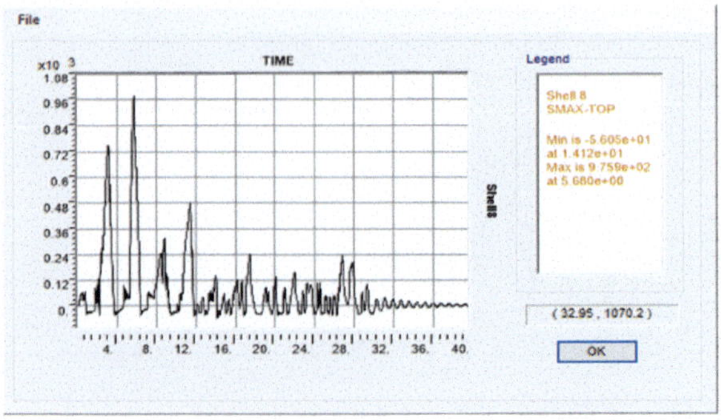

(d) Shell stress at top of the of LRB-isolated dome

Fig. 7.5 (continued)

Table 7.7 Results of LRB-isolated dome under different earthquakes

Parameters	El Centro	Chamoli	Bhuj
Maximum shell stresses (MPa)	0.975	0.6	0.243
Base shear (kN)	2386	2850	1845
Top displacement (mm)	0.3	0.79	0.47

7.9 Conclusions

Response of fixed base and base-isolated dome is studied for three real earthquake ground motions. LRBs and FPSs are employed at the base of the dome. From the nonlinear time history analysis performed in SAP2000, the following conclusions are drawn.

1. LRB and FPS isolators significantly reduce the tensile stresses, base shear and displacement of the top of the dome, compared to the fixed base dome, thereby reducing the structural damages during strong ground shaking.
2. Base isolation increases the time period of the structure resulting in reduced earthquake-induced forces on the structure.
3. When subjected to low seismic forces, the FPS is comparatively more efficient in the reduction of the base shear than the LRB. Further, it also reduces the tensile stresses and displacements of the apex of the dome than LRB does and thus resulting in comparatively low structural damage.
4. When subjected to high PGA earthquake ground motions, the LRB system is more efficient in the reduction of tensile stresses, displacements and the base shear of the masonry dome structure.

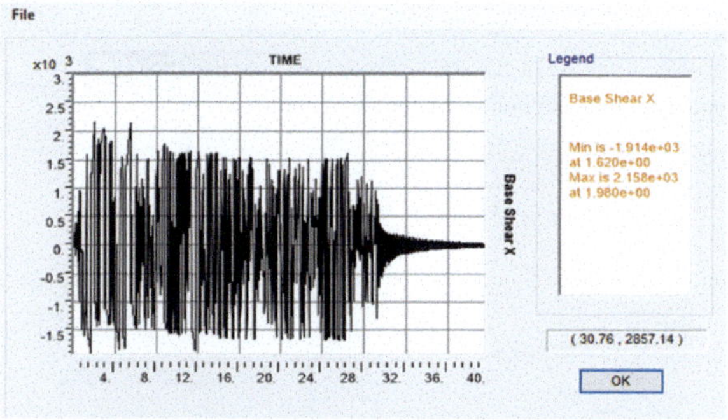

(a) Base shear response of the FPS-isolated dome

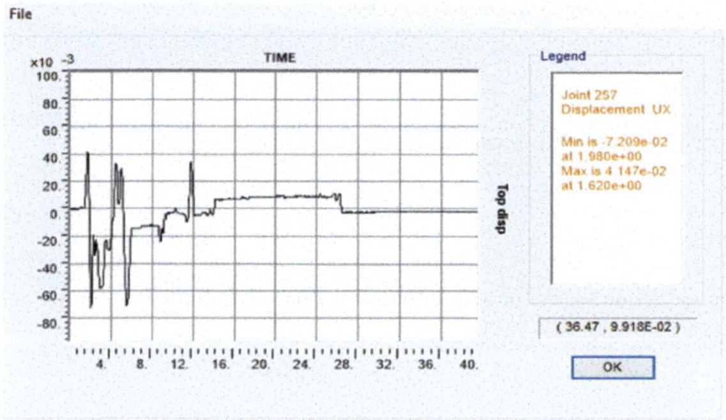

(b) Displacement at top joint of the FPS-isolated dome

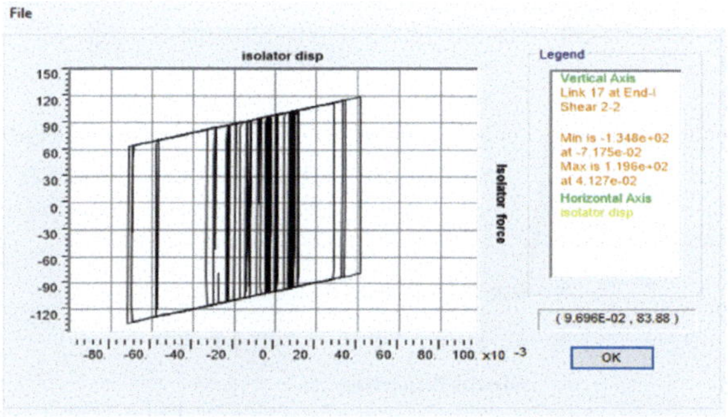

(c) Hysteresis loop of the FPS isolator

Fig. 7.6 Response of FPS-isolated dome: El Centro earthquake (1940)

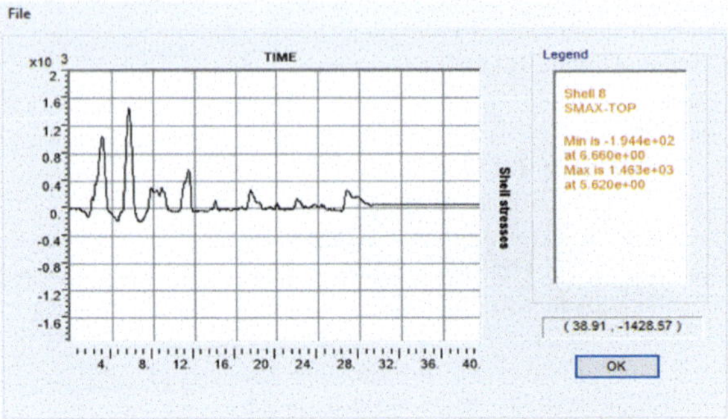

(d) Shell stress at top of the of FPS-isolated dome

Fig. 7.6 (continued)

Table 7.8 Results of FPS-isolated dome under different earthquakes

Parameter	El Centro	Chamoli	Bhuj
Maximum shell stresses (MPa)	1.463	0.971	0.2
Base shear (kN)	2156	3091	1642
Top displacement (mm)	0.34	0.9	0.09

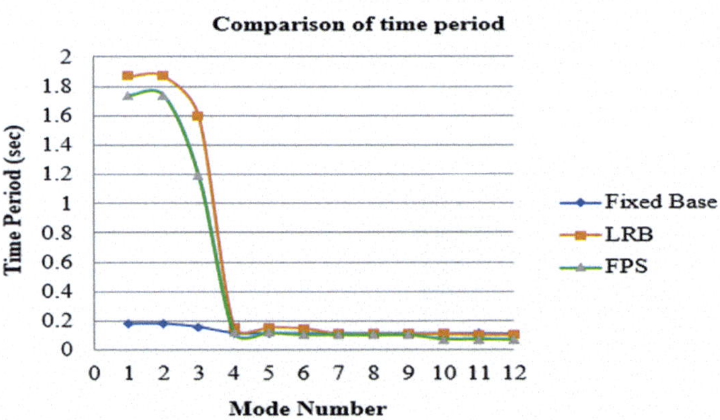

Fig. 7.7 Comparison of fundamental time period

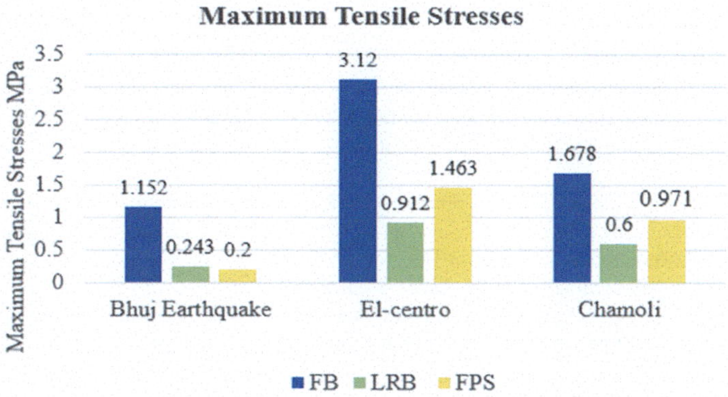

Fig. 7.8 Comparison of maximum tensile stress

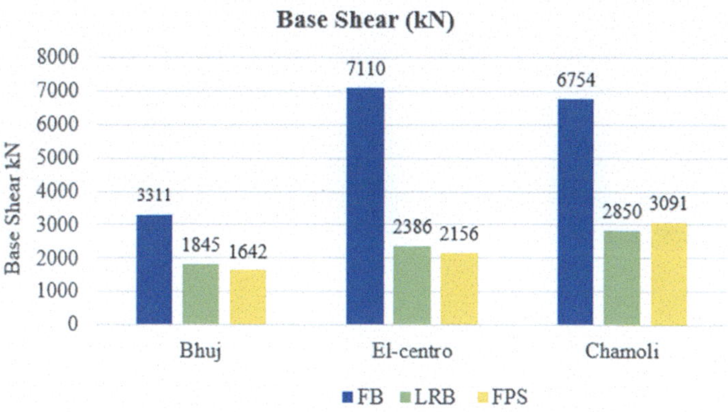

Fig. 7.9 Comparison of base shear

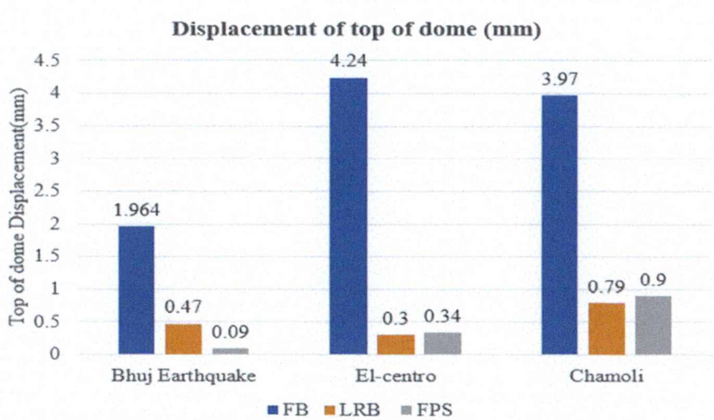

Fig. 7.10 Comparison displacement at the top of the dome

References

1. Croci G (2006), Seismic behavior of masonry domes and vaults of hagia sophia in instanbul and St. Francis in Assisi. In: First European conference on earthquake engineering, K8
2. Matsagar VA, Jangid RS (2008) Base isolation for seismic retrofitting of structures. ASCE 13:4(175)
3. Narayanan P, Sirajuddin M (2013) Properties of brick masonry for FE modelling. Am J Eng Res (AJER) 1:6–11
4. Michiels T, Adriaenssens S, Jorquera-Lucerga JJ (2017) Parametric study of masonry shells form-found for seismic loading. J Int Assoc Shell Spat Struct, 267–275
5. Lupasteanu V, Soveja L, Lupasteanu R, Chingalata C (2019) A study of the seismic protection techniques for the Byzantine churches in Macedonia. J Eng Struct 34(2/3):63–69

Chapter 8
Rapid Retrofitting of RC Columns Using Fe-SMA for Enhanced Seismic Performance

M. Sarmah, S. K. Deb, and A. Dutta

Abstract Collapse of many buildings and bridges in past earthquakes can be primarily attributed to damage in RC columns due to insufficient ductility. Studies have shown that addition of lateral confinement to concrete increases its strength and ductility significantly. Prestrained strips/wires (2–8%) made of shape memory alloy (SMA), a smart material, have recently gained popularity for their ability to provide in-situ active confinement of RC columns without the use of mechanical anchorages and in much lesser time. Prestrained SMAs can regain their original (undeformed) shape when heated above their transformation temperature. Thermal activation of prestrained SMA while restrained leads to the development of a large recovery stress, which is utilized to exert active confinement pressure externally on RC column. A low-cost iron-based SMA (Fe-SMA) is adopted for rapid retrofitting of RC column. A parametric study on behavior of Fe-SMA-confined concrete was carried out. It was found that the residual stress of Fe-SMA confined concrete is independent of concrete strength and only function of active confinement pressure on concrete. This material-level study was further expanded to the component-level for retrofitting of RC columns using Fe-SMA strips. Design of rapid retrofitting strategies of RC column using Fe-SMA strips in its plastic hinge region is discussed in this chapter.

Keywords Fe-SMA · Retrofitting · Shape memory effect · RC column

8.1 Introduction

There is a growing emphasis on the sustainability of existing structures. Due to this, a lot of research has been done on seismic repair and retrofitting of RC structures, which aims at improving their performance against seismic forces. Necessity for seismic retrofitting arises in non-ductile columns of RC structures, which were designed prior to the introduction of ductile detailing guidelines. These techniques may also be used

M. Sarmah · S. K. Deb (✉) · A. Dutta
Civil Engineering Department, IIT Guwahati, Assam 781039, India
e-mail: skdeb@iitg.ac.in

T. G. Sitharam et al. (eds.), *Theory and Practice in Earthquake Engineering and Technology*, Springer Tracts in Civil Engineering,
https://doi.org/10.1007/978-981-19-2324-1_8

for upgradation of structures so as to fulfill the requirements of the latest reforms in codal practices and/or hazard levels. Other issues that may necessitate strengthening include aging of the structure, change in its usage and loading requirements, and deterioration of structural strength at local or global level due to poor maintenance or post-earthquake damage. Seismic retrofit strategies can broadly be divided into global and local retrofit strategies. Global retrofit strategies target the overall performance of the structure when subjected to lateral loads by addition of shear walls, infill walls, bracings, supplemental damping, and base isolation among others. Local retrofit strategies aim at enhancing the seismic resistance of a structural member without having a substantial impact on the overall resistance of the structure. These include jacketing of beams, columns, beam-column joints, strengthening of individual footings, etc. Retrofitting is a better alternative to replacement because it not only saves time and money but also helps in cutting CO_2 emissions in construction.

RC columns constructed prior to the introduction of ductile detailing code of practice are termed as substandard as they had insufficient ductility arising from inadequate transverse reinforcement and/or lap splice length at its plastic hinge region. Research studies have demonstrated that lack of flexural ductility and/or poor shear resistance are the key factors that cause the collapse of RC columns [1–4]. Over the last few decades, a significant number of studies have been carried out by researchers, which focused on the development of appropriate strengthening and repair techniques for RC columns. These techniques aim to fulfill the requirement of improved strength and ductility and lower the cost of implementation while ensuring minimum disruption to building occupants during retrofitting activities keeping the structure's aesthetics in mind. Another critical consideration is repair time, particularly for lifeline structures such as bridges and hospitals. Hence, there is a growing emphasis for the need of rapid strengthening and repair techniques.

The addition of lateral confinement to concrete has been proven to significantly increase the strength and ductility of the concrete. Richart et al. [5] first investigated the effect of lateral confinement on ductile failure of concrete. The findings of this study encouraged numerous other researchers to come up with alternative strategies to apply lateral confinement and access the behavior of concrete under these confinement methods. Concrete confinement methods can be divided into two types: passive and active. In case of passive confinement, the confining pressure develops gradually after dilation of the concrete during loading. Several researchers [6, 7] have studied the behavior of concrete confined by transverse reinforcement. For existing structures with poorly designed RC columns, additional (external) confinement for concrete is provided at the potential plastic hinge regions. This passive confinement is commonly applied using external reinforced concrete (RC) jackets [1, 8, 9], steel jackets [2, 10, 11], and fiber-reinforced polymer (FRP) [12–18].

In case of active confinement, the lateral confining pressure is applied to the concrete prior to its dilation. Although the methodology adopted by several researchers to apply active confinement on-site differed, majority of them sought to employ prestressed steel strands or FRP belts [19, 20]. The results demonstrated efficacy of retrofitting using active confinement strategies in enhancing shear capacity and ductility of the columns. This is because prior application of confinement pressure causes a delay in the onset of damage in concrete. Even though active confinement is widely acknowledged to be superior to passive confinement, its implementation in the retrofitting strategies has been hindered. This is because such techniques result in many practical problems on-site related to use of specialized equipment and excessive amount of time and labor. As such, conventional passive confinement methods have become more viable in the construction industry. In light of the above, a smart material known as shape memory alloy (SMA) has lately gained popularity for use in active confinement of civil engineering structures.

By virtue of its unique thermo-mechanical property, shape memory effect (SME), prestrained SMAs can regain their original (undeformed) shape when heated above their transformation temperature. If the SMA is restrained while heating, large recovery stresses are developed in it, which in turn aids in prestressing the structural component against which it is restrained. Unlike conventional prestressing techniques, this method doesn't require heavy machinery on-site. Also, the time required for generation of recovery stress is much lesser.

The idea of employing SMA spirals for active confinement of concrete through thermal prestressing was initially presented by Andrawes and Shin [21]. Subsequently, several researchers [22–28] carried out experiments to study the behavior of concrete confined by NiTiNb SMA spirals. Results indicated an improvement in ultimate stress and strain of concrete. However, its widespread application was hindered owing to its high cost and complex production procedure. Hence, for prestressing structural elements at a low cost, a Fe-SMA with composition Fe–17Mn–5Si–10Cr–4Ni–1(V, C) (mass %) was developed in 2009 at the Swiss Federal Laboratories for Materials Science and Technology (Empa) [28].

Several studies have been carried out for characterization of this material [29–31]. In addition, several experimental investigations have shown that Fe-SMAs have a high potential for strengthening of structural components [32–34]. Recovery stresses of 300–350 MPa were developed after heating the Fe-SMA to 160 °C. Prestressing using Fe-SMA strips was found to be successful in improving the strength, ductility, and energy dissipation capacity of RC columns. In this chapter, design of rapid retrofitting strategies of RC column using Fe-SMA strips in its plastic hinge region will be discussed. Five SMA configurations, namely hoop, end-anchored (EA), near-surface mounted (NSM), combination of hoop and EA, and combination of hoop and NSM are highlighted.

8.2 Retrofitting Techniques Using Confinement Approach

Poor performance of structures during the past seismic events has drawn researchers in the field of structural retrofitting. One of the most common retrofitting techniques that are used to increase the strength and ductility of RC columns is provision of additional external confinement. These concrete confinement techniques can be categorized into two categories: (1) Passive confinement and (2) Active confinement. These are further elaborated in the following subsections.

8.2.1 Comparison of Stress–Strain Behavior of Passively and Actively Confined Concrete

The fundamental difference between passive and active confinement is the instant mobilization of lateral confining pressure. Figure 8.1a shows cross-sections of passively and actively confined concrete cylinders. As seen in the figure, lateral confining pressure is applied to the concrete section prior to axial loading in case of active confinement. However, in passive confinement technique, concrete has to dilate, i.e., deform laterally for mobilization of the confining pressure. Hence, early application of confinement pressure causes a delay in the onset of damage in actively confined concrete.

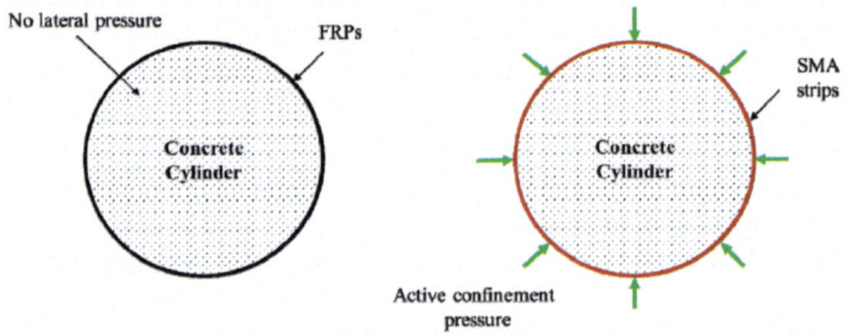

(a)

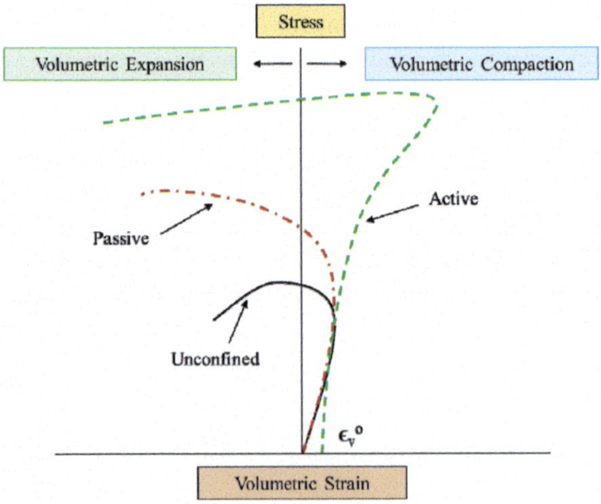

(b)

Fig. 8.1 Passive versus active confinement (**a**) cross-sections of passively and actively confined concrete before loading (**b**) stress versus volumetric strain curves of unconfined, passively, and actively confined concrete

Figure 8.1b illustrates the typical axial stress versus volumetric strain responses of unconfined, passively confined, and actively confined concrete. In the elastic region, the unconfined concrete undergoes volumetric compaction, following which there is rapid expansion leading to failure. Similarly, when passively confined concrete is subjected to axial stress, its volume decreases in the elastic region. As the axial stress increases, passive confining pressure aids in delaying the instant at which the concrete begins to expand volumetrically. However, in the active confinement case, the lateral confining pressure is mobilized on the concrete as a prestress before application of service loading. This leads to the development of an initial volumetric strain $\varepsilon_v{}^o$ resulting from compaction. Additional axial stress is required in order to counteract the effect of this strain, thus delaying the failure of the concrete as compared to passively confined concrete.

8.2.1.1 RC Jacketing

RC jackets have been used to enhance the performance of vulnerable columns for several decades. In traditional RC jacketing, the original column cross-section is enlarged over a portion or its entire length using reinforced concrete. This new section is connected to the original section using dowel bars. Although this retrofitting technique improves the strength and ductility of the columns [1, 8, 9], it is labor intensive with a high operation cost owing to roughening, cleaning, and dowelling of the existing damaged concrete section so as to ensure composite deformation between the jacket and the column surface. It increases the size of the structural element, thereby changing its dynamic properties, which is likely to impose a higher seismic demand on the structure.

8.2.1.2 Steel Jacketing

Steel jacketing involves installation of a steel jacket by welding the components and using cement-based grout to fill the space between the jacket and the original or enlarged column. This technique was able to enhance the strength and ductility of the retrofitted columns resulting in considerably larger lateral loads and energy dissipation capacities as compared to the as-built columns [2, 9, 14]. Furthermore, similar to RC jacketing, this approach alters the stiffness of the structure by changing the cross-sectional dimension of the RC column section. This approach, however, has certain downsides, corrosion, and debonding being the common problems with the bonded sheets. There are also challenges involved in its transportation and installation in remote areas.

8.2.1.3 FRP Jacketing

In the recent decades, strengthening of RC columns by FRP composites has become increasingly popular. FRP comprises fibers (e.g., carbon, aramid, glass, etc.) connected effectively with the help of resins, which provide the mechanism for load transfer between the fibers and are responsible for composite strength. In FRP jackets, fibers are oriented in different directions to achieve different objectives. Fibers aligned along the longitudinal axis of the column increase its flexural strength, whereas those aligned in the transverse direction improve its shear capacity and ductility [12–17]. FRP composites have been more prevalent for retrofitting of RC columns as compared to RC and steel jackets. This is because of its high strength-to-weight ratio, which results in minimal modification of the geometry of RC columns while enhancing its capacity. Several disadvantages associated with the use of FRP confinement are its need for experienced workmanship, high cost, ineffective functioning in the compression part of cyclic loads and delamination.

8.2.2 Active Confinement Techniques

Active confinement, as compared to passive confinement, effectively slows down the lateral expansion of concrete under compression because the pressure is applied before the concrete dilates; hence increasing the ductility and strength of concrete. These attractive features encouraged many researchers to consider applying active confinement by the following techniques:

8.2.2.1 Triaxial Pressure Vessel/Triaxial Testing Machine

Most of the early studies on active confinement used triaxial testing devices to investigate the behavior of concrete. The study by Richart et al. [5] was one of the first studies on active confinement of concrete, which used fluids in a triaxial pressure vessel to exert a constant lateral confining pressure. The following equations were proposed to determine the peak strength f_{cc} and strain ϵ_{cc} of confined concrete under a lateral confining pressure f_l:

$$\begin{aligned} f_{cc} &= f_{co} + k_1 f_l \\ \varepsilon_{cc} &= \varepsilon_{co}\left(1 + k_2 \frac{f_l}{f_{co}}\right) \end{aligned} \tag{8.1}$$

where f_{co} and ε_{co} are the peak strength and corresponding strain of the unconfined concrete and k_1 and k_2 are coefficient values that take into account the effect of active confining pressure. The suggested values for k_1 and k_2 are 4.1 and $5k_1$, respectively. Many analytical models have been built on the basis of these fundamental equations.

After the pioneering study by Richart et al. [5], several researchers conducted experiments on concrete blocks or cylinders under biaxial/triaxial stress using a triaxial testing machine. These studies demonstrated that the strength and the strain of concrete are dramatically improved by increasing the confining pressure.

8.2.2.2 Prestressing Steel Reinforcement/FRP Jackets

Active confinement was applied in the field by researchers using various methodologies. Some researchers adopted lateral prestressed steel strands and special anchorage devices for confinement of RC columns [19]. Others examined the feasibility of applying confining pressures with the help of prestressed FRP belts [20]. A key difference between using triaxial pressure vessel/triaxial testing machine and prestressed steel strands/FRP belts to apply active confining pressure is that the active confining pressures applied through the former remain constant throughout the test, whereas in the latter case, active confining pressure is exerted prior to loading and additional passive confining pressure keeps developing with concrete expansion during loading. Since it incorporates the advantages of both active and passive confinement techniques, it is found to effectively improve the axial and lateral capacity of RC columns. However, this technique requires complex maneuvering to apply the required external pressure. Hence, researchers began exploring for a simpler yet effective approach to apply active confinement on-site with minimal hardware and labor. This objective is fulfilled by the use of SMAs, details of which are elaborated in the next section.

8.2.2.3 SMA Confinement

The concept of employing SMAs to provide active confinement of concrete is based on utilizing the recovery stress generated upon thermal activation of SMAs. SMAs can regain their original shape even after undergoing excessive deformations up to 8% strain. By virtue of its unique thermo-mechanical property, shape memory effect (SME), prestrained SMAs can regain their original (undeformed) shape when heated above their transformation temperature. If the SMA is restrained while heating, large recovery stresses are developed in it, which in turn aids in prestressing the structural component against which it is restrained. The recovery stress is largely dependent on the SMA composition, manufacturing technique, and level of prestrain prior to shape recovery. The idea of employing SMA spirals for active confinement of concrete through thermal prestressing was initially presented by [21] Andrawes and Shin (2008). Prestrained SMA wires were wrapped around the plastic hinge region of the column and anchored at both ends. These were then heated using electric current or a fire torch. The induced shrinkage of the SMA wires causes a squeezing effect on the concrete column, which in turn provides active confinement to the column. Unlike conventional prestressing techniques, this method doesn't require heavy machinery on-site. The effectiveness of this technique can be attributed to the thermo-mechanical properties of the SMAs, which will be discussed in the next section.

8.2.3 Shape Memory Alloys

Shape Memory Alloys (SMAs) are metallic alloys that deform significantly under external load at low temperatures and recover when heated over a particular temperature. This ability to remember its original undeformed shape lends this metallic alloy group its name, i.e., shape memory alloy and this phenomenon is referred to as shape memory effect (SME). Another thermo-mechanical phenomenon that is unique to these groups of alloys is known as superelasticity (SE). By virtue of SE, if this material undergoes substantial deformation by the action of external load, above a particular temperature, it is able to recover the same upon unloading. These distinctive properties of the SMAs, SME, and SE are related to the specific characteristics at the microstructural level that permits reversible solid–solid phase transformation among variants of its constituent phases, namely Martensite [M], and Austenite [A]. This was first discovered in 1951 in Au-Cd alloy, followed by the discovery of Ni–Ti alloy. These are divided into three major groups: NiTi-based, Cu-based, and Fe-based. SMAs are widely used in actuator applications, valves, pipe couplings, dampers, and medical applications. Majority of uses of SMAs are confined to military, aerospace, and medical applications. However, there has lately been a surge in demand for SMAs in civil engineering applications as well.

8.2.3.1 Thermo-Mechanical Properties of SMAs

SMAs have two main crystallographic phases, the low-temperature phase is called Martensite [M], and the high-temperature phase is called Austenite [A]. The regulations governing the transition from one phase to the next are highly dependent on the alloy's temperature in terms of four transformation temperatures, which are unique to each alloy as shown in Fig. 8.2(a). These transformation temperatures are (1) the martensite start temperature (M_s), where the martensitic transformation of austenite to martensite starts, (2) the martensite finish temperature (M_f), where the martensitic transformation of austenite to martensite finishes, and below Mf the SMAs are completely in martensite phase (3) the austenite start temperature (A_s), where the reverse transformation of martensite to austenite starts, (4) the austenite finish temperature (A_f), where the reverse transformation of martensite to austenite finishes, and beyond A_f, the SMAs are completely in austenite phase.

If an SMA is excessively deformed in the austenite phase, stress-induced martensite is formed. However, once the stress is removed, reverse transformation from martensite to austenite starts and the SMA returns back to its original shape. This phenomenon is known as superelasticity. A typical flag-shaped stress–strain behavior

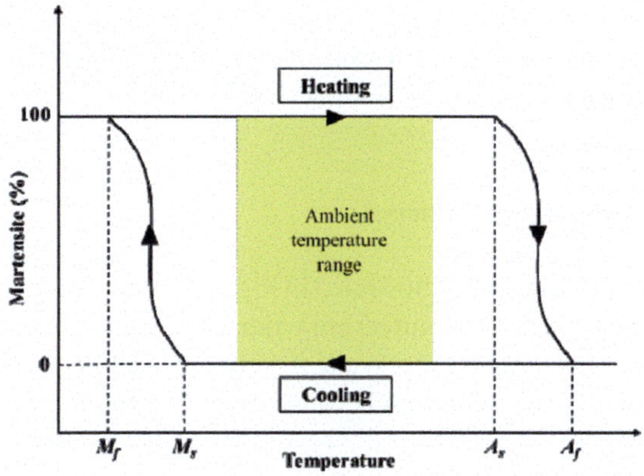

(a)

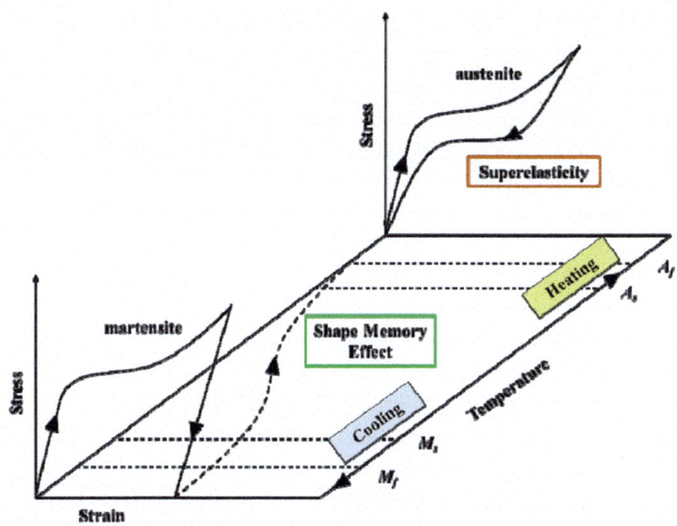

(b)

Fig. 8.2 Thermo-mechanical properties of SMA (**a**) phases (**b**) superelasticity and shape memory effect

of SMA loaded in the austenite phase is shown in Fig. 8.2b. As the loading and unloading paths are different, superelastic SMAs exhibit large hysteresis, which are utilized in civil structures as dampers and base isolation devices. When an SMA is deformed in its martensite phase, it is subjected to significant residual strains even after removal of loading. Upon heating up to above A_f, it recovers its original undeformed shape as shown in Fig. 8.2b. This phenomenon is known as shape memory effect (SME). However, if the SMA is restrained, it will not be able to recover its original shape and a recovery stress develops in the SMA. Researchers have made efforts to utilize this recovery stress of SMAs in the application of active confinement technique for concrete.

8.2.3.2 SMA Confinement Concept

The SMA confinement concept is illustrated in Fig. 8.3. Prestrained SMA wires/strips (2–8% prestrained by the manufacturer) are wrapped around concrete members, and heated to their activation temperature, while restrained, to activate SME. As the spiral is unable to restore its original length because of the restraining action of concrete, a large recovery stress is generated in it. This recovery stress squeezes the concrete member, exerting confining pressure on it.

Andrawes et al. [21] initially investigated the possibility of using SMA spirals for active confinement of concrete by thermal activation. Even at low confining pressures, uniaxial compression tests of concrete cylinders wrapped with NiTiNb SMA spirals revealed a considerable increase in concrete strength and ductility. Later, Shin and Andrawes [22–24] incorporated this study to experimentally investigate the retrofitting of RC columns using prestrained NiTiNb SMA spirals. The approach was found to be beneficial in improving the ductility and energy dissipation capacity of the RC columns. A retrofitting approach was proposed which employed SMA confinement to improve the ductility of concrete more efficiently as compared to conventional passive confinement techniques. Although the NiTi-based alloys are the most widely used SMAs, these are very expensive for large-scale applications. Hence, for prestressing structural elements at a low cost, an Fe-SMA with composition Fe–17Mn–5Si–10Cr–4Ni–1(V, C) (mass %) was developed in 2009 at the Swiss Federal Laboratories for Materials Science and Technology (Empa) [28]. For mobilization of active confinement pressure, this Fe-SMA strip is heated to 160 °C in less than 60 s. Because of the advantages offered by Fe-SMA for prestressing applications, a number of investigations have been conducted to further characterize this alloy [29–31]. Most of the experimental studies are conducted on strengthening and repair applications of RC beams [32–34]. To numerically analyze the performance of

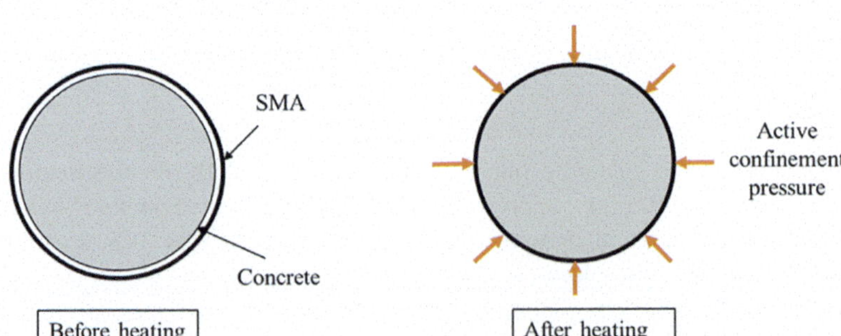

Fig. 8.3 Active confinement concept of SMA

RC beams with Fe-SMA as NSM reinforcement, Abouali et al. [35] developed a 3D nonlinear finite element model in ABAQUS. Results of these studies showed significant strengthening effect as compared to that of reference beam and that Fe-SMAs have a high potential for retrofit of structural components.

8.2.4 Fe-Based Shape Memory Alloys

The use of Fe-SMA strips for prestressing applications in RC members comprises three steps, which are shown in Fig. 8.4a:

1. Fe-SMA strips are initially prestrained to a specific strain level. After attainment of this strain level, it is fully released (Segment 1).
2. Fe-SMA strips are thermally activated by heating to their activation temperature while affixed on concrete surface or embedded in concrete (Segment 2).
3. When the RC member is loaded, the behavior of the Fe-SMA strips is as shown in Fig. 8.4a (Segment 3).

The recovery stress generation in the Fe-SMA strip during the above-mentioned steps is shown in Fig. 8.4b. At the beginning of the heating step, due to thermal expansion, the stresses in the Fe-SMA strip decrease. Upon attainment of the transition temperature, tensile stresses develop in the strips due to reverse transformation of the alloy from martensite state to austenite. Once the strip is cooled down to the room temperature, there is a further increase in its tensile stress due to thermal contraction. Since the Fe-SMA strips are restrained against concrete, compressive stresses develop in the concrete member. Phase change temperatures for the Fe-SMA

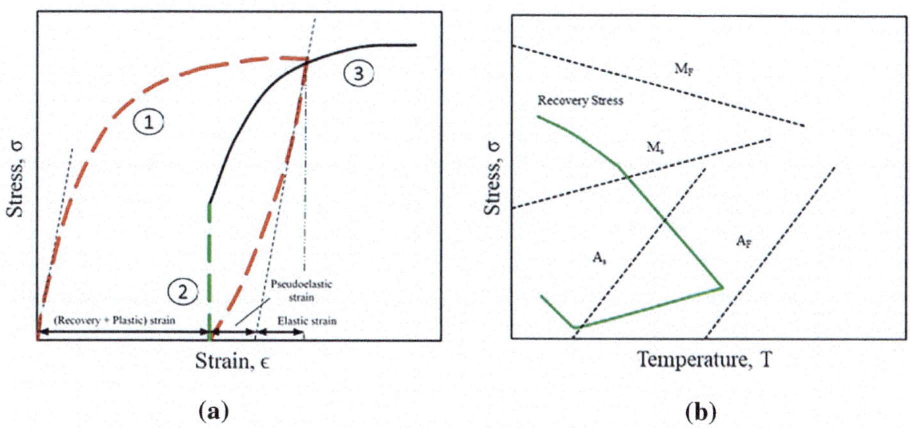

Fig. 8.4 Prestressing application of Fe-SMA (**a**) stress–strain behavior (**b**) recovery stress generation stress development

have been reported by Lee et al. [30] as follows: $M_f = -64$ °C, $M_s = -60$ °C, $A_s = 103$ °C, and $A_f = 163$ °C.

Shahverdi et al. [33] carried out material characterization of the Fe-SMA strips for the purpose of prestressing applications in civil engineering structures. The ultimate tensile strength and corresponding failure strain of the Fe-SMA strips was found to be 1000 MPa and 40%, respectively. This indicated that the strips have high strength as well as ductility. After heating the Fe-SMA strips to 160 °C, recovery stresses of 300–350 MPa were recorded. It was found that the recovery stresses generated after thermal activation depend on the prestrain level of the strips as well as the temperature of activation. A 2% prestrain level was found to be optimum for achieving maximum recovery stress at an activation temperature of 160 °C. These results are used for structural design purposes.

8.3 Parametric Study on Concrete Confined by Fe-SMA Strips

Abouali et al. [35] developed a 3D nonlinear finite element (FE) model in ABAQUS to numerically explore the strengthening of RC beams in flexure with NSM Fe-SMA strips. This modeling approach is used to numerically predict and analyze the behavior of concrete confined by Fe-SMA strips using FE method. Parametric studies are carried out to study the effect of two design variables: concrete strength and Fe-SMA strip spacing on the constitutive behavior of concrete. For this purpose, standard concrete cylinders of 150 mm diameter and 300 mm height were modeled for FE simulation. Three different concrete grades of characteristic strength 25, 30, and 40 MPa, which are designated as A, B, and C, respectively, are considered in this study.

To understand the effect of Fe-SMA strip spacing on the constitutive behavior of concrete, four different SMA strip spacing are adopted in the study. The dimensions of Fe-SMA strips used are 24 mm width and 1.5 mm thickness. Four different strip spacing of 92, 69, 55.2, and 46 mm, which are designated as S1, S2, S3, and S4, are considered in this parametric study. Hence specimen A-S1 implies M25 grade concrete cylinder with Fe-SMA spacing of 92 mm.

Previous studies on active confinement have shown that the axial stress–strain response of concrete under monotonic loading appears to be the envelope of axial stress–strain response of cyclically loaded concrete [36]. Hence in this study, the cylinders were subjected to monotonic loading only.

Active confining pressures corresponding to different strip spacing were calculated based on the effective confining pressure proposed by Mander et al. [7]. According to Mander et al. [7], for a circular section, the effective confining pressure $f_l{}'$ can be calculated as

$$f_l' = \frac{2k_e f_y A_{sp}}{sD} \tag{8.2}$$

where k_e is the confinement effectiveness ratio, f_h is the SMA recovery stress [36], A_{sp} and s are the cross-sectional area and center-to-center spacing of the Fe-SMA strip and D is the diameter of the specimen. Details of modeling in ABAQUS are discussed in the following sub-section.

8.3.1 Material Models

8.3.1.1 Concrete Constitutive Model

To model the behavior of concrete, the CDP model available in ABAQUS was used [35]. In this model, damage is incorporated both in tension as well as compression. Concrete having cylindrical compressive strength, f_c' of 50 MPa and direct tensile strength, f_t of 2.4 MPa is used [16]. The stress–strain curve of concrete under uniaxial compression was defined using the Hognestad parabola [37], as described in Eq. (8.3):

$$\begin{aligned}
\sigma_c^{(1)} &= E_{c0}\varepsilon_c \text{ for } \sigma_c < 0.4f_c' \\
\sigma_c^{(2)} &= f_c'\left[2\left(\frac{\varepsilon_c}{\varepsilon_c'}\right) - \left(\frac{\varepsilon_c}{\varepsilon_c'}\right)^2\right] \text{ for } \frac{\varepsilon_c}{\varepsilon_c'} \le 1 \\
\sigma_c^{(3)} &= f_c' - \frac{f_c'\left(\varepsilon_c - \varepsilon_c'\right)^2}{\left(\varepsilon_{c,\max} - \varepsilon_c'\right)^2} \text{ for } \frac{\varepsilon_c}{\varepsilon_c'} > 1
\end{aligned} \tag{8.3}$$

where σ_c and ε_c are the compressive stress and strain, respectively, $\varepsilon_{cm,ax}$ is the maximum strain, ε_c' is the strain corresponding to the peak stress, E_{co} is the modulus of elasticity of concrete. In tension, the stress-crack opening displacement curve post-cracking of concrete was defined in the ABAQUS CDP model by an exponential function described by Eq. (8.4) [38]:

$$\begin{aligned}
&\frac{\sigma}{f_t} = f(\omega) - \frac{\omega}{\omega_c} f(\omega_c) \\
&\text{where, } f(\omega) = \left[1 + \left(\frac{c_1\omega}{\omega_c}\right)^3\right]\exp\left(-\frac{c_2\omega}{\omega_c}\right) \text{ and } \omega_c = 5.14\frac{G_f}{f_t}
\end{aligned} \tag{8.4}$$

where ω is the crack opening displacement, ω_c is the crack opening displacement at which no stresses can't be transferred, $f(\omega)$ is a displacement function, G_f is the concrete fracture energy, and c_1 and $c_2$are material constants.

The values of the plasticity parameters used in this study are summarized in Table 8.1. Compressive and tensile damage variables in the CDP model are defined as per Birtel et al. [39].

Table 8.1 Plasticity parameters for CDP model [39]

Dilation angle, Ψ (°)	50
Plastic potential eccentricity, ε	0.1
Stress ratio, σ_{b0}/σ_{c0}	1.16
Shape of the yielding surface, K_c	0.667
Viscosity parameter, μ	0.0001

8.3.1.2 Iron-Based Shape Memory Alloy (Fe-SMA) Strips

The Fe-SMA strips were modeled with the stress–strain curve obtained in the tension test [33], using isotropic hardening plasticity as shown in Fig. 8.5. For the concrete cylinders confined with activated Fe-SMAs, the loading to failure segment of the stress–strain diagram was implemented in the FE model.

8.3.2 Finite Element Model

Continuum eight-node linear brick elements with reduced integration and hourglass control (C3D8R) were used to model the concrete as it provides a solution of comparable accuracy as compared to second-order tetrahedral elements at less computational cost. Fe-SMA strips, as shown in Fig. 8.6, were modeled with two-node Timoshenko beam elements (B31) with linear interpolation as they can be subjected to large axial strains.

Assuming no slip between the SMA strips and concrete surface, a tie constraint was used to model the interaction between the two. A fixed boundary condition was applied to the bottom surface of the model during loading phase. Using the initial

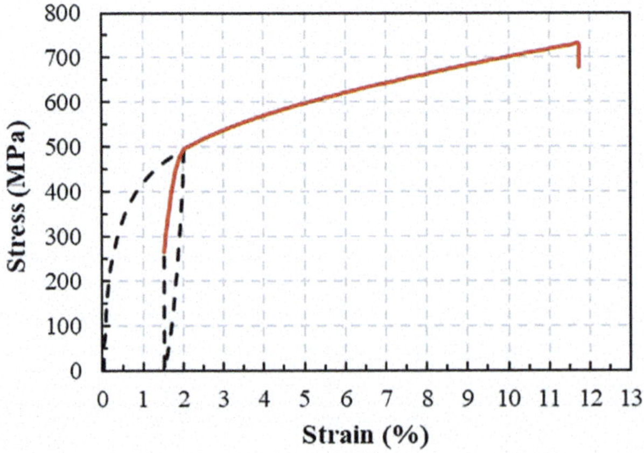

Fig. 8.5 Stress–strain curve of Fe-SMA

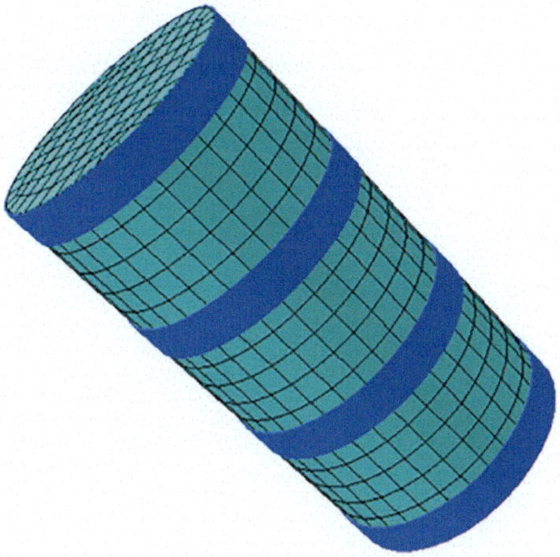

Fig. 8.6 FE model of concrete cylinder confined by Fe-SMA strips

stress feature in ABAQUS, the recovery stress was enforced as initial stress in the Fe-SMA strips in the axial direction. To allow establishment of self-equilibrium after assigning the initial stress, an idle step without any loading was defined.

Displacement-controlled loading was applied to the models by applying a monotonic vertical displacement with a loading strain rate of 0.5% per min until failure of the cylinder. The nonlinear solution was then obtained by configuring an implicit static general analysis method.

8.3.3 Results and Discussion

A parametric study is carried out to understand the effect of concrete strength and Fe-SMA strip spacing on the constitutive behavior of concrete. The results of all the specimens are summarized in Table 8.2.

8.3.3.1 Effect of Grade of Concrete

Four spacing values were considered in the study, namely, 92, 69, 55.2, and 46 mm. Figure 8.7 compares the axial stress–strain relationships of SMA confined concrete having same strip spacing with different concrete grades.

As the grade of concrete is increased from M25 to M40, it is observed that the peak stresses of A-S1, B-S1, and C-S1 increased by 51%, 45.3%, and 38.9%, respectively, as compared to that of their respective unconfined concrete specimens A, B, and C. However, residual stress of Fe-SMA confined concrete is found to be independent of concrete grade and is only a function of active confinement level for normal strength concrete. A similar trend is observed for specimens with SMA spacing of S2, S3, and S4. This observation is in agreement with the experimental results obtained by [27] for NiTiNb-SMA spirals confined concrete.

Table 8.2 Summary of results for all specimens

Specimen	Active confinement pressure (MPa)	f'_{cc} (MPa)	f_{res} (MPa)
A	0	19.6	–
A-S1	1.09	29.7	15.2
A-S2	1.76	33.9	21.9
A-S3	2.44	35.6	26.8
A-S4	3.14	45.2	32.7
B	0	23.5	–
B-S1	1.09	34.1	15.4
B-S2	1.76	38.3	22.1
B-S3	2.44	40.8	26.9
B-S4	3.14	49.5	32.8
C	0	30.9	–
C-S1	1.09	42.9	15.1
C-S2	1.76	47.1	22.1
C-S3	2.44	49.2	27.1
C-S4	3.14	58.1	32.9

8.3.3.2 Effect of Spacing of Fe-SMA Strips

As the spacing of Fe-SMA strips is decreased from S1 to S4, the effective confinement pressure exerted by the strips on concrete increases. With increasing value of confinement pressure, it is observed (Fig. 8.8) that the peak stresses of A-S1 to A-S4 increased by 51%, 72.9%, 81.6%, and 130%, respectively, as compared to that of the unconfined concrete specimen, A. From Fig. 8.8, it is evident that the ductility of the specimens also increases as compared to that of the unconfined specimen. Also, as established in the earlier section, residual stress is a function of confinement level of concrete. Thus, with increase in the confinement value of Fe-SMA strips the residual stress increases. A similar trend is observed for specimens of grade of concrete B, C when the spacing is varied from S2 to S4.

8.3.4 Observations Based on Parametric Study

This study focused on analyzing the behavior of SMA confined concrete using FEM within the framework of damaged plasticity model in ABAQUS. Active confinement of concrete is provided by prestrained Fe-SMA strips, which are restrained during heating. This is due to induction of large recovery stresses in the SMA, as it tries to recover its original undeformed configuration. This study is an initial step towards understanding the behavior of Fe-SMA confined concrete by numerically investigating the change in the constitutive behavior of concrete with changes in grade

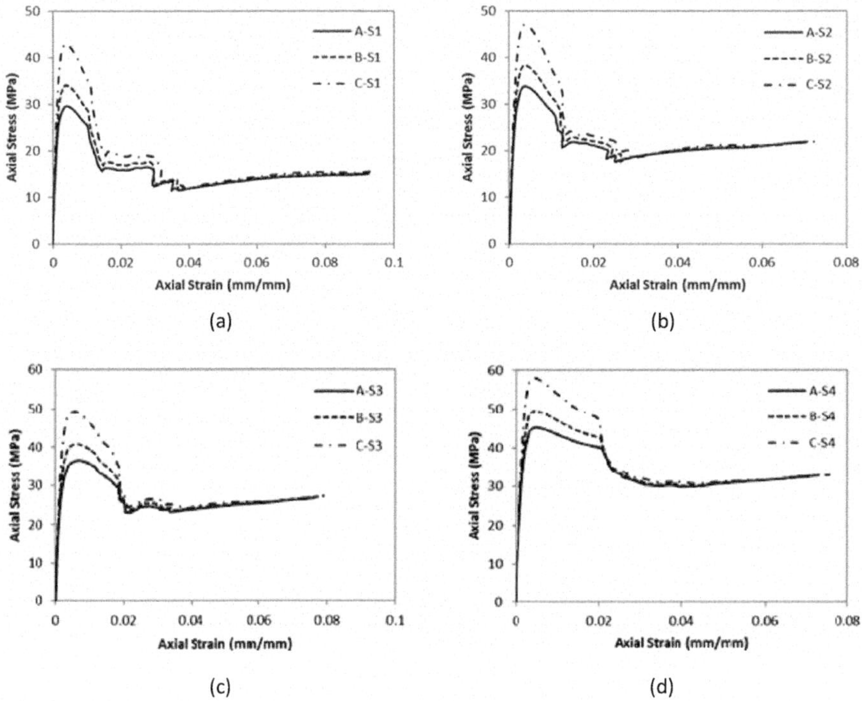

Fig. 8.7 Comparison of axial stress–strain relationships of Fe-SMA confined concrete of different grades (A = 25 MPa, B = 30 MPa and C = 40 MPa) for strip spacing (**a**) 92 mm (**b**) 69 mm (**c**) 55.2 mm (**d**) 46 mm

of concrete and SMA spacing. Concrete cylinders, 150 mm diameter and 300 mm height, having three different concrete grades of 25, 30, and 40 MPa and four different SMA strip-spacing of 92, 69, 55.2, and 46 mm were modeled and simulated under uniaxial compressive monotonic loading. The results show that there is an increase in the strength as well as ductility of the concrete specimens due to this active confinement.

The following conclusions could be drawn from the study:

1. The strength and ductility of Fe-SMA confined concrete increases as the active confinement pressure increases.
2. The residual stress of Fe-SMA confined concrete is independent of concrete strength and only function of active confinement pressure on concrete.
3. Therefore, it may be concluded that active confinement would provide enhanced seismic performance by increasing both strength and ductility of concrete.

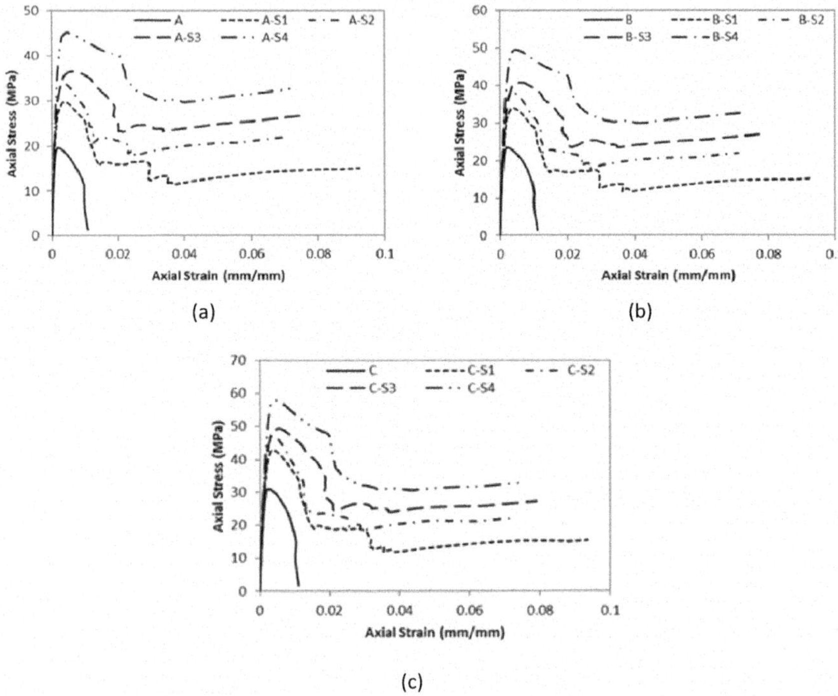

Fig. 8.8 Comparison of axial stress–strain relationships of Fe-SMA confined concrete having different strip spacing (S1 = 92 mm, S2 = 69 mm, S3 = 55.2 mm, and S4 = 46 mm) for concrete grades (**a**) 25 MPa (**b**) 30 MPa (**c**) 40 MPa

8.4 Design Methodology Adopted for Rapid Retrofitting Strategies of RC Columns Using Fe-SMA Strips

The parametric study presented in the previous section demonstrated the effectiveness of the active confinement technique using Fe-SMA strips in enhancing the strength and ultimate strain of concrete. This material-level study is further expanded to the component-level for retrofitting of RC columns using Fe-SMA strips.

Inelastic response of RC columns subjected to axial compression and bending is characterized by plastic deformations localized in small regions, namely the plastic hinge region. Under seismic events, the performance of plastic hinge regions of RC columns is extremely important as it governs its load-carrying and deformation capacities.

From a design perspective, additional confinement in RC columns should be provided along the length of its plastic hinge zones as shown in Fig. 8.9 so that the retrofitted RC columns can attain the desired ductility characteristics in order for them to survive severe seismic events. Hence, RC columns are retrofitted with thermally prestressed Fe-SMA spirals at the plastic hinge region as shown in Fig. 8.8.

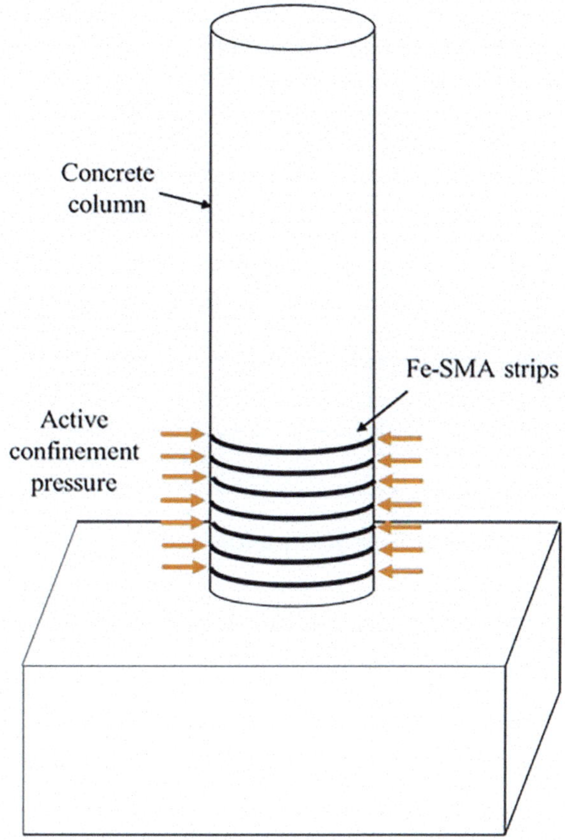

Fig. 8.9 Retrofitting of RC column using Fe-SMA strips

The length of the plastic hinge region where the SMA strips were applied is estimated as (1) $l_p = 1.5$ times the column diameter [40] or (2) $l_p = 0.08L + 0.022d_b f_y$ [41] where f_{sy} and d_b = yield strength and bar diameter of the main column reinforcement, respectively.

When SMA strips are used for confinement of RC columns, the total confining pressure acting on it comprises two components: an active component and a passive component. According to Mander et al. [7], for a circular section, the effective confining pressure $f_l{}'$ can be calculated from Eq. (8.2). In this equation, f_h is the hoop stress along the transverse reinforcement, which Mander et al. [7] assumed to be the yield strength of transverse steel for internal transverse steel confined concrete.

The above-mentioned equation can be applied for SMA confinement to calculate the effective active component of the total confining pressure from Fe-SMA strips by using SMA recovery stress as the hoop stress. The passive component is calculated using hoop stress that develops in the Fe-SMA strips during loading. Tensile test data of Fe-SMA strips from Shahverdi et al. [33] can be utilized to calculate the hoop stress along the strips when concrete dilates. When the level of active confinement decreases to a certain level, it's efficiency in improving the strength and ductility of RC columns decreases. Chen et al. [36] recommended this limiting active confinement pressure to be 0.91 MPa.

The use of Fe-SMA strips for active confinement of RC columns facilitates rotation of the plastic hinge, thus preventing failure of the column-footing connection. Owing to early application of confinement pressure, prior to loading, Fe-SMA strips protect the concrete in the plastic hinge region of the RC column from early spalling hence delaying its damage. It also improves the energy dissipation of the RC column. However, this depends on the effective active confinement pressure, which in turn depends on the spacing of the Fe-SMA strips in the plastic hinge region. Specimens with higher confinement pressures perform better as compared to those with lower confinement pressures. The height of application of Fe-SMA strips is advised to be slightly higher than the height of the plastic hinge to minimize uncertainties for estimation of plastic hinge length.

8.5 Concluding Remarks

The present available retrofitting methods for RC columns (e.g., concrete or FRP jackets) cannot be implemented immediately after the occurrence of an earthquake. As a result, there is an urgent requirement for an efficient retrofitting method that can be executed in the field in less time. This can be fulfilled by the use of Fe-SMA strips for active confinement of RC columns. The time and labor required in this rapid retrofitting technique are minimal. Unlike using prestressed steel strands or FRP belts, installation of thermally prestressed Fe-SMA strips doesn't require heavy machinery. Thermal activation of prestrained SMA leads to the development of a large recovery stress, which is utilized to exert active confinement pressure externally on RC column on which it is fixed. A recovery stress of 350 MPa is generated in the Fe-SMA strips when it is heated to 160 °C. A parametric study was carried out to understand the behavior of Fe-SMA confined concrete by numerically investigating the change in the constitutive behavior of concrete with changes in grade of concrete and SMA spacing. It was found that the residual stress of Fe-SMA confined concrete is independent of concrete strength and only function of active confinement pressure on concrete. Active confinement using Fe-SMA strips would provide enhanced seismic performance by increasing both strength and ductility of concrete. From a design perspective, additional confinement in form of Fe-SMA strips provided along the plastic hinge length of RC columns can help in attainment of the desired ductility characteristics in order for it to survive severe earthquakes. The use of Fe-SMA strips for rapid repair takes only a few hours, which makes them very suitable for situations where performing emergency retrofit is required.

References

1. Bett B, Klingner R, Jirsa J (1988) Lateral load response of strengthened and repaired reinforced concrete columns. ACI Struct J 85:499–508
2. Chail YH, Priestley MN, Seible F (1991) Seismic retrofit of circular bridge columns for enhanced flexural performance. Struct J 88(5):572–584
3. Priestley MN, Seible F, Xiao Y, Verma R (1994) Steel jacket retrofitting of reinforced concrete bridge columns for enhanced shear strength-part 1: theoretical considerations and test design. Struct J 91(4):394–405
4. Priestley MN, Seible F, Xiao Y (1994) Steel jacket retrofitting of reinforced concrete bridge columns for enhanced shear strength–Part 2: test results and comparison with theory. Struct J 91(5):537–551
5. Richart FE, Brandtzæg A, Brown RL (1928) A study of the failure of concrete under combined compressive stresses. University of Illinois at Urbana Champaign, College of Engineering
6. Scott B, Park R, Priestley M (1982) Stress-strain behavior of concrete confined by overlapping hoops at low and high strain rates. J Proc 79(1):13–27
7. Mander JB, Priestley MJ, Park R (1988) Theoretical stress-strain model for confined concrete. J Struct Eng 114(8):1804–1826
8. Rodriguez M, Park R (1994) Seismic load tests on reinforced concrete columns strengthened by jacketing. Struct J 91(2):150–159
9. Fukuyama K, Higashibata Y, Miyauchi Y (2000) Studies on repair and strengthening methods of damaged reinforced concrete columns. Cement Concr Compos 22(1):81–88
10. Daudey X, Filiatrault A (2000) Seismic evaluation and retrofit with steel jackets of reinforced concrete bridge piers detailed with lap-splices. Can J Civ Eng 27(1):1–16
11. Saiidi MS, Wehbe NI, Sanders DH, Caywood CJ (2001) Shear retrofit of flared RC bridge columns subjected to earthquakes. J Bridg Eng 6(3):189–197
12. Saadatmanesh H, Ehsani MR, Li M-W (1994) Strength and ductility of concrete columns externally reinforced with fiber composite straps. Struct J 91(4):434–447
13. Sheikh SA, Yau G (2002) Seismic behavior of concrete columns confined with steel and fiber-reinforced polymers. Struct J 99(1):72–80
14. ElSouri AM, Harajli MH (2011) Seismic repair and strengthening of lap splices in RC columns: carbon fiber–reinforced polymer versus steel confinement. J Compos Constr 15(5):721–731
15. He R, Grelle S, Sneed LH, Belarbi A (2013) Rapid repair of a severely damaged RC column having fractured bars using externally bonded CFRP. Compos Struct 101:225–242
16. He R, Sneed LH, Belarbi A (2013) Rapid repair of severely damaged RC columns with different damage conditions: an experimental study. Int J Concr Struct Mater 7(1):35–50
17. He R, Sneed LH, Belarbi A (2014) Torsional repair of severely damaged column using carbon fiber-reinforced polymer. ACI Struct J 111(3)
18. Rutledge ST, Kowalsky MJ, Seracino R, Nau JM (2014) Repair of reinforced concrete bridge columns containing buckled and fractured reinforcement by plastic hinge relocation. J Bridg Eng 19(8):A4013001
19. Saatcioglu M, Yalcin C (2003) External prestressing concrete columns for improved seismic shear resistance. J Struct Eng 129(8):1057–1070
20. Nesheli KN, Meguro K (2006) Seismic retrofitting of earthquake-damaged concrete columns by lateral pre-tensioning of FRP belts. In: Proceedings of the eighth US national conference on earthquake engineering
21. Andrawes B, Shin M (2008) Seismic retrofitting of bridge columns using shape memory alloys. In: Active and passive smart structures and integrated systems 2008, vol 6928. International Society for Optics and Photonics, p 69281K
22. Shin M, Andrawes B (2010) Experimental investigation of actively confined concrete using shape memory alloys. Eng Struct 32(3):656–664
23. Shin M, Andrawes B (2011) Lateral cyclic behavior of reinforced concrete columns retrofitted with shape memory spirals and FRP wraps. J Struct Eng 137(11):1282–1290

24. Shin M, Andrawes B (2011) Emergency repair of severely damaged reinforced concrete columns using active confinement with shape memory alloys. Smart Mater Struct 20(6):065018
25. Choi E, Chung Y-S, Choi J-H, Kim H-T, Lee H (2010) The confining effectiveness of NiTiNb and NiTi SMA wire jackets for concrete. Smart Mater Struct 19(3):035024
26. Dommer K, Andrawes B (2012) Thermomechanical characterization of NiTiNb shape memory alloy for concrete active confinement applications. J Mater Civ Eng 24(10):1274–1282
27. Chen Q, Andrawes B (2017) Cyclic stress–strain behavior of concrete confined with NiTiNb-shape memory alloy spirals. J Struct Eng 143(5):04017008
28. Dong Z, Klotz UE, Leinenbach C, Bergamini A, Czaderski C, Motavalli M (2009) A novel Fe–Mn–Si shape memory alloy with improved shape recovery properties by VC precipitation. Adv Eng Mater 11(1–2):40–44
29. Leinenbach C, Kramer H, Bernhard C, Eifler D (2012) Thermo-mechanical properties of an Fe–Mn–Si–Cr–Ni–VC shape memory alloy with low transformation temperature. Adv Eng Mater 14(1–2):62–67
30. Lee W, Weber B, Feltrin G, Czaderski C, Motavalli M, Leinenbach C (2013) Stress recovery behaviour of an Fe–Mn–Si–Cr–Ni–VC shape memory alloy used for prestressing. Smart Mater Struct 22(12):125037
31. Ghafoori E, Hosseini E, Leinenbach C, Michels J, Motavalli M (2017) Fatigue behavior of a Fe–Mn–Si shape memory alloy used for prestressed strengthening. Mater Des 133:349–362
32. Soroushian P, Ostowari K, Nossoni A, Chowdhury H (2001) Repair and strengthening of concrete structures through application of corrective posttensioning forces with shape memory alloys. Transp Res Rec 1770(1):20–26
33. Shahverdi M, Czaderski C, Motavalli M (2016) Iron-based shape memory alloys for prestressed near-surface mounted strengthening of reinforced concrete beams. Constr Build Mater 112:28–38
34. Michels J, Shahverdi M, Czaderski C, Schranz B, Motavalli M (2017) Iron based shape memory alloy strips, part 2: flexural strengthening of RC beams. Proc SMAR
35. Abouali S, Shahverdi M, Ghassemieh M, Motavalli M (2019) Nonlinear simulation of reinforced concrete beams retrofitted by near-surface mounted iron-based shape memory alloys. Eng Struct 187:133–148
36. Chen Q, Andrawes B (2017) Plasticity modeling of concrete confined with NiTiNb shape memory alloy spirals. In: Structures, vol 11. Elsevier, pp 1–10
37. Hognestad E (1951) Study of combined bending and axial load in reinforced concrete members. University of Illinois at Urbana Champaign, College of Engineering
38. Cornelissen H, Hordijk D, Reinhardt H (1986) Experimental determination of crack softening characteristics of normalweight and lightweight. Heron 31(2):45–46
39. Birtel V, Mark P (2006) Parameterised finite element modelling of RC beam shear failure. In: ABAQUS users' conference, vol 14
40. Caltrans C (2013) Seismic design criteria (SDC), v 1.7. California Department of Transportation, Sacramento, CA
41. Priestley MN, Seible F, Calvi GM (1996) Seismic design and retrofit of bridges. Wiley

Chapter 9
Earthquake Early Warning System: Its Relevance for India

Ashok Kumar, B. P. Chamoli, Bhavesh Pandey, Pankaj Kumar, and Govind Rathore

Abstract Earthquake early-warning system (EEWS) uses modern communication infrastructure and real-time seismology to estimate the size of the earthquake and gives warning to target cities well before arrival of damaging waves. For earthquakes originating from the central Himalayas, such a system can provide tens of seconds of warning to the adjoining plains and Delhi can get as much as 70 s of warning time. A successful EEWS, in the event of 7+ magnitude earthquake from the central Himalayas, can save millions of lives. This paper provides concept, methodology, algorithms, and other details of EEWS and its importance for India. The paper also presents details and performance of a pilot project of EEWS operational in Uttarakhand, India.

9.1 Introduction

More than 50% land area of India is prone to big earthquakes. Through their technical papers and presentations, several prominent seismologists from India and abroad have concluded that tectonic movement between Indian and Eurasian plates may result in large earthquakes (greater than M7) in the Himalayas [33]. Scientists have also concluded that the risk of such earthquakes striking soon in central Himalayas (Uttarakhand and part of Himachal Pradesh) is relatively high because stresses in this part of the Himalayas have not been released for the last several centuries. It may be mentioned here that plains adjoining the central Himalayas (Punjab, Haryana, Delhi, western UP, and Eastern UP) are most vulnerable to a disaster due to earthquakes having epicenter in the Himalayas. This is because the large part of this region is sitting on soft soil, which amplifies the ground motion several-fold. Unfortunately, most dwellings of this region do not have earthquake-resistant features and are likely to face total collapse in an M7 and above earthquake having an epicenter in the Himalayas. There may be more than twenty million houses in this region, which are

A. Kumar (✉) · B. P. Chamoli · B. Pandey · P. Kumar · G. Rathore
Indian Institute of Technology Roorkee, Roorkee, India
e-mail: ashokeq@gmail.com

T. G. Sitharam et al. (eds.), *Theory and Practice in Earthquake Engineering and Technology*, Springer Tracts in Civil Engineering,
https://doi.org/10.1007/978-981-19-2324-1_9

vulnerable to have a total collapse. Similarly, other areas covering the Himalayas and adjoining Himalayas are susceptible to earthquake disasters. Thus, earthquakes in the Himalayas have a well-established risk of having a casualty number that may be more than any other curse to humanity in the known history.

Using the earthquake early warning (EEW) system, people living in the ground and first floors more than 100 km from the epicenter can be provided sufficient warning time to save their lives. Those living on higher floors can safeguard themselves from falling material and debris. An EEW system can provide 30 s or more warnings to the populated region. In fact, Delhi can get as much as 70 s of warning time for earthquakes originating from the central Himalayas. Today every city, town, and village of this region have mobile connectivity, and soon every village will have broadband connectivity. Almost all houses in rural areas and many houses in urban areas are single or two-storied houses. Thus, the entire region can get the benefit from the EEW system. Therefore, it is apparent that a successful EEW system in the event of M7+ earthquake in the Himalayas can save millions of lives.

Such systems are in operation in Japan, Mexico, Taiwan, Turkey and are under development in the USA and few European countries. There are several success stories of these systems. The success of such a system will be heard more frequently in the future due to fast advancements in communication technology. However, it is essential to understand that usefulness, relevance, and possibility of getting full advantage of the regional EEW system is much more in India than in any other place in the world.

9.2 Brief Background

An EEW system aims to provide a quick and substantially reliable estimate of the magnitude and location of a damaging earthquake before it hits the target site. The concept of EEW was first put forth by Dr. Cooper around 145 years ago. The basis behind his approach was that the severe damage due to an earthquake is caused by the S-waves, which travels at about half the speed of P-waves and are much slower than electromagnetic waves [13]. This idea is the basis of most EEW systems that detect the disaster using a P-wave and issues warning through modern communication systems that use electromagnetic waves. Most of the modern algorithms of EEW assume earthquakes to be a point source for estimating magnitude [6]. This is because the least delay in the issue of warning is a crucial requirement of EEW and therefore EEW cannot wait to estimate the rupture area, which cannot be determined within few seconds of nucleation of the earthquake. Assuming point source does make under-estimation of earthquakes having a magnitude greater than M7, but the determination of the exact magnitude of an earthquake is not the objective of EEW. EEW systems aim to issue a warning if the estimated magnitude exceeds the threshold magnitude value (generally set at 6), which can be achieved by assuming an earthquake as a point source.

For issuing the warning, all EEW systems generally use the initial portion of the P-wave to determine the magnitude of the impending earthquake and time onset difference between stations to estimate the earthquake's location [6]. Various EEW parameters have been used in the past by analyzing the initial portion of the P-wave to estimate the event's magnitude. Some of the parameters are described below.

$\tau_p{}^{max}$ represents the predominant period of the first few seconds of the earthquake [5] and is defined as

$$\tau_i^p = 2\pi\sqrt{\frac{X_i}{D_i}}$$

$$X_i = \alpha X_{i-1} + x_i^2$$

$$D_i = \alpha\ D_{i-1} + \left(dx/dt\right)_i^2$$

where

- τ_i^p is the predominant period at the time i
- x_i is the recorded ground velocity
- X_i is the smoothed ground velocity squared
- D_i is the smoothed velocity derivative squared
- α is the smoothing constant

The maximum value of the calculated period during the first three seconds of the event is called $\tau_p^{\max}$.

T_c represents the characteristic period parameter of the time window of t_0 (which is taken as 3 s) and is given by Wu and Kanamori [36]

$$T_c = \frac{2\pi}{\sqrt{\left(\int_0^{t_0} \dot{u}^2(t)dt\right)/\left(\int_0^{t_0} u^2(t)dt\right)}}$$

where

- $\dot{u}$ is the high pass filtered velocity
- u is the high pass filtered displacement
- t_0 is the fixed time window

A small magnitude earthquake results in a slip over a small area of the fault and thus results in high-frequency energy. In contrast, in larger magnitude earthquakes, a larger fault area gets ruptured, resulting in lower frequency energy. This difference is used to differentiate between small and large magnitude earthquakes. Regression laws based on past earthquake records have been established between $\tau_p^{\max}$ and M and between T_c and M, which can be used to determine the magnitude of the impending earthquake.

The most commonly used EEW parameter is peak displacement (P_d). P_d is peak displacement during the first three seconds after the P onset. The streamed acceleration time history is first high-pass filtered in real-time with cut-off frequency of 0.075 Hz and then double integrated to calculate displacement. The maximum value of displacement in the first three seconds is taken as P_d. Using many past earthquake records, regression equations have been developed with magnitude on the left-hand side and P_d and hypo-central distance on the right-hand side. This parameter is being used in almost all EEW systems operational/under development in the world.

Another EEW parameter is Cumulative Absolute Velocity (CAV) [15]. It is defined as the area under the absolute accelerogram and includes the cumulative effects of ground motion duration. The velocity content present in the ground velocity record is associated with the earthquake energy content of the recording site and used to determine the damage threshold for engineered structures related to the impending earthquake. However, its use in the warning system is limited because CAV represents the intensity of shaking, which may not provide desired information about the size of the earthquake.

Several other attributes like bracketed CAV [16], windowed bracketed CAV [2], Root sum of squares of cumulative velocity [8], etc. have been reported in the literature, which estimates the size of an earthquake using the first few seconds of the vertical component of ground motion records. However, their uses are limited and are in the process of research.

9.3 Details of EEW Systems

The earthquake early warning (EEW) system is comprised of four segments. A prerequisite of an EEW system is dense seismic instrumentation in the epi-central region which is the first segment. The second segment of the EEW system is real-time, 24 × 7, streaming of data from sensors to servers of the central monitoring station (CMS). Data received in real-time at CMS is processed for deciding to issue or not to issue warnings—it is the third segment of the EEW system. The fourth segment is the dissemination of warning through mobile apps, radio, TV, and sirens. Figure 9.1 gives a flow diagram of a typical regional EEW system.

Various steps involved in the above segments of EEW are given below and will be discussed in this paper.

9.3.1 Network of Sensors

9.3.1.1 Accelerometer Versus Seismometer

The combined requirements of seismologists and earthquake engineers vary from measuring the weakest possible ground motion to measuring the strongest possible

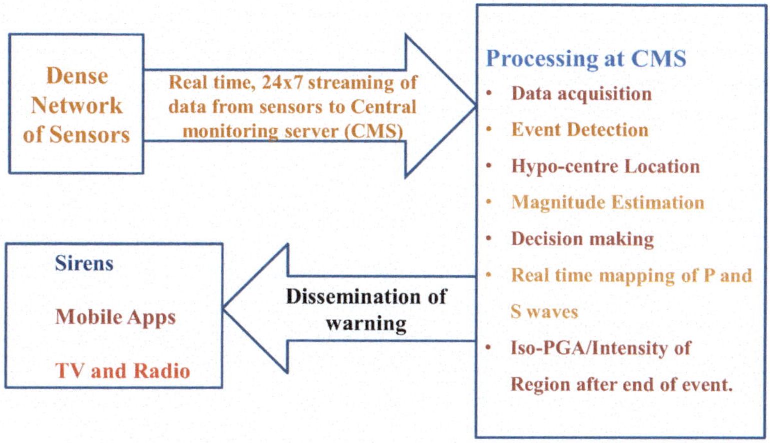

Fig. 9.1 Flow chart showing segments of EEW

ground motion. At one end, seismologists require recordings of ground motion caused by very small earthquakes and earthquakes occurring several thousand kilometers away from sensors. On another end, earthquake engineers need recordings of the largest possible shaking from even the biggest size earthquake and as close as possible to the epicenter, resulting in shaking levels of 2 g. As of today, no single sensor can cater to this wide range of recording to satisfy seismologists and earthquake engineers' needs. Thus, seismic sensors have been divided into two broad categories—weak motion recording sensors (seismometers) and strong motion recording sensors (accelerometers).

Seismometers have a very small natural frequency (soft system) and therefore give an output that is proportional to ground velocity. Seismometers having a natural frequency as small as 1/360 Hz are now available to record ground motion caused by an even drop of a pin. However, records of seismometers saturate and may even get damaged if subjected to strong shaking. Seismometers are much costlier than accelerometers and require a special noise-free installation facility.

Accelerometers have a high natural frequency of the order of 200 Hz or more (stiff system) and therefore give an output which is proportional to ground acceleration. Unlike seismometers, records of accelerometers are mainly required for engineering applications. Therefore, most strong motion sensor installations worldwide are within the city, inside small buildings, near existing structures, and within an environment with anthropological noise. Since accelerometers are very stiff systems, such noises have a negligible effect on records. Therefore, records of these sensors at these locations are considered as required ground acceleration for all practical purposes. Fundamental research on EEW and seismic hazard are based on records of such sensors.

For an EEW system, sensors are necessarily required to be installed near to epicenter and EEW systems are primarily used to save life and infrastructure during big earthquakes (say M.6). It is, therefore, quite evident that the primary sensor of the EEW system should be an accelerometer.

9.3.1.2 Force Balance Versus MEMS Accelerometers

Low-cost MEMS accelerometers are being manufactured with advanced technology and are widely used in automobiles, aviation, space applications, etc. But these accelerometers have a high noise floor and thus have a low dynamic range. Typically, the dynamic range of 72 dB is good enough for industrial applications, but a dynamic range of 120 dB or more is required for most earthquake engineering and seismology applications. MEMS accelerometer with a dynamic range of more than 120 dB is not commonly manufactured and may become costly for the low volume requirement of seismic applications. Regarding EEW, a MEMS accelerometer with a dynamic range of 72 dB is good enough for estimating P_d, but attributes like T_c and τ_p^{max} need a high dynamic range. Also, records of low dynamic range accelerometers limit most research carried out in seismology and earthquake engineering.

As of now, force balance accelerometers (FBA) are most commonly used to record acceleration in the field and during laboratory experiments. Most seismic instrumentation manufacturers of the world make FBAs having a dynamic range >120 dB. However, FBA costs slightly more than MEMS accelerometers, but almost all field installations worldwide for recording strong ground motion use FBAs.

As long as a low-cost MEMS accelerometer with a dynamic range >120 dB does not hit the market, FBA will remain the most preferred accelerometer for recording acceleration for most seismic applications, including EEW systems.

9.3.1.3 Density of Sensor Network

For an effective earthquake early warning system, the density of instrumentation is an essential factor. A sound EEW system requires that about 8–10 sensors should be available within 50 km hypo-central distance. It may be mentioned that the epicenter of the earthquakes may lie anywhere around the fault line; hence the instrumentation should cover the entire active seismic region. For example, it is well known that the main central thrust is the main seismogenic source of the Himalayas, but the earthquake's rupture can happen from anywhere in the Himalayas. Thus whole (or as much as possible) of the Himalayan region should be densely instrumented for the EEW system. Considering this requirement of the EEW system, availability of the logistics, redundancy needed in rugged terrain, and experience, an average station-to-station distance between sensors must be 8 km.

9.3.2 Location and Communication Between Sensors and CMS

Sensors should be installed only at those locations which fulfill the following logistic requirements:

a. Sensors require the availability of a reliable power supply. Most sensors operate on ±15 V DC. Thus, at the site, a UPS with good battery back-up or a battery bank that provides DC output with a generator back-up (in case of shut down of mains) is required. If required, the site may also be supplied with solar cell backup.
b. Sensors require ready access for connectivity with a network for sending data to CMS.
c. Installation should be safe against vagaries of weather and human/animal vandalism.

Through National Informatics Centre (NIC), the government of India has networked all district headquarters with NIC-NET. In addition, this network has further been branched to the level of Tehsils and blocks through the state-wide area network (SWAN) of each state. These are the most appropriate location for sensor installation as these offices satisfy all three logistic requirements mentioned above.

However, the number of blocks, Tehsils, and districts will not meet the desired density requirement of an array in different regions. Thus, sensors must be installed up to the village level. The best location to install sensors at these places is in the base room of mobile towers, which have come up in vast numbers. Base rooms of mobile towers satisfy all three logistic requirements for the installation of sensors. BSNL is the most prominent owner of mobile towers in remote locations of the Himalayas. Thus, suitable mobile towers of BSNL can be chosen for the installation of sensors.

Transmission between the sensor and Central Monitoring Station (CMS) is crucial, as it is the backbone of the EEW system. There are no issues as far as sensors installed in NICNET/SWAN are concerned as the network, which is entirely based on OFC cable, is available. However, the decision for connectivity is required for sensors, which are not installed at NIC/SWAN. Options for modes of communication available are VSAT, mobile technology, leased line, and virtual private network.

VPN on broadband is most suitable for the present purpose. VSAT may be ruled out due to antenna displacement during strong shaking of an earthquake, which will disconnect the antenna from the satellite. In addition, VSAT takes more than 1 s for data to reach CMS from the sensor. Mobile phone technology (using SIM card-based modem for each sensor) was experimentally found to have massive data drop. The leased line between each station and CMS is too costly. Whereas VPN on broadband was found to have very little or no data drop, it takes <100 ms for data to reach CMS from sensors, it is entirely on OFC cable, and OFC cable is available at mobile towers. Figure 9.2 given depicts the proposed transmission between sensor and CMS using VPNoBB.

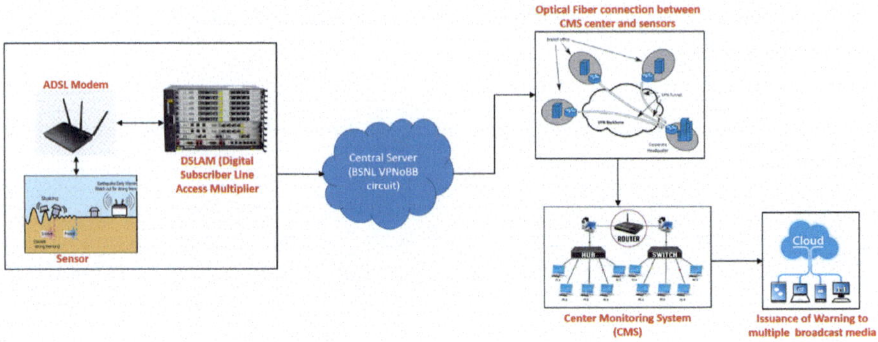

Fig. 9.2 Flow chart of transmission of data from sensor to CMS

9.3.3 Central Monitoring Station

The central monitoring station (CMS) is the heart of the EEW system and comprises a high-performance computational facility and advanced networking hardware. There should be sufficient redundancy in power supply, computers, servers, and networking hardware to achieve zero-second failure of CMS. The actions to be taken at CMS are (1) data acquisition (2) Event detection (3) Hypo-center location (4) Magnitude Estimation (5) Decision-making (6) Real-time mapping of P- and S-wave (7) Development of shake map at the end of the event.

9.3.3.1 Data Acquisition

Data from all sensors are streamed using TCP/IP protocol through BSNL OFC and NIC SWAN network. Data acquisition must be done using an open-source platform to remain transparent and open to be checked, modified, retrieved, or corrected in the future by permitted authorities. The Earthworm is a commonly used open-source platform for seismological applications and data processing, which may be used.

The Earthworm has been developed by USGS [18] and is widely used for the real-time processing of seismic data. It is an open-source program where several seismologists worldwide have developed various modules, and since the program is open source, need-based custom modules can be developed in-house. Using this functionality of Earthworm, a central server can be configured, and the EEW system can be developed by using and modifying the already available program as per our needs and requirement. The decision to choose Earthworm was taken considering the following main design criterion.

The modules developed for performing each task are independent and can function in isolation in both hardware and software. This implies that different modules can be modified and more functionality can be added without affecting the overall system configuration and processing.

The modules in particular and Earthworm as a whole are system independent, which implies that different modules can be run in various types of computer systems, and all these systems can be connected, where they can act as a part of a whole single system, despite having different software hardware and geographical location.

Another added advantage is that data acquisition modules, formatting modules, and sharing modules are readily available. So even if a variety of other sensors are needed to be added, the module performing the EEW task could run without any significant change in the future. A data acquisition program to receive data in real time from the sensors is already available in Earthworm version 7.7. For the processing of data, a high-pass filter with a cut-off frequency of 0.075 Hz should be implemented to filter the data in real time.

9.3.3.2 Event Detection

One of the primary tasks for any successful earthquake early warning system is to correctly pick the onset of the P-phase of the earthquake from otherwise streaming noise in real-time. There are several algorithms available for picking of the start of P-phase of earthquake record [1, 12, 17, 29, 30, 35, 38, 39]. The most popular and used algorithm out of these many algorithms works on the concept of change in the ratio of short-time average (STA) and long-time average (LTA) of the data stream [3, 4]. This algorithm is found to be quite robust and readily used in most EEW networks as the STA shows a sharp rise as soon as P-phase arrives in the time history while the change in LTA is much slower [11, 31]. Due to this reason, the ratio also shows a sharp change with the arrival of the P-phase, which helps detect the start of seismic record by the computer program.

EEW system should use modified Allen's method [11]. The modified form has two additional parameters for acceleration and velocity threshold. These new parameters are beneficial in removing spikes created by noise. Threshold values for these parameters can be fixed for each of the sensors in the configuration files.

9.3.3.3 Hypo-Center Location

The grid search method should be used to find the location of the hypocenter of the earthquake. This method needs a P-wave velocity profile of the instrumented region. For the Uttarakhand region, the velocity profile as given by Kanaujia et al. [19] may be used. Other methods like the back azimuth method [14] using principal component analysis have an advantage in finding the location of long distant earthquakes but will not be suitable for near-source recordings. In the grid search method, Δts of trigger of at least four stations are used, and the search point that gives the least error is picked as focus. The search region is generally kept as $\pm$100 km from a position of the first trigger in both horizontal directions and 50 km deep. On the surface, most of this search region will be within the instrumented area. However, for earthquakes originating from outside instrumented sites, the search region will also be beyond instrumented area (±100 km from the first trigger position). This method will provide the focus location with about 5 km accuracy for earthquakes having epicenters within the search region. Thus, locations can also be found accurately for earthquakes, originating outside but close to the instrumented area. For EEW applications, this method is recommended as it provides good accuracy.

9.3.3.4 Magnitude Estimation and Decision-Making

Once the earthquake is picked, EEW parameters will be calculated, and then regression models should be used to estimate the magnitude and take the decision to issue or not to issue a warning.

The regression model for P_d will be of the following form.

$$M = A * \log(P_d) + B * \log(R) + C \tag{9.1}$$

where R is Hypocentral distance and $R = \sqrt{d^2 + h^2}$.

where d is epi-central distance and h is the depth of focus.

The regression model for T_c will be of the following form.

$$M = A\ \log(T_c) + B \tag{9.2}$$

A, B, and C are regression constants in the above equations.

Initially, the work of [37] for Eq. (9.1) and the work of Wu and Zhao [37] for Eq. (9.2) should be used for regression coefficients of equations given above. However, later, these coefficients can be improved using past Indian earthquake data sets and future records through developed EEW system. Generic estimation parameters as described by Lior and Ziev [22, 23] should also be determined. This estimation should be used as a backup plan, i.e., if a warning is missed using Tc, Pd approach, then a few seconds later, Lior and Ziev approach will decide to issue the warning. It may be mentioned that such situations may arise only if the magnitude of an earthquake is close to the threshold magnitude, which should be 6.0. Figure 9.3 gives a flow chart for the task carried out for magnitude estimation.

9.3.3.5 Decision-Making

Each triggered sensor will estimate two magnitudes; one from P_d and one from T_c. Decision-making should be initiated only after at least four sensors have been triggered. One of the ways is to take an average of all calculated magnitudes. However, another way is to give more weight to sensors that are located with less error while finding the average. It is recommended that the second method be used to estimate the magnitude for the issue of warning, which is explained below.

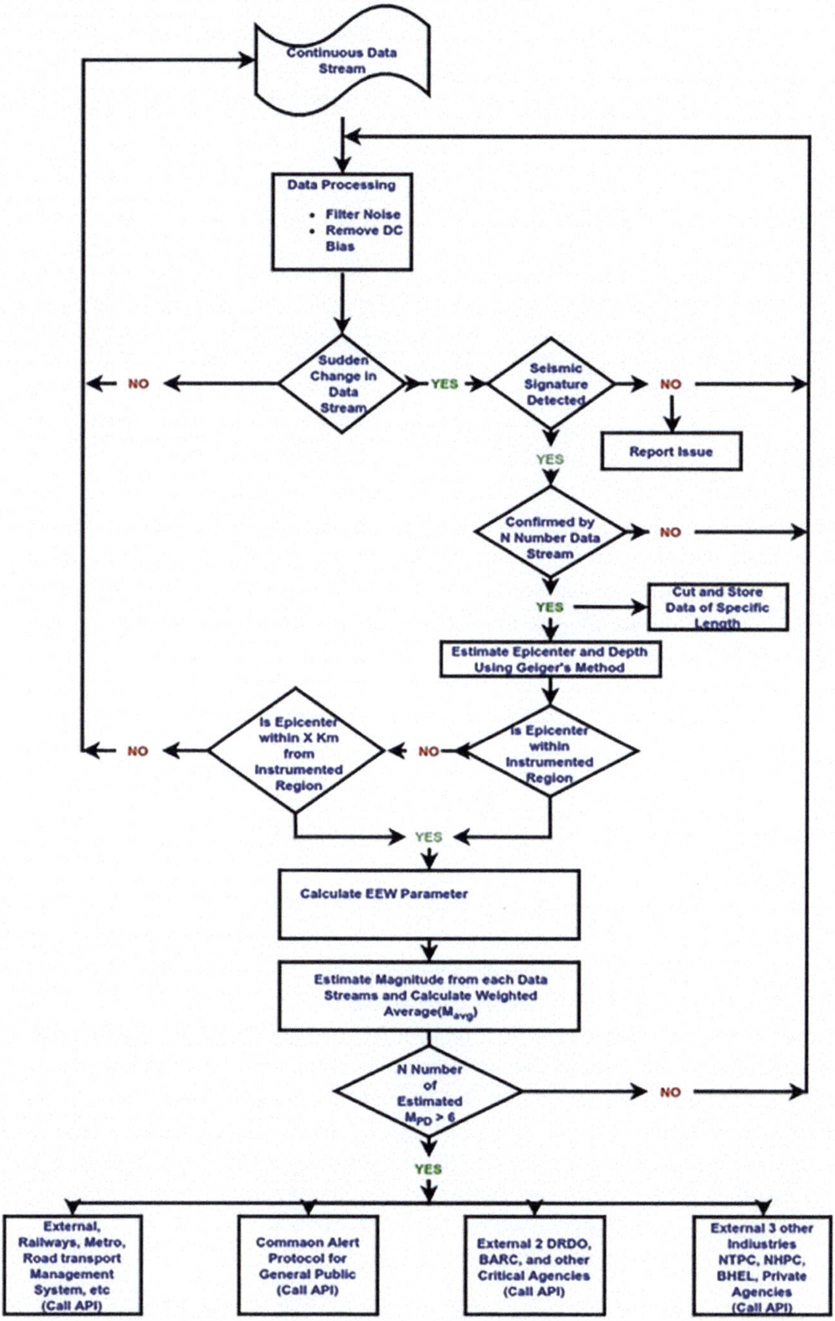

Fig. 9.3 Flow chart showing steps involved in magnitude estimation

$$W_i = (1/(1 + |R_i|))^2 \quad (9.3)$$

where W_i is the weight and R_i is the travel time residual respectively of the station number 'i'. The average magnitude is then found.

$$M = \sum\left(\left(W_i/\sum W_i\right) * M_i\right) \quad (9.4)$$

The warning should then be issued if this value exceeds the threshold value of 6.0. It should be clearly understood that warning only means that an earthquake is likely to be of magnitude greater than 6.0. The estimated value of magnitude will not be part of the warning to the general public.

Evaluation of Decision-Making

Past earthquake records can be used to evaluate the performance of decision-making. Evaluation should be performed on a template with actual magnitude on the horizontal axis and estimated magnitude on the vertical axis and have four quadrangles, namely, 'Correct alarm', 'False alarm', 'Correct all clear', and 'Missed alarm'. It may be mentioned here that if an EEW alarm is issued for earthquakes of magnitude between 5.5 and 6.0, which will cause substantial ground shaking and be felt by everyone, then academically, this scenario may be a false alarm. Still, for all practical purposes, it is a correct alarm. Therefore, in the template, the region between magnitude 5.5 and 6.0 is considered a win–win zone: implying if the warning is issued, it will be considered as 'correct alarm'. If it is not issued, it will be regarded as 'correct all clear'. The template to be used for the evaluation of earthquakes is shown in Fig. 9.4.

Twenty-one earthquakes, each having four or more records within a 50 km hypocenter from KNET (Japan) data set, were used for this evaluation. However, in this evaluation, only the regression model of Wu and Zhao [37] was used. Figure 9.5 gives the performance of EEW for 21 earthquakes.

From the Fig. 9.5, it is pretty clear that for 21 earthquakes used for this work, the EEW system performed with almost 100% correct results. Two earthquakes had a magnitude between 5.5 and 6.0, but the estimated magnitude was greater than 6.0; hence alert will be issued, but these earthquakes would have created a substantial amount of ground motion and therefore has been plotted in the 'correct alarm' region. Also, two other earthquakes were between 5.5 and 6.0 and the estimated value was also between 5.5 and 6.0; hence are in the region of 'correct all clear'.

9.3.3.6 Real-Time Mapping of P- and S-Waves

Once a warning is triggered, it will be communicated to various media, discussed in the subsequent section. However, there is a need to develop a Graphical User

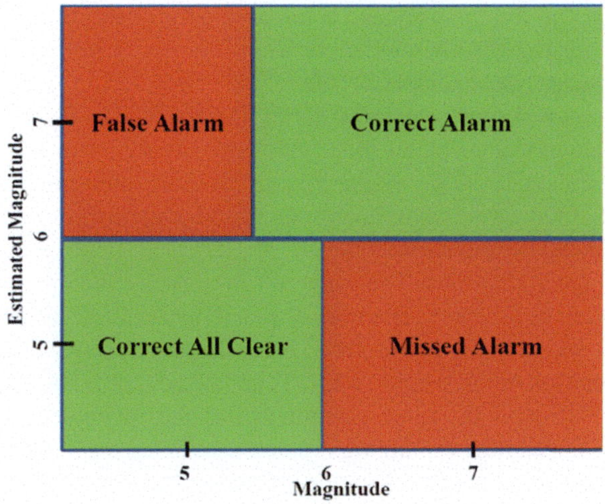

Fig. 9.4 Template to be used to evaluate decision-making of EEW

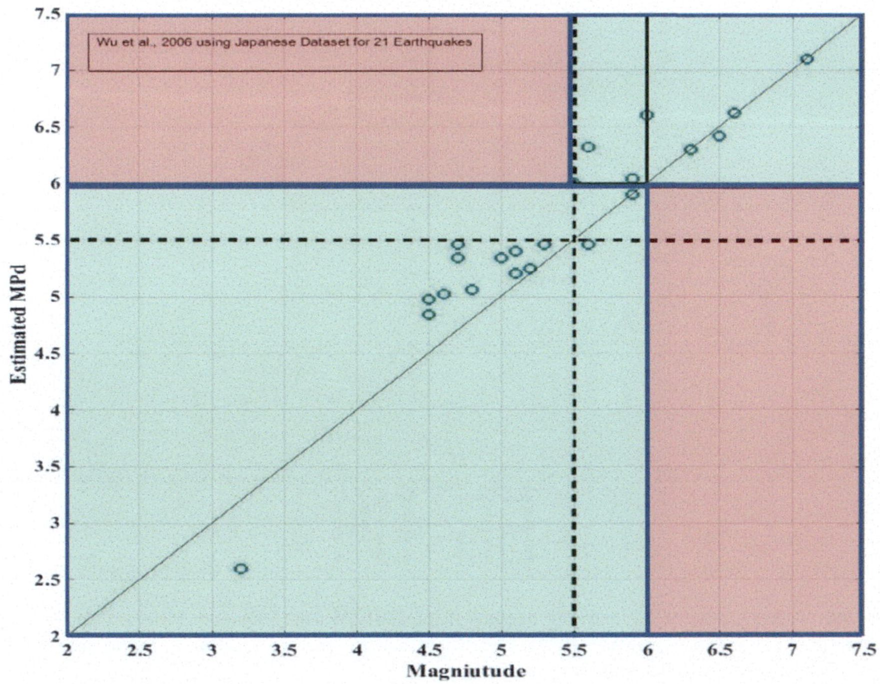

Fig. 9.5 Performance of EEW system for 21 earthquakes recorded by KNET (Japan)

Interface (GUI) for mobile phones and applications for other gadgets at CMS. This GUI is developed using PyQt4, a Qt application framework in Python and contains two main components: a map display of the region monitored and a countdown timer to display the estimated time remaining before the arrival of a detected earthquake. The map is displayed using Leaflet JS, a JavaScript library for interactive maps.

The GUI continuously monitors the system drive for the report files created in the event of a detected earthquake. Each such report contains information about the sensors triggered due to the earthquake and the estimated epicenter location with respect to the triggered stations. The triggered stations are then plotted on the map display 1after every new report. The map displays the current estimated position of the P-wave and S-wave as concentric circles based on the velocity model of the region. The estimated position of the seismic waves is updated every 1/10th of a second. The countdown timer displays the time remaining till the estimated earthquake reaches the user's location. The time remaining is calculated using the distance between the user's location and the epicenter location of the earthquake. After an earthquake reaches the user's location, the timer is reset to zero, and the GUI goes back to monitoring the system drive for any new event. Figure 9.6 shows one example of the GUI.

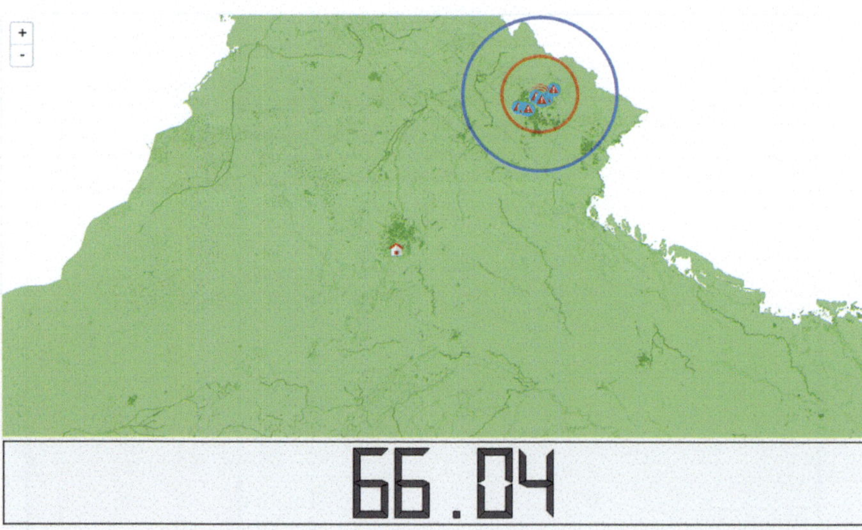

Fig. 9.6 The outcome of the warning display of the Early Warning Display Program during the simulation of the 29th November 2015 Chamoli earthquake of magnitude 4. In the picture, blue dots are the sensors picked; the innermost red circle is the estimated epicenter, the red circle is S-wave envelop, the blue circle is P-wave envelop, the hut is in the center of the map is the target location, in this case, New Delhi. 66.04 is the expected time of S-wave arrival at the target location in seconds

9.3.3.7 Development of Shakemap

In addition to issuing a warning, EEW systems should also develop an online shake map immediately after the earthquake is over. This shake map will provide important information about ground motion shaking in the region during the earthquake. Disaster management authorities for rescue and relief can use this information immediately after the earthquake. The shake map being developed uses a dense grid (say 1 km × 1 km). Data used will be recorded and generated using modern interpolation methods and the applicable attenuation relationship of the region [21]. A typical shake map developed for an earthquake is shown in Fig. 9.7. Such map will be uploaded automatically on the website and so that any disaster management authority can use this shake map for its search, rescue, and response operations.

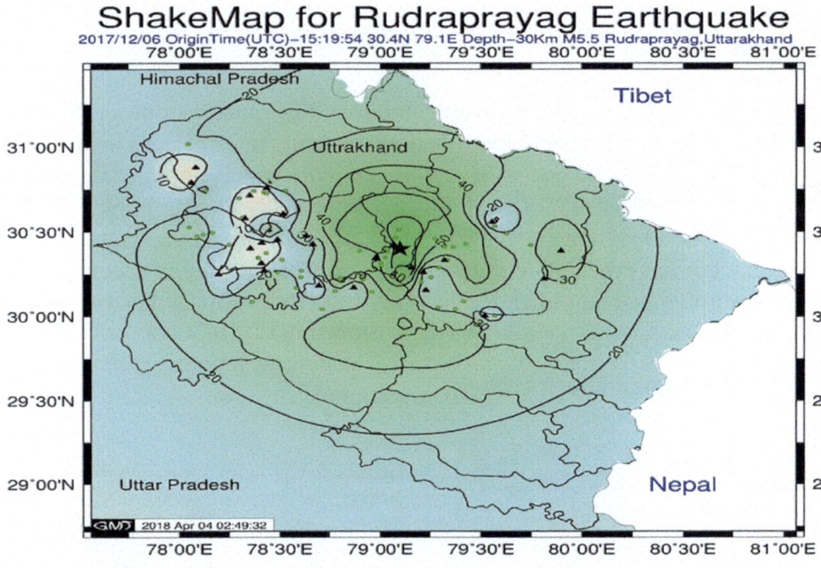

This Map is developed by EEW System Laboratory, IIT Roorkee

Perceived Shaking	Not Felt	Weak	Light	Moderate	Strong	VeryStrong	Severe	Violent	Extreme
Potential Damage	None	None	None	Very Light	Light	Moderate	Moderate /heavy	Heavy	Very Heavy
Peak Acc(gal)	"<1.7"	1.7–14	14–39	39–92	92–180	180–380	380–650	650–1240	">1240"
Instrumental Intensity	I	II–III	IV	V	VI	VII	VIII	IX	X+

Fig. 9.7 A typical shake map developed immediately after an earthquake. Acceleration values of the contours are in gals

9.3.4 Dissemination of Warning

Once the warning is issued by decision-making software, it needs to be communicated to the public in the shortest possible time (within few seconds). Since earthquake warning does not have any scope of any human intervention, it should be ensured that no hacker or unknown device or unauthorized person should hack the dissemination system. It is, therefore, essential that whichever mode for dissemination is used should provide adequate safety and use dynamically encrypted warning messages. This implies that the receiving end (siren, radio, TV, mobile broadcasts, etc.) should also have an intelligent device (like a microcomputer) that activates the alert to receive the coded message. Various modes of dissemination of warning are discussed in this section.

9.3.4.1 Sirens

The design of the EEW siren can be divided into four significant modules [28]. The first is the controller board, which controls the siren and stores various sounds for warning purposes. The second one is the amplifier circuit, which amplifies the sound signal coming from the controller board. The third one is the speaker/hooter, which converts these sound electrical signals into loud sounds. The last one is the power supply unit that converts the 230 V A.C. current into D.C., which is necessary for working the controller and amplifier. Figure 9.8 gives a flow chart of these modules.

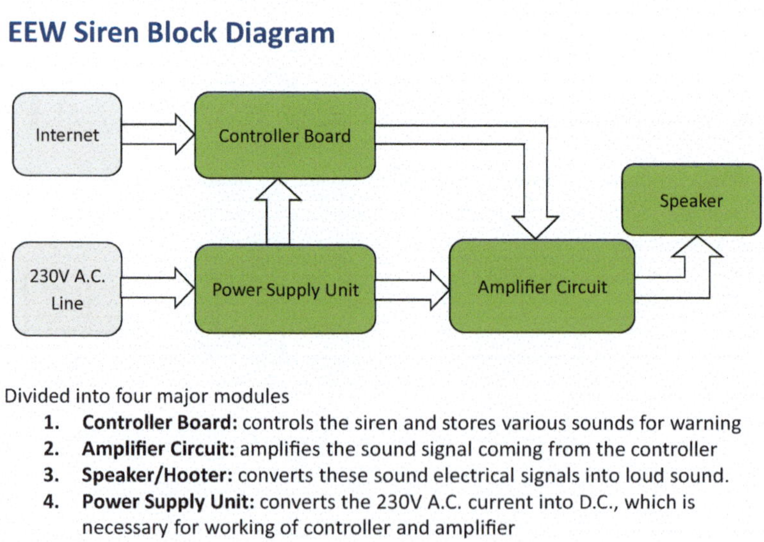

Fig. 9.8 Flow chart of EEW siren modules

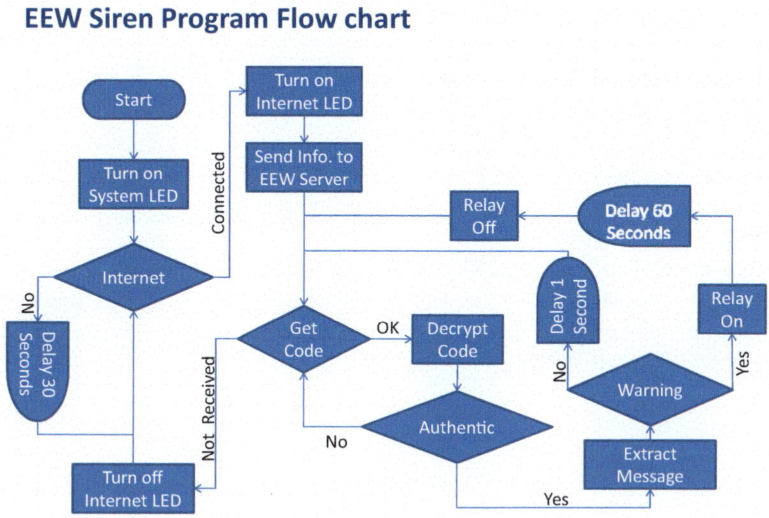

Fig. 9.9 Flow diagram of computer program of siren

The Controller board is a microcomputer (like Raspberry Pi) connected through the Internet via a LAN port. It will also have General Purpose Input Output (GPIO) pins for interacting with other things. A computer program will communicate with the warning server to receive real-time warning messages through the Internet and control hooter/speaker relays with the help of GPIO pins. The flow diagram of the computer program is given in Fig. 9.9.

9.3.4.2 Mobile Apps

Today Android systems mobile phones have become very common and are being owned by a substantial percentage of the population. It is, therefore, important that warning messages should also be disseminated through mobile Apps. The messaging app should be programmed to produce a specific sound on the mobile phone to alert users on receiving a warning. The app should also contain information about the steps needed to be followed on the occurrence of an earthquake and information about how to make and find safe places in a house for hiding during the earthquake. Android Studio can be used to develop this app, which is the official integrated development environment for Google's Android operating system. This should also provide graphic details of approximate real-time locations of P- and S-waves and time left for the S-wave to hit the location of the mobile phone, as shown in Fig. 9.6.

9.3.4.3 Other Methods of Dissemination

EEW warnings should also be issued through radio and TV. Warnings may be provided to All India Radio and FM channels to broadcast a pre-recorded earthquake alert message automatically. All TV stations should be fed with earthquake alerts, and predesigned messages with a specific sound should automatically overlap over the transmitted program. Again, safeguarding the system from hacking should be ensured.

Another method to issue an alert should be through Multimedia Broadcast and Multicast Services (MBMS) of the mobile phone service provider. A ready to use text should be automatically broadcasted to all its users by the service provider of mobile phones instantly on receiving the alert.

9.4 Need for EEW System in India

People living more than 120 km from the epicenter can be provided a lead time of more than 30 s using the EEW system. A lead time of 30 s means availability of 30 s before the arrival of the S-wave. However, most structures take additional tens of seconds for the build-up of vibration and their sequential collapse. Therefore, effectively more than a minute will be available for saving lives. More than one minute will generally be enough for those living on the ground or first floors to get out of buildings fast. Those living on higher floors or those sitting in a group in the classroom or a hall will have enough time to perform drop, cover, and hold to save them effectively.

A glimpse of the region of adjoining plains (West UP) to Uttarakhand in terms of vulnerability is shown in Table 9.1. It can be seen that about 10 million houses (in which about 40 million people live) in this region are not earthquake resistant (BMTPC, 2019) and are likely to have a total collapse in the event of a 7+ magnitude earthquake. Plains adjacent to the Himalayas are most vulnerable to earthquakes because most of this region is sitting on soft soil where ground motion amplifies tremors several times.

An EEW system can provide several tens of seconds of warning as a lead time to the populated region. The estimated lead time includes the time to reach S-wave after subtraction of the sum of P-wave arrival time at the station, time taken by EEW algorithm, transmission, and processing delay. For example, if an earthquake originates from the central Himalayas with an epicenter at Chamoli, Delhi can get 70s of warning time. Similar estimated lead times to other densely populated cities in the region show that EEW can successfully work as one of the crucial real-time earthquake disaster reduction measures [8].

It is important to note that India has a communication network and connectivity in the remotest parts and even to the most economically disadvantaged population. Today every city, town, and village in this region has mobile connectivity, and

Table 9.1 Number of houses having high and moderate risk in region adjacent to Uttarakhand

Based on vulnerability Atlas of India (first revision) 2008 by BMTPC					
	No. of houses at high risk		No. of houses at moderate risk		Total
Delhi	146,960	4.30%	3,101,055	91.70%	3,379,956
Total	**146,960**		**3,101,055**		**3,379,956**
Uttar Pradesh					
District	No. of houses		No. of houses		Total
	High risk	Percentage	Moderate risk	Percentage	
Baghpat	8967	3.5	242,415	95.6	253,469
Bijnor	103,464	15.3	545,325	80.6	676,443
Bulandshahar	55,485	8.3	596,709	89.0	670,224
Gautam Budh Nagar (NOI	16,505	5.8	257,296	90.4	284,519
Ghaziabad	38,463	5.3	663,712	91.8	722,691
Meerut	37,933	6.0	575,197	91.3	629,693
Muzzaffarnagar	62,274	8.6	648,804	89.4	725,764
Saharanpur	119,297	18.2	510,279	77.9	654,889
Total	**442,388**		**4,039,737**		**4,617,692**
Uttarakhand					
District	No. of houses		No. of houses		Total
	High risk	Percentage	Moderate risk	Percentage	
Dehradun	81,880	24.4	234,502	69.8	336,012
Haridwar	42,110	12.6	274,763	82.5	333,008
Total	**123,990**		**509,265**		**669,020**
Haryana					
District	No. of houses		No. of houses		Total
	High risk	Percentage	Moderate risk	Percentage	
Yamuna Nagar	24,052	9.3	220,851	85.1	259,609
Kurukshetra	10,456	5	189,954	91.3	208,099
Karnal	13,435	4.3	291,624	93.1	313,344
Panipat	8863	3.8	217,408	93.1	233,547
Sonipat	5681	1.8	311,050	96.6	321,869
Rohtak	7535	3.1	227,507	94.9	239,806
Gurgaon	100,205	26.2	269,018	70.3	382,839
Faridabad	53,782	10.6	437,813	86.4	506,539
Total	**224,009**		**2,165,225**		**2,465,652**

soon every village will have broadband connectivity. Therefore, it is apparent that a successful EEW system for a 7+ magnitude earthquake in the central Himalayas can save millions of lives.

9.5 EEW System for Northern India

As a first step toward establishing EEW System for Northern India, a network comprising 84 low-cost MEMS accelerometers has been installed in Central Himalayas (Uttarakhand) to provide earthquake early warning [10]. This system was established by the funding provided by the Ministry of Earth Sciences (MoES), Government of India, and is currently being operated by funds provided by Disaster Mitigation and Management Center (DMMC), Government of Uttarakhand, India. The objective of the project granted by MoES was to test the feasibility and establish a prototype EEW system for Northern India, which was successfully concluded in March 2017. With the successful completion of the first installation phase, another 85 accelerographs were installed in the Kumaon region of Uttarakhand. The Government of Uttrakhand funded this extended network. This system, which is the first EEW system in India, can give a warning of about 76 s [8] to Delhi in the event of an earthquake having a magnitude greater than six and having an epicenter in the instrumented region. Similarly, other important cities such as Dehradun, Haridwar, Roorkee, Muzaffarnagar, and Meerut can have a lead time of 20, 22, 31, 44, and 57 s, respectively [8].

This system uses sensor connectivity as described in Sect. 9.3.2, works on the Earthworm platform and uses the P_d attribute to issue a warning. As described in the above sections, capabilities to issue a warning through sirens and mobile apps have been developed. About 80 sirens have been installed at several places and connected with CMS through NICNET or SWAN or Internet. A mobile app in similar lines as described in the above sections has been developed, available in the play store for Android and Apple phones. Figure 9.10 gives locations and types of connectivity used for this network.

9.5.1 Region for EEW Sensor Network

For the initial phase of installation, a region in central Himalaya which lies between Yamunotri in East (31.0140° N, 78.4600° E) and Badrinath in the west (30.7433° N, 79.4938° E) was selected. This region was chosen because of the following two main reasons.

Primarily this region has been identified as one of the seismically most active regions in the central Himalaya.

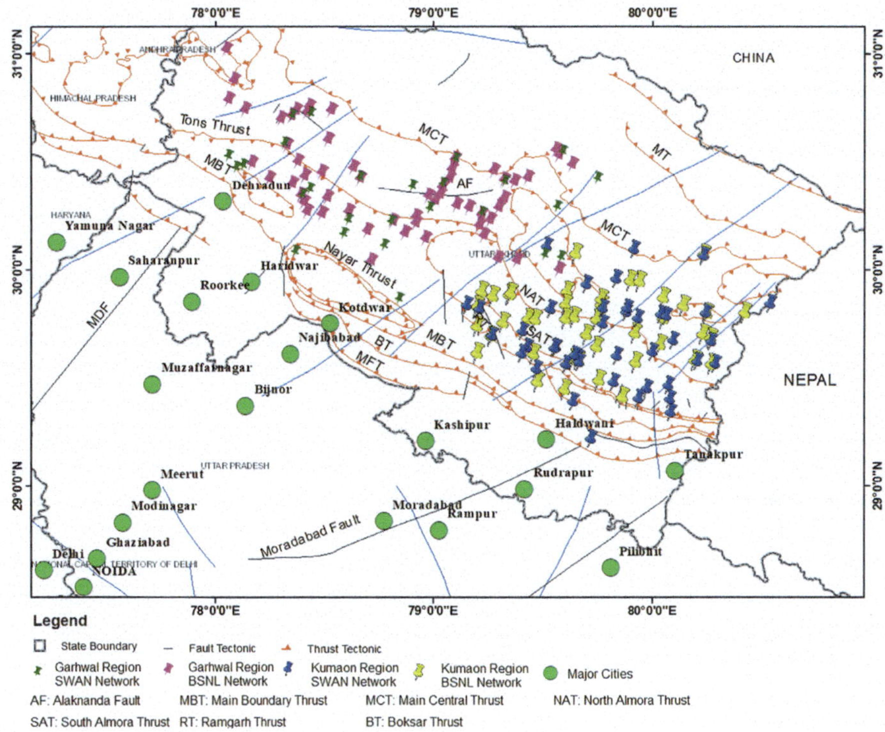

Fig. 9.10 Sensor network created for the EEW for Uttarakhand, India [9]

Secondly, the project's scope was to establish the feasibility to issue warnings to cities falling in western Uttar Pradesh and NCR, which would be most affected by an earthquake having epicenter in the selected region.

During the last 25 years, this region has seen two major earthquakes, viz—Uttarkashi earthquake of 1991 (Magnitude 6.8) and Chamoli earthquake of 1999 (Magnitude 6.5). The selection of a region for installing sensors for the EEW network was done after considering the suggestions of many eminent scientists. Most of them have identified this region as a seismic gap where a large magnitude earthquake (Magnitude >7) can occur. The studies have shown that a large earthquake with a magnitude >7 in Central Himalayas could generate significant ground shaking up to the National capital region Delhi, which is situated at a distance of nearly 200 km from Main Boundary Thrust (MBT) and 300 km from Main Central Thrust (MCT). The MCT and MBT are the two most active thrust planes of the Himalayas. The study conducted by Mittal et al. [24] has also estimated that for magnitude 8.0 and 8.5 earthquakes from the Himalayas, the postulated time history's response spectrum will exceed the design response spectrum for the zone IV at several frequencies on all soft soil sites for Delhi. Most of the thickly populated cities and industrial hubs in and around NCR have been mapped in zone IV as per the seismic zoning map of India provided in IS 1893:2002. Thus a major earthquake in the central Himalayas could prove to be disastrous for NCR and adjoining cities. In the later stages of the research, the region was further extended up to Dharchula so that the Uttarakhand-Dharchula seismic gap can be exhaustively covered.

Characterized by its thickly populated cities, large population, and proximity with Himalayas, western UP and Delhi are among the most vulnerable regions in India for disasters related to earthquakes originating in the Himalayas. Because of this, it was first decided that the first-ever EEW network in India must be installed in Kumaun-Garhwal Himalayas so that warning could be given to the whole of Western UP and the national capital region (NCR).

Kumaun and Garhwal Himalaya are part of central Himalaya and lie in the state of Uttarakhand. This region is surrounded by the international boundaries of Tibet (China) in the northeast and Nepal in the east. This region consists of all four Litho-tectonic subdivisions of Himalaya [34], viz., Tethys Himalaya, Higher Himalaya, Lesser Himalaya, and sub-Himalaya (Shivalik). These subdivisions are further separated from each other by North dipping intra-crustal boundary thrust. Shivalik, which comprises Cenozoic sediments, is the youngest of all and lies at the southern end; it reached a maximum height of 1200 m. Himalayan Frontal Thrust marks South of Shivalik, and the north boundary is delimited by Main Boundary Thrust (MBT). The region between the South of Greater Himalaya and north of Shivalik is known as the Lesser Himalaya. This region has its boundaries marked by MBT in the south and Main Central Thrust in the north and comprises mainly Meso- and Neo-proterozoic period rocks. The region between MCT and south of the Tethys Himalaya is called Greater Himalaya or Himadri. It primarily consists of Archaean-Palaeoproterozoic sequences, constituting the Central Crystalline and Mesoproterozoic, also known as the Dar formation; this also forms the basement for the Phanerozoic succession of the Tethys Himalayas. The MCT marks the southern boundary. In contrast, it imperceptibly passes into the Tethys Himalaya in the north; however, at few places, the Dar-Martoli (Himadri) fault also delimits its boundaries. Greater Himalayas are invariably covered with snow and consist of famous peaks like Kedarnath, Nanda Devi, Trishul, Gangotri Panchachuli, etc.

The extreme north subdivision of Himalaya is known as Tethys Himalaya; it is separated from the trans-Himalayan zone in the north by Indu-Tsangpo Suture. It is mainly comprised of Neo-proterozoic to late Mesozoic aged sedimentary sequence, which is predominantly fossiliferous.

The region between MBT and MCT is Garhwal-Kumaun Himalaya is further characterized with many secondary thrusts. The most commonly known among those are Ramgarh Thrust (RT), North Almora thrust (NAT), South Almora Thrust (SAT), and Munsiary Thrust (MT). Seismologists have also indicated the activation of many faults/thrust oblique or transverse to the major thrust, marking the tectonic boundaries in the above-mentioned region.

9.5.2 *Target Location*

The target cities in which warning is planned to be issued lie in Western Uttar Pradesh and the national capital region of Delhi, which are approximately 100–300 Kms from the instrumented area. Geologically this region mainly comprises Indo-Gangetic plains. The possibility and relevance of getting full advantage of the EEW system in India are better in terms of possible lead time than in most of the world's areas. This is because, for Northern India, potential sources of large earthquakes are located in the Himalayas. In contrast, the centers of large populations and major industrial hubs (including the capital city of Delhi) are in the plains adjoining the Himalayas, which are at least 150 km away from the expected potential source. As mentioned earlier, thick population density and poor adherence to earthquake-resistant practices have substantially increased the seismic vulnerability of this region. However, in the case of a large earthquake in the Himalayas, most of these places, between the central Himalayas and Delhi, can have a lead time varying from 30 to 70 s before the damaging S-waves arrive. If this real-time seismological information is adequately tuned to the operational requirements of technical systems, life and industrial loss could be significantly reduced.

9.5.2.1 Plains Adjoining Kumaun—Garhwal Himalaya

Morpho-stratigraphically speaking, the Indo-Gangetic plains comprise two units: Bangar, an older upland, which is mostly free from floods, and Khadar, a comparatively younger lowland, which is mostly free from having high vulnerability for floods. Based on sediment characteristics and gradient, Bangar is further divided into the Piedmont zone or Bhabhar and the Varanasi plain. Bhabhar is upland adjoining Shivalik and primarily consists of coarse clastic sediments of upper Pleistocene age to middle age and consist of alluvial and colluvial. Most of the seasonal rivers from Shivalik moves under the Bhabhar and disappear. However, they again emerge at the Bhabhar zone in Varanasi plains, resulting in a swampy region known as the Tarai region. Compared to the Varanasi plain, the Bhabhar zone has a more significant gradient and varies from 10 to 0.4 m/km from North to south and its boundary is demarked by HFT. The important cities in which warning is to be issued in this zone are Vikasnagar, Dehradun, Haridwar, Rishikesh, Kotdwar, Haldwani, Khatima, Tanakpur, Roorkee, Laksar, Kashipur, and Udham Singh Nagar. Geophysical site investigation performed by Pandey et al. [25–27] also points toward a stiff, deep alluvial deposit consisting of a mixture of sand and boulders.

9.5.2.2 Western UP and Eastern Haryana

This region is thickly populated and is bounded mainly by Yamuna and Ganga rivers; sometimes, it is also referred to as the Doab region. This region is characterized by deep alluvial deposits of soft soil mainly composed of sand and silt, thus making it highly vulnerable to seismic hazards. Similar inference can be made from the geophysical site investigation results presented by Pandey et al. [26, 27]. The Varanasi plains have a mostly negligible gradient. A great boundary fault demarks this region in the south-east and the Mahendragarh-Dehradun fault in the north-west. Several minor faults like the great boundary fault, Moradabad fault, and several other smaller faults and lineaments also characterize this region.

The Ganga basin is one of India's most prominent sedimentary basins, which forms the northern province of India and is mainly covered with quaternary alluvial deposits.

9.5.2.3 National Capital Region, Delhi

Delhi, the national capital of India, is highly vulnerable to seismic induced hazards not only because of its proximity to the highly active Himalaya but also due to complex nearby tectonics and local site effects. The most important features around Delhi are Main Boundary Thrust, the Main Central Thrust in Himalaya, the Delhi-Haridwar ridge, the Delhi-Lahore ridge, the Aravalli-Delhi fold, The Sohna fault, the Mathura fault, the Rajasthan great Boundary fault, and the Moradabad Fault, along with lots of other lineaments. This highly and thickly populated capital territory of India lies in zone IV as per the Bureau of Indian standard, 2002, and thus is most crucial in terms of the target location for issuing earthquake warnings.

Many studies have concluded that a major or great earthquake in the Himalayas could significantly impact Delhi and surrounding regions. In an analysis performed [32], a scenario for two earthquakes with magnitude 8 and 8.5 in the Himalayas was created. It was concluded that for such events in Himalaya with a distance of more than 300 Kms, Delhi could experience peak ground acceleration of order 96–140 gals for rocks sites and 174–218 gals for soft soils sites [7].

9.5.3 Performance of EEW System

The first phase of the EEW system with about 85 sensors has become fully operational in 2015, and the second phase with additional about 85 sensors was completed in 2018. Since its inception in 2015, this EEW system has recorded about 40 earthquakes and has performed satisfactorily in most situations. The system has not issued an earthquake alert since it was not subjected to any earthquake of M > 6.0 having an epicenter in the instrumented region. Table 9.2 gives details of the earthquake recorded by this system till September 2019.

Table 9.2 Details of earthquakes recorded by EEW system

S. no	Date	Lat	Long	Depth	Mag	Region
1	18 July 2015	30.5 N	79.1 E	13	4.3	Uttarakhand
2	6 August 2015	30.3 N	79.5 E	10	2.9	Chamoli
3	29 November 2015	30.6 N	79.6 E	10	4	Chamoli
4	25 September 2016	30.5 N	78.9 E	10	3.7	Rudraprayag
5	23 November 2016	30.3 N	78.0 E	10	3.4	Distt. Dehradun
6	1 December 2016	29.8 N	80.6 E	10	5.2	Nepal–India Border Region
7	13 December 2016	30.9 N	78.0 E	5	3.4	Uttarkashi
8	19 December 2016	30.9 N	78.0 E	10	3.4	Uttarkashi
9	26 December 2016	30.8 N	77.9 E	10	3.5	Distt. Dehradun
10	10 January 2017	30.3 N	79.4 E	5	3.2	Distt. Chamoli
11	23 January 2017	30.8 N	78.2 E	10	3.5	Distt. Uttarkashi
12	3 February 2017	30.5 N	79.2 E	10	3.6	Chamoli
13	6 February 2017	30.6 N	79.0 E	10	3.6	Distt. Rudraprayag
14	11 February 2017	30.5 N	79.1 E	5	3.2	Rudraprayag
15	10 April 2017	30.7 N	78.6 E	10	3.8	Uttarkashi
16	16 April 2017	30.5 N	79.1 E	10	3.5	Rudraprayag
17	6 December 2017	30.4 N	79.1 E	10	5.5	Distt. Rudraprayag
18	14 June 2018	30.8 N	78.2 E	10	4	Uttarkashi
19	11 November 2018	29.7 N	80.6 E	10	5	Nepal–India Border
20	13 April 2019	30.9 N	78.2 E	10	2.9	Distt. Uttarkashi
21	17 May 2019	30.5 N	79.3 E	10	3.8	Distt. Chamoli
22	14 June 2019	29.6 N	80.4 E	10	3.8	Distt. Pithoragarh
23	6 July 2019	30.7 N	78.4 E	5	3.1	Distt. Uttarkashi
24	14 July 2019	30.9 N	78.2 E	10	3	Distt. Uttarkashi
25	11 September 2019	30.4 N	79.7 E	14	3.6	Distt. Chamoli
26	16 September 2019	29.6 N	80.7 E	10	4.3	Nepal-India Border Region
27	1 January 2019	30.4 N	79.3 E	10	3.3	Distt. Chamoli
28	12 November 2019	29.9 N	80.2 E	10	4.5	Distt-Pithoragarh
29	19 November 2019	29.4 N	81.2 E	33	5.2	Western Nepal
30	24 November 2019	30.4 N	79.3 E	10	3.4	Distt. Chamoli
31	7 December 2019	30.5 N	79.3 E	5	3.2	Joshimath
32	8 February 2020	29.9 N	79.7 E	20	4.7	Pithoragarh
33	26 October 2015	36.5 N	70.8 E	190	7.5	Afganistan
34	25 December 2015	36.5 N	71.2 E	186	6.5	Hindukush-Afghanistan
35	2 January 2016	36.5 N	70.9 E	170	5.8	Hindukush, Afghanistan

(continued)

Table 9.2 (continued)

S. no	Date	Lat	Long	Depth	Mag	Region
36	31 January 2018	37.4 N	69.6 E	190	6.2	Afghanistan–Tajikistan Border
37	9 May 2018	36.9 N	71.3 E	96	6.1	Afghanistan–Tajikistan Border
38	24 September 2019	33.1 N	73.7 E	10	6	India–Pakistan (J & K) Border

9.6 Concluding Remarks

The most significant limitation of the EEW system is that it takes about 12–15 s to issue a warning after its first sensor picks the motion. This means that a distance of about 40–50 km around the epi-center will be a blind zone. Also, for the next 50 km, the warning time will be <15–20 s, too short for people to respond and save a life. Thus, the regional EEW system is effective at a distance of more than 100 km from the epi-center, where the warning will save several lives. It is, therefore, evident that the EEW system is not relevant for moderate-sized earthquakes of magnitude <6.0 as such earthquakes will not cause serious ground shaking at distances of 100 km or more. However, this limitation of the EEW system will be more than compensated as major earthquakes ($M > 6.0$) having epi-center in the Himalayas will benefit thickly populated places in plains adjacent to the Himalayas.

The developed EEW system will not estimate the intensity of shaking in real time at target cities. To achieve reasonably correct real-time intensity, each site's local site properties and site amplification factors are required, which is not available now. In due course of time, when tens of small earthquakes will be recorded at each sensor location, local site conditions and site amplification can be estimated from these records. Subsequently, the wave-front approach, commonly called the PLUM method, can be used to estimate real-time intensity at desired target locations.

The developed EEW system uses low-cost MEMS sensors with a high noise floor (low dynamic range). However, records of these sensors provide P_d quite accurately but will give erratic results for calculating other parameters like T_c and τ_p^{max}, etc.

EEW is a multi-disciplinary system, and the developed EEW system was only a pilot project which has more than adequately served its purpose. The development of this system has given enough confidence to scientists and decision-makers of India for rolling out the most modern EEW system for the entire Himalayan region.

Acknowledgements Authors are thankful to the Ministry of Earth science, Government of India and Uttarakhand Government for providing funds for the development and operation of earthquake early warning system. Authors are thankful to Prof. Ajay Gairola, Prof. R.S. Jakka, Prof. Kamal, and Prof. M.L. Sharma of IIT Roorkee for their support and active involvement in developing, operating, and maintaining the EEW system. This project was closely monitored by a committee of experts under the chairmanship of Dr. B.K. Gairola. This committee provided support and suggestions during crucial phases of implementation, which proved to be fruitful. The authors are also thankful to this project's laboratory and field staff, who worked tirelessly to implement and maintain the EEW system.

References

1. Ait Laasri EH, Akhouayri ES, Agliz D et al (2014) Automatic detection and picking of P-wave arrival in locally stationarynoise using cross-correlation. Digit Sign Process 26:87–100. https://doi.org/10.1016/j.dsp.2013.12.009
2. Alcik H, Ozel O, Apaydin N, Erdik M (2009) A study on warning algorithm forIstanbul early warning system. Geophys Res Lett 36(5):1–3
3. Allen R (1982) Automatic phase pickers: their present use and future prospects. Bull Seism Soc Am 72:S225–S242
4. Allen RV (1978) Automatic earthquake recognition and timing from single traces. Bull Seismol Soc Am 68:1521–1532
5. Allen RM, Kanamori H (2003). The potential for earthquake early warning in Southern California. Science (80–) 300:786–789. https://doi.org/10.1126/science.1080912
6. Allen RM, Melgar D (2019) Earthquake early warning: advances, scientific challenges, and societal needs. Ann Rev Earth Planet Sci 2019(47):361–388. https://doi.org/10.1146/annurev-earth-053018-060457
7. Bansal BK, Singh SK, Dharmaraju R et al (2009) Source study of two small earthquakes of Delhi, India, and estimation of ground motion from future moderate, local events. J Seismol 13:89–105. https://doi.org/10.1007/s10950-008-9118-y
8. Bhardwaj R (2013) Algorithm for earthquake early warning system. PhD thesis, Department of Earthquake Engineering, Indian Institute of Technology Roorkee, India
9. Chamoli BP (2021) Development of algorithm for earthquake early warning system. PhD thesis, Centre of Excellence in Disaster Mitigation and Mangement, Indian Institute of Technology Roorkee, India
10. Chamoli BP, Kumar A, Chen D-Y, Gairola A, Jakka RS, Pandey B, Kumar P, Rathore G (2019) A prototype earthquake early warning system for Northern India. J Earthq Eng. https://doi.org/10.1080/13632469.2019.1625828
11. Chen DY, Hsiao NC, Wu YM (2015) The earthworm based earthquake alarm reporting system in Taiwan. Bull Seismol Soc Am 105:568–579. https://doi.org/10.1785/0120140147
12. Chi-Durán R, Comte D, Díaz M, Silva JF (2017). Automatic detection of P- and S-wave arrival times: new strategies based on the modified fractal method and basic matching pursuit. J Seismol 21:1171–1184. https://doi.org/10.1007/s10950-017-9658-0
13. Cooper JD (1868) Letter to editor. San Francisco Daily Evening Bulletin, November 3
14. Eisermann AS, Ziv A, Wust-Bloch GH (2015) Real-time back Azimuth for earthquake early warning. Bull Seismol Soc Am 105(4), August 2015. https://doi.org/10.1785/0120140298
15. EPRI-Electric Power Research Institute (1988) A criterion for determining exceedance of the operating basis earthquake: Report of Electric Power Research Institute, EPRI NP-5930, Palo Alto, California
16. EPRI-Electric Power Research Institute (1991) Standardization of the cumulative absolute velocity: Report of Electric Power Research Institute, EPRI TR-100082, Palo Alto, California
17. Gentili S, Michelini A (2006) Automatic picking of P and S phases using a neural tree. J Seismol 10:39–63. https://doi.org/10.1007/s10950-006-2296-6
18. Johnson CE, Bittenbinder A, Bogaert B et al (1995) Earthworm: a flexible approach to seismic networkprocessing. IRIS Newsl 14:4
19. Kanaujia J, Kumar A, Gupta SC (2015) Velocity structure and characteristics of contemporary local seismicity around the Tehri region, Garhwal Himalaya. Bull Seismol Soc Am 105:1852–1869

20. Kodera Y, Yamada Y, Hirano K, Tamaribuchi K, Adachi S, Hayashimoto N, Morimoto M, Nakamura M, MitsuyukiHoshiba (2018) The Propagation of Local Undamped Motion (PLUM) Method: a simple and robust seismic wavefield estimation approach for earthquake early warning. Bull Seismol Soc Am 108(2):983–1003, April 2018. https://doi.org/10.1785/0120170085
21. Kumar P (2020) Earthquake early warning system: site classification, intensity map and attributes. PhD thesis, Centre of Excellence in Disaster Mitigation & Management of the Indian Institute of Technology Roorkee
22. Lior I, Ziv A (2018) The relation between ground motion, earthquake source parameters, and attenuation: implications for source parameter inversion and ground motion prediction equations. J Geophys Res: Solid Earth 123:5886–5901. https://doi.org/10.1029/2018JB015504
23. Lior I, Ziv A (2020) Generic source parameter determination and ground-motion prediction for earthquake early warning. Bull Seismol Soc Am 110:345–356. https://doi.org/10.1785/0120190140
24. Mittal H, Kumar A, Kumar A (2013) Site effects estimation in Delhi from the Indian strong motion instrumentation network. Seismol Res Lett 84:33–41. https://doi.org/10.1785/0220120058
25. Pandey B, Jakka RS, Kumar A (2016) Influence of local site conditions on strong ground motion characteristics at Tarai region of Uttarakhand, India. Nat Hazards 81:1073–1089. https://doi.org/10.1007/s11069-015-2120-0
26. Pandey B (2018) Site characterization, local site effects and development of attenuation relationships for Northern India. PhD thesis, Department of Earthquake Engineering, Indian Institute of Technology, Roorkee, Roorkee, India
27. Pandey B, Jakka RS, Kumar A, Sharma ML (2021) Site characterization of strong-motion stations of Himalaya and adjoining plains. Arab J Geosci 14:879. https://doi.org/10.1007/s12517-021-07231-y
28. Rathore G, Kumar A, Jakka RS, Chamoli BP (2018) Development of earthquake early warning siren for regional earthquake early warning system in India. In: Proceedings 16th symposium on earthquake engineering, IIT Roorkee, paper no. 154
29. Ross ZE, Ben-Zion Y (2014) Automatic picking of direct P, S seismic phases and fault zone head waves. Geophys J Int 199:368–381. https://doi.org/10.1093/gji/ggu267
30. Saragiotis CD, Hadjileontiadis LJ, Panas SM (2002) PAI-S/K: a robust automatic seismic P phase arrival identification scheme. IEEE Trans Geosci Remote Sens 40:1395–1404. https://doi.org/10.1109/TGRS.2002.800438
31. Satriano C, Elia L, Martino C et al (2011) PRESTo, the earthquake early warning system for Southern Italy: concepts, capabilities and future perspectives. Soil Dyn Earthq Eng 31:137–153. https://doi.org/10.1016/j.soildyn.2010.06.008
32. Singh SK (2002) Ground motion in Delhi from future large/great earthquakes in the central seismic gap of the Himalayan Arc. Bull Seismol Soc Am 92:555–569. https://doi.org/10.1785/0120010139
33. Sreejith KM, Sunil PS, Agarwal R, Saji AP, Rajawat AS, Ramesh DS (2018) Audit of stored strain energy and extent of future earthquake rupture in central Himalaya. Sci Rep 8:16697. https://doi.org/10.1038/s41598-018-35025-y
34. Valdiya KS (1980) Geology of the Kumaun Lesser Himalaya. Dehradun, p 289
35. Wang Z, Zhao B (2017) Automatic event detection and picking of P, S seismic phases for earthquake early warning and application for the 2008 Wenchuan earthquake. Soil Dyn Earthq Eng 97:172–181. https://doi.org/10.1016/j.soildyn.2017.03.017
36. Wu Y-M, Kanamori H (2008) Development of an earthquake early warningsystem using real-time strong motion signals. Sensors 8(1):1–9
37. Wu Y-M, Zhao L (2006) Magnitude estimation using the first three seconds P-wave amplitude in earthquake early warning. Geophys Res Lett 33:L16312. https://doi.org/10.1029/2006GL026871

38. Zhang Y, Chen Q, Liu X (2018) Adaptive and automatic P- and S-phase pickers based on frequency spectrum variation of sliding time windows. Geophys J Int 215:2172–2182. https://doi.org/10.1093/gji/ggy400
39. Zhu M, Wang L, Liu X et al (2018) Accurate identification of microseismic P- and S-phase arrivals using the multi-step AIC algorithm. J Appl Geophys 150:284–293. https://doi.org/10.1016/j.jappgeo.2018.01.007

Chapter 10
Earthquake Loss Information System for the City of Guwahati, Assam, India

Abdelghani Meslem, Jayanta Pathak, Conrad Lindholm, Dominik Lang, Yogendra Singh, and Kiron Mazumdar

Abstract The city of Guwahati is one of the most rapidly growing cities in India, at the same time being the most important hub of Northeast (NE) India. According to the reported seismic activity in India, the entire Northeastern region, where Guwahati City falls into, is among the most seismically active parts of the Indian subcontinent and even of entire South Asia. The seismicity of NE India has been proven by many damaging earthquakes in the past and is also reflected by the seismic zoning map of India's current seismic building code which classifies the entire region into Zone V, i.e. the country's highest seismic zone. The present chapter describes the different technical aspects that were implemented for the development of an earthquake loss information system for the city of Guwahati. The development of the system has involved both extensive fieldwork and computational efforts including ground shaking modelling considering soil amplification effects, defining ground shaking scenarios for earthquakes on local active faults and significant historical earthquakes, demarcation of the Guwahati City area and its subdivision into geographical units, the definition of building typology classes and generating their vulnerability functions, collection of building inventory data and socio-economic information throughout the city and finally the computation of damage and loss scenarios. The developed earthquake loss information model for Guwahati City can be used as guidance to local authorities on future city planning and earthquake mitigation and response actions.

A. Meslem (✉) · C. Lindholm · D. Lang
Stiftelsen NORSAR, Kjeller, Norway
e-mail: abdelghani.meslem@norsar.no

J. Pathak
Assam Engineering College, Guwahati, India

Y. Singh
Indian Institute of Technology Roorkee, Roorkee, India

K. Mazumdar
Geological Survey of India, Kolkata, India

A. Meslem
Norwegian University of Life Sciences, Ås, Norway

T. G. Sitharam et al. (eds.), *Theory and Practice in Earthquake Engineering and Technology*, Springer Tracts in Civil Engineering,
https://doi.org/10.1007/978-981-19-2324-1_10

10.1 Introduction

Guwahati City, the capital of the state of Assam, is one of the most rapidly growing cities in India, at the same time being the most important hub of Northeast (NE) India.

According to the reported seismic activity in India, the entire Northeastern region, where Guwahati City falls into, is among the most seismically active parts of the Indian subcontinent and even of entire South Asia. The seismicity of NE India has been proven by a large number of damaging earthquakes in the past and is also reflected by the seismic zoning map of India's current seismic building code [1] which classifies the entire region into Zone V, i.e. the country's highest seismic zone (Fig. 10.1). Due to the state of Assam's elongated slender shape, and also since it covers almost 30% of NE India, the state of Assam has been affected by most larger earthquakes that have occurred in both NE India and neighbouring regions, often

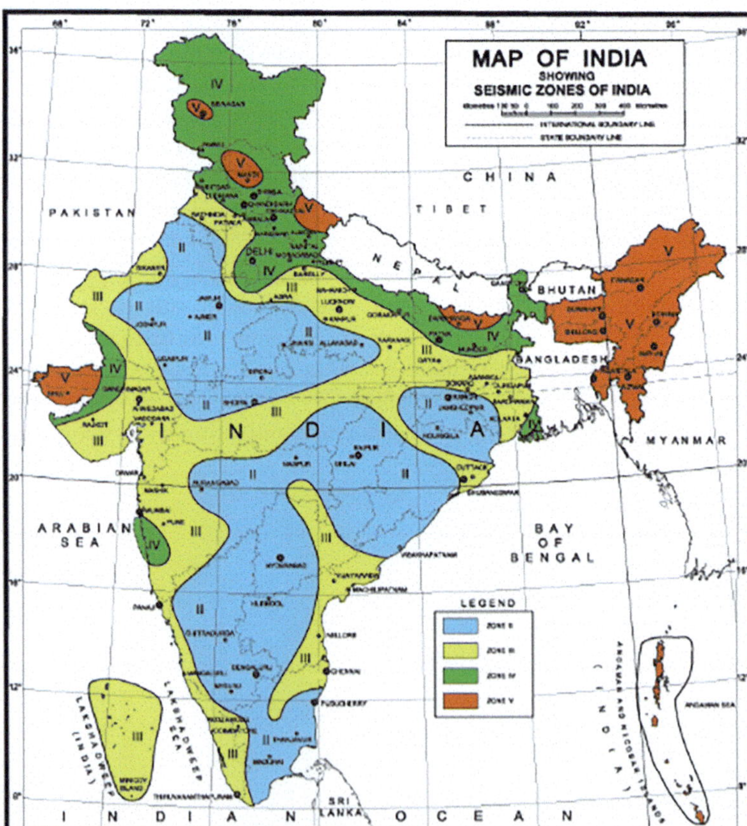

Fig. 10.1 Seismic zonation map subdividing India into four hazard zones [1, 5]

severely. Two of India's five largest historical earthquakes caused huge losses in the state of Assam, i.e. the 1897 Shillong (Great Assam) earthquake [2] and the 1950 Assam earthquake [3]. A compendium of historical earthquakes in NE India and the state of Assam is given elsewhere, e.g. AMTRON [4].

In addition to the region's seismic hazard and risk situation, unplanned land use patterns of surrounding hills, as well as the number of hillocks that are located within the city, contribute to additional landslide risk. The development of the herein presented earthquake damage and loss model has involved both extensive field-work and computational efforts including ground shaking modelling considering soil amplification effects, defining ground shaking scenarios for earthquakes on local active faults and for significant historical earthquakes, demarcation of the Guwahati City area and its subdivision into 258 geographical units, definition of building typology classes (structural typologies) and generating their vulnerability functions, collection of building inventory data and socio-economic information throughout the city and finally the computation of damage and loss scenarios.

The main scope of this article is to describe the different technical aspects that were implemented for the development of earthquake damage and loss model for the city of Guwahati [6].

10.2 Seismic Hazard Situation of Guwahati

Historically, the city of Guwahati has been affected by a large number of earthquakes that had been generated by various faults either directly surrounding the city or from larger distances. Being located in a highly seismic active zone flanked by the Shillong Plateau to the south and the Mikir Plateau to the east, the city has been affected by both crustal and interplate earthquakes during the last several hundred years. The region is associated with complex geological and tectonic features (Fig. 10.2).

The area of Guwahati City, located on the banks of the Brahmaputra River in the Assam valley, falls in a wedged tectonic block between the Himalayan collision zone to the north (bounded by the Main Boundary Thrust (MBT) and the Main Central Thrust (MCT)) and the Indo-Myanmar subduction zone to the southeast. The Shillong Plateau to the south of Assam valley represents a 'pop-up' tectonic block [8] between two E–W trending boundary faults, i.e. the Dauki Fault to the south and the Oldham (and/or the Brahmaputra Fault) to the north [9].

With respect to the more ancient earthquakes, large uncertainties exist with respect to locations and magnitudes. An overview of the larger historical earthquakes that had been felt in Guwahati City and that even produced damage to its building stock is provided in Erteleva et al. [9]. However, with the tremendous growth in population density and built area over the past 50 years, the historical damage reports only provide vague ideas on how today's Guwahati may suffer under a future large earthquake.

Table 10.1 lists historical earthquakes and hypothetical scenario earthquakes that have been chosen to serve as the basis for scenario events for the present study.

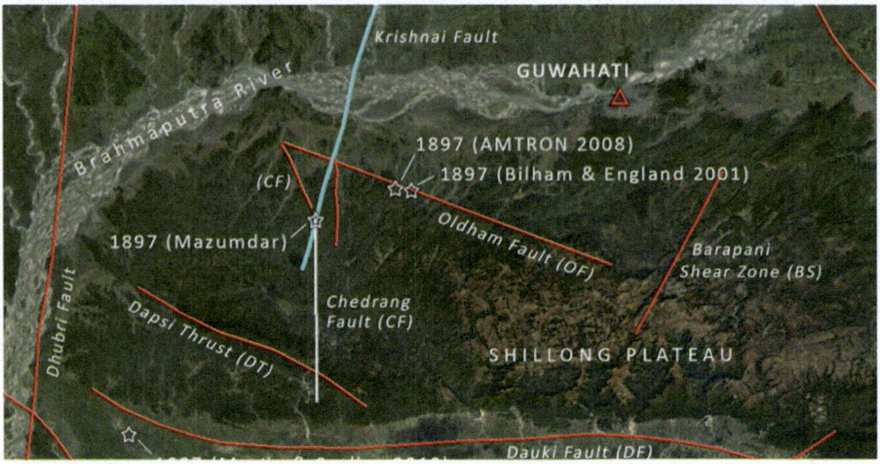

Fig. 10.2 Major tectonic features of the region around the city of Guwahati. Locations of the fault lines taken from Baruah and Hazarika [7]

Table 10.1 Overview of scenario earthquakes with decided focal parameters

Scenario event	Lat [N]	Lon [E]	Focal depth	Mw	Fault orientation	Dip angle	Mechanism
1897 Chedrang	26.00°	92.67°	10 km	8.7	155°	45°	Reverse
1897 Oldham	26.80°	91.44°	10 km	8.7	110°	45°	Reverse
Kopili	27.34°	91.41°	30 km	7.7	148°	90°	Strike-slip
USLRP–MFT	26.80°	91.44°	10 km	7.7	265°	45°	Reverse
Kulsi	25.87°	91.47°	10 km	6.0	90°	60°	Reverse

Judging from their location and hence distance to the city as well as magnitude, the selected earthquake events have been evaluated to be able to cause damage to the building stock of Guwahati. The Table 10.1 also provides information on these events which was collected from various sources.

10.3 Earthquake Loss Estimates: Methodology and Tool

The earthquake risk study of Guwahati City was carried out using SELENA, an open-source tool developed jointly by NORSAR (Norway) and the University of Alicante (Spain). The software can provide local, state and regional institutions with a state-of-the-art decision-support tool for estimating possible losses from future earthquakes [10]. The methodology for the present study is based on 'scenario earthquakes' and a suite of empirical ground motion prediction models (GMPE) to evaluate the seismic ground motion distribution and corresponding losses for the city. The user needs to provide earthquake sources, GMPEs, soil information along with soil amplification factors, built-up areas or number of buildings distributed over the various building typologies, corresponding capacity curves and fragility curves and finally cost estimates (i.e. building repair and replacement values). Using this information, SELENA computes the probabilities for each of the four levels of damage (i.e. Slight, Moderate, Extensive and Complete) and provides the mean damage ratio for each Geo-Code and building typology. Economic losses and casualty numbers are also estimated.

10.4 Ground Shaking, Exposure and Vulnerability for Guwahati

10.4.1 Ground Shaking Modelling

10.4.1.1 Ground Motion Attenuation

For the current study in NE India, it is necessary to select ground motion prediction equations (GMPEs) that represent both the present seismotectonic situation (seismogenic faults and zones) and the seismic hazard level (distance to active faults, magnitude and depths range of expected earthquakes). So far, customized GMPEs have not yet been developed for NE India. It is therefore essential to choose alternative GMPEs that are considered representative of the considered seismotectonics conditions and scenario earthquakes. For the present study, it has been decided to select the GMPE model by Ambraseys et al. [11], which has been developed using data from Europe and the Middle East with a magnitude range Mw 5.0–7.6 (lack of data for Mw > 6.5) and a distance range rjB < 100 km. In total, five earthquake scenarios are defined for Guwahati City and for which damage and economic losses have been estimated. The corresponding site-specific design response spectra that are considered local subsoil amplification are computed following provisions of the current Indian seismic building code [1].

10.4.1.2 Soil Model

Guwahati City falls into the Lower Assam valley which is located between the Eastern Himalayas to the north and the Shillong Plateau to the south. The Lower Assam valley mainly consists of crystalline rocks covered by gently dipping tertiary and younger sediments [4]. Thicknesses of overlying sediments vary from ten to a few hundred meters. The study area is located in the valley of the Brahmaputra River with large areas characterized by very soft water-saturated sediments. In contrast to the low-altitude sedimentary valleys, many relatively steep-sided granite hillocks are distributed over the city. In the course of Guwahati's Microzonation project conducted in 2008 [4], large efforts were made in order to identify the near-surface geology of

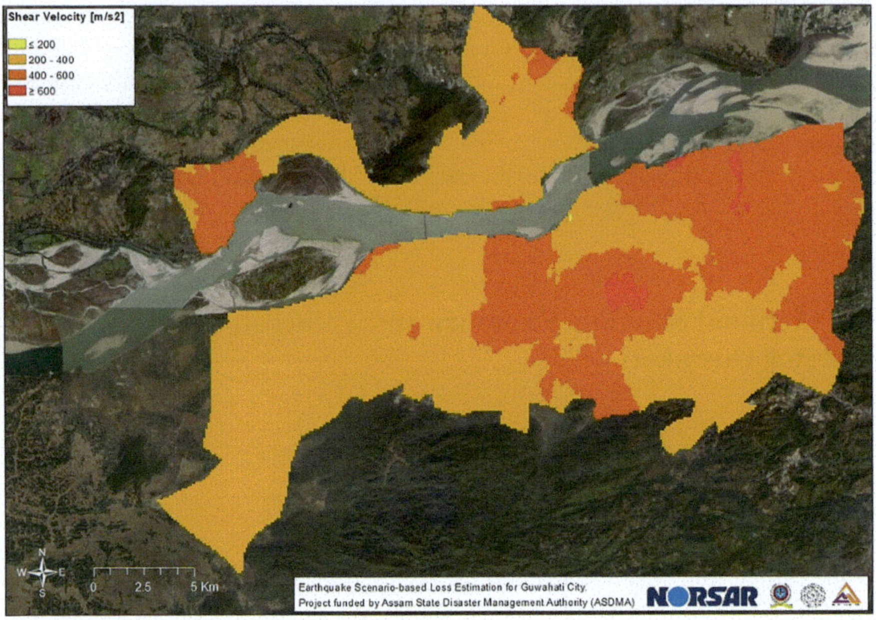

Fig. 10.3 Soil characteristics represented by average shear wave velocities for Guwahati City

the area and to identify the main geotechnical parameters of the various soil types. Figure 10.3 shows the soil characteristics profile and associated average shear wave velocities (VS, 30) values that have been developed and used as soil model input data for the present study.

10.4.2 Exposure Modelling

10.4.2.1 Scale and Resolution of Risk Modelling

The scale of an earthquake loss estimation may range from an entire country to cities or even a city district. The decision on the size of each Geo-Code and how to demarcate it has to be made considering different variables such as geological conditions, constant surface topography or level of infrastructure quality within the demarcated area (socio-economic aspects). The larger the study area, the more likely it is that ground shaking-induced damage to buildings will dominate the overall losses, and that any secondary hazards affecting a particular zone, such as earthquake-induced liquefaction or landslides, will be of less significance. The smaller the size of the Geo-Codes, the higher will be both the level of detail and the reliability of results for an individual infrastructure. However, this could also result in an increased number of Geo-Codes, which will increase the efforts required to set up the inventory data model and perform an assessment of a given region.

For the present study, the derived demarcation of the Guwahati City into Geo-Codes (a total number of the 258 geographical units is done considering various aspects (Fig. 10.4):

- In order to provide Geo-Codes of homogenous spatial size, wards of larger dimensions were subdivided into sub-units, i.e. Geo-Codes.
- Since the study area is surrounded by hilly terrain (with rock soil conditions) and further characterized by a number of steep rocky hillocks that are protruding from the sedimentary plains (with very soft soil conditions), wards were subdivided taking into account these topographic–geological differences.
- Those (outer) parts of the study area that are not covered by the original ward map are subdivided into Geo-Codes with equal topographical-geological conditions as well as numbers of buildings. The Geo-Code limits are assigned following major streets or natural borders (i.e. rivers, streams, topography contours) as well as equal building inventory patterns.

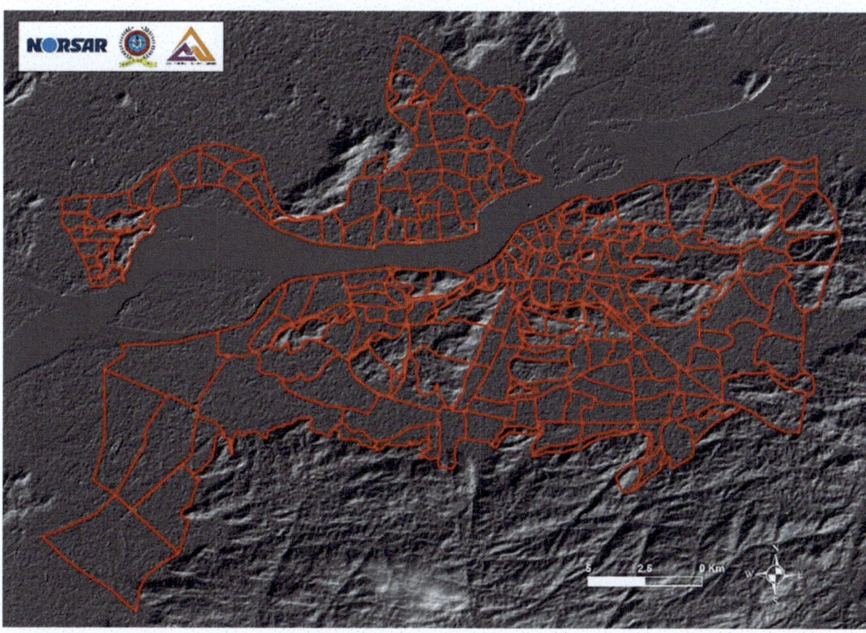

Fig. 10.4 Demarcation of the 2025 Guwahati Master Plan area into 258 geographical units, illustrated with respect to the varying topographical-geological conditions

10.4.2.2 Building Inventory Model

A building stock inventory of 4003 individual buildings was initially compiled by Assam Engineering College [12] and was considered to be satisfactory as it guaranteed a certain level of reliability. However, the collected samples did not cover the entire study area, rather they covered only certain pockets of certain wards. Since the distribution of the building stock can vary greatly not only between urban, rural and suburban areas but also between different neighbouring wards, it was decided to generate a larger inventory database through the conduct of additional walk-down surveys. During a 1-month field trip in September–October 2013, additional walk-down surveys were conducted in each Geo-Code of the study area. During this time, data from additional 11,531 individual buildings was collected.

The results of the walk-down surveys, i.e. the Geo-Code-wise building stock distributions, are utilized to set up the Geo-Code-based inventory model for the entire study area. In doing so, two steps were accomplished: Step-1: identification of the building typology distributions in each of the 258 Geo-Codes that were investigated during the field; Step-2: identification of the total amount of individual buildings in each of the 258 Geo-Codes using available satellite imagery (e.g. Google Earth images). In doing so, the database used in this study consists of 117,484 individual buildings. Linear interpolation has been implemented to estimate the distribution of the buildings according to their typologies and occupancy types. Based on these typology distributions as well as the estimated number of buildings in each Geo-Codes, the building-wise inventory for the entire study area was generated.

10.4.3 Vulnerability Modelling

10.4.3.1 Building Classification Scheme

The building classification scheme that is described herein resulted from various inventory surveys in Guwahati conducted by Assam Engineering College (AEC) in recent years as well as more recent investigations on the current building stock. It reflects the building typologies and materials identified in the Guwahati urban area as well as in several revenue villages around Guwahati City. The building classification may thus be only representative of the municipal and suburban area around the city of Guwahati and may require adjustment or amendment if applied to surrounding regions or other Indian cities. However, the identified typologies can be considered representative of the entire building stock in the urban and rural areas of Assam. Accordingly, the building stock in Guwahati City is considered to be dominated by three groups of building typologies (see Figs. 10.5 and 10.6): traditional Assam-type houses (generally denoted as "Ikra houses"), confined clay brick masonry houses, as well as ductile and non-ductile RC frame structures. The description of available building typologies in Guwahati is summarized in Table 10.2.

Traditional Assam-type (Ikra) house | Confined masonry building | Older (non-ductile) RC frame building | Ductile RC building under construction

Fig. 10.5 Example of existing building typology classes in Guwahati City

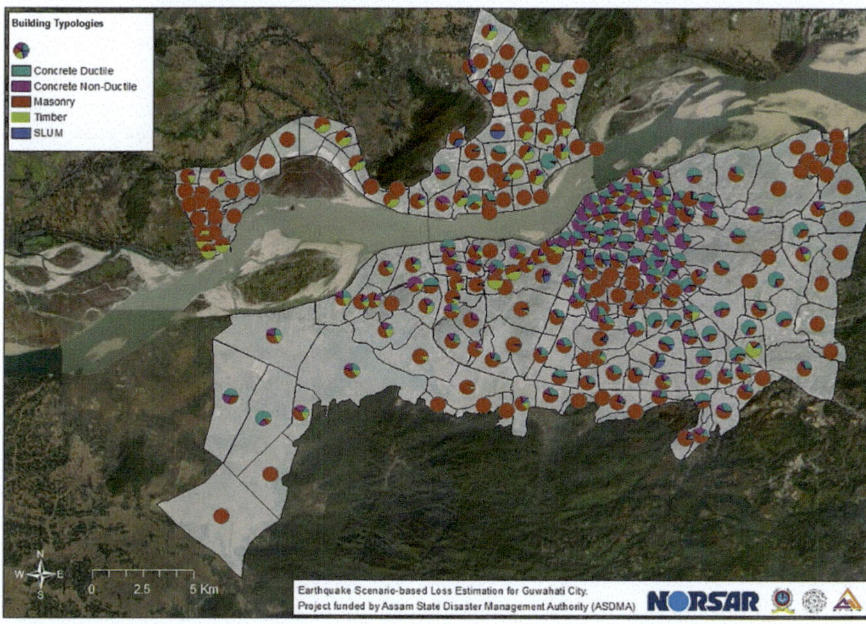

Fig. 10.6 Building typology class distributions in each geographical unit (Geo-Code)

10.4.3.2 Structural Vulnerability Model

Vulnerability functions, which are one of the major components of earthquake loss estimation studies, represent the structural capacity and behaviour of a given building typology and define the probability of suffering a certain level of damage along a given ground motion intensity parameter. In this respect, a thorough vulnerability model (i.e. capacity curves and fragility functions) for the building stock in Guwahati was developed considering the building classification scheme presented earlier. The vulnerability functions were generated based on the use of nonlinear static-based approaches, taking into account the dispersion due to the uncertainty

Table 10.2 Building typology classes, observed in Guwahati City, and seismic vulnerability classification

Classification			Quality Construction; State of Preservation	MBT Code	Vulnerability Classification Scheme			
					Low	Mid	High	Very High
Wood	Load-Bearing Timber Frame	Ikra (Assam-type, wattle and daub)	Low	WFlq				
			Poor	WFpq				
Masonry	Unreinforced Masonry	Load-Bearing Wall	Low	UMlq				
			Poor	UMpq				
	Confined Masonry	Load-Bearing Wall	Low	CMlq				
			Poor	CMpq				
Reinforced Concrete	Frame System (Beams and Columns)	Nonductile Moment Resisting Frames	Low	NDRCFlq				
			Poor	NDRCFpq				
		Nonductile Moment Resisting Frames with Open Ground Floor	Low	ONDRCFlq				
			Poor	ONDRCFpq				
		Ductile Moment Resisting Frames	Good	DRCFgq				
			Medium	DRCFmq				
		Ductile Moment Resisting Frames with Open Ground Floor	Medium	ODRCFmq				
Steel	Light Metal Frame	Steel metal frames	Medium	SMFmq				
			Low	SMFlq				
	Moment Resisting Frame	Moment Resisting Frame	Medium	SFmq				
			Low	SFlq				
Other	Slum	Slum - Mixed materials	Very Poor	SLM				

in structural characteristics-related parameters, building-to-building variability, as well as the record-to-record dispersion in ground motion. For instance, the structural characteristics-related parameters that have been considered in creating the numerical models for the analysis of RC buildings are those associated to mechanical properties and structural details; i.e. compressive strength of concrete, yield strength of reinforcing steel bars and transverse reinforcement spacing. In terms of dispersion, the choice of expected mean and standard deviation for each of these parameters is based on the results of structural characteristics assessment as well as the requirements from different versions of earlier seismic codes in India (Fig. 10.7).

Fragility curves are defined as the conditional probability of being in or exceeding a particular damage state ds_i given the spectral displacement S_d:

$$P[ds \geq ds_i | S_d] = \Phi\left[\frac{1}{\beta_{ds_i}} \ln\left(\frac{S_d}{\overline{S_{d,ds_i}}}\right)\right] \quad (10.1)$$

where $\overline{S_{d,ds_i}}$ is the median value of spectral displacement at which the building reaches the threshold of damage state ds_i; β_{ds_i} is the standard deviation of the natural logarithm of spectral displacement for damage state ds_i; Φ is the standard normal cumulative distribution function (ds_1 represents the attainment of slight damage level—SD; ds_2 represents the attainment of moderate damage level—MD; ds_3 represents the attainment of extensive damage level—ED; ds_4 represents the attainment of complete damage level—CD).

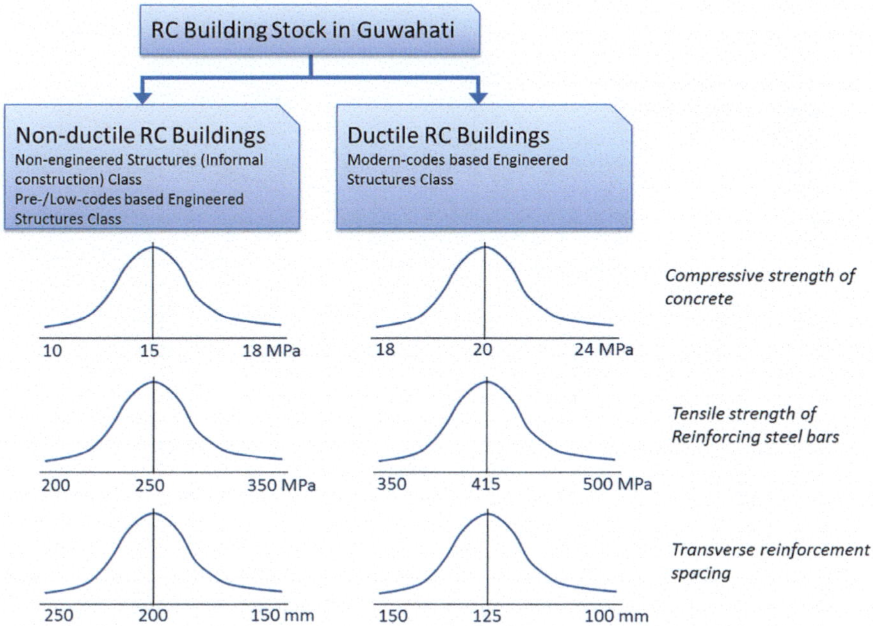

Fig. 10.7 Range of values of the structural characteristics-related parameters for the existing RC buildings in Guwahati

The different values of median capacity, $\overline{S_{d,ds_i}}$, corresponding to different damage states, are obtained as the 50% of all the spectral displacements from the capacity curves recorded at each damage state threshold. The total dispersion parameter, $\boldsymbol{\beta}_{ds_i}$, includes the dispersion in structural capacity (which represents the variability at the level of single building and variability at the level of building to building ($\boldsymbol{\beta}_{C,ds_i}$), uncertainties in capturing the real behaviour of the modelled buildings (β_{DM}) in the overall estimation of the total dispersion [13], test data variability (β_{BTD}) representing available test data (knowledge) related to material properties [14] and the record-to-record variability characterizing the randomness in the demand imposed on the structure by the earthquake ground motion. The following equation could be used for the calculation of total dispersion:

$$\beta_{ds_i} = \sqrt{\beta_{C,ds_i}^2 + \beta_M^2 + \beta_{BTD}^2 + \beta_D^2} \tag{10.2}$$

β_{C,ds_i}^2 is the dispersion in structural capacity and which is estimated directly as the lognormal standard deviation for the selected capacity curves. The following formula is used to calculate the dispersion in the structural capacity associated to each damage threshold [13]:

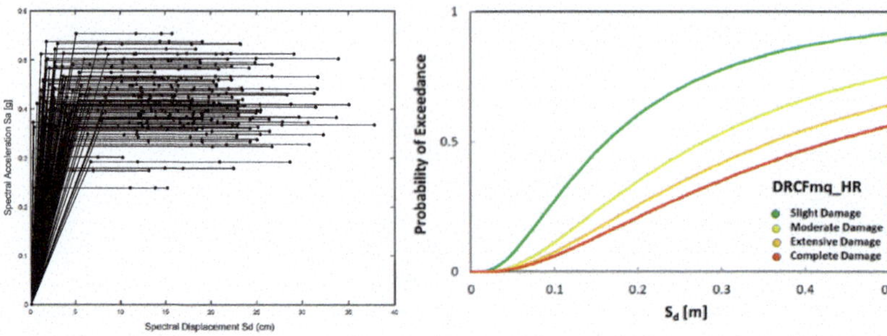

Fig. 10.8 Pushover analysis and example plot of structural capacity and fragility curves for RC building

$$\beta_{C,ds_i} = \sqrt{ln\left(1 + CoV^2\right)} \quad \text{where} \quad CoV = \frac{STDEV}{Mean} \tag{10.3}$$

The derivation of a capacity curve by use of pushover analysis does not directly account for the specific seismic motion, as the dynamic characteristics of demand and response system are not taken into account in the analysis. Hence, dispersion for record-to-record variability β_D is assumed through default values suggested by ATC-58 [15]. Given the fundamental period of the structure and a strength ratio, the values of β_D can be chosen in a range between 0.05 and 0.45. If data on the above two parameters is lacking or uncertain, a maximum default value of 0.45 can be used.

In addition, it has been estimated that a value of 0.65 can be used for a combined record and modelling uncertainties [16]. And also, a value of 0.20 can be added for test data variability (β_{BTD}) representing available test data (knowledge) related to material properties [17]. Hence, from Eq. (10.2), we obtain the following:

$$\beta_{ds_i} = \sqrt{\beta^2_{C,ds_i} + 0.65^2 + 0.20^2} \tag{10.4}$$

Example of generated fragility curves from pushover analysis is illustrated in Fig. 10.8.

10.5 Damage and Loss Computation: Results and Discussion

In the following, the results of the scenario-based earthquake risk assessment are illustrated in terms of damage distribution. The computations are implemented for each selected historical earthquake generated by the various faults as per the results of hazard assessment presented in the previous section. Figure 10.9 exemplarily

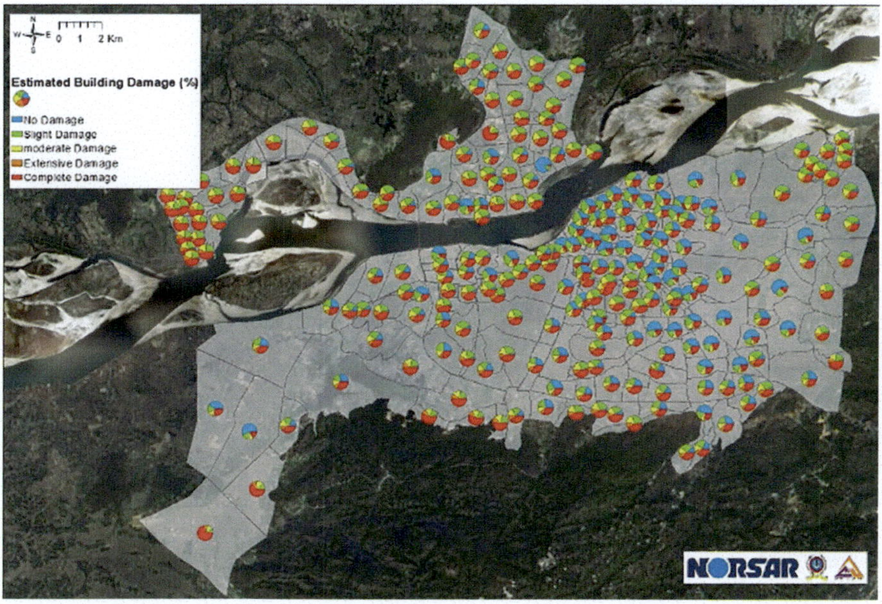

Fig. 10.9 Predicted earthquake damage for a repeat of the 1897 Shillong earthquake (Magnitude 8.7, located at 80 km distance to the city centre)

illustrates results for a scenario representing the 8.7 magnitude Shillong earthquake of 1897, generated by the Oldham fault approximately at 80 km distance to Guwahati's city centre.

Figures 10.10 and 10.11 graphically summarize results for the different earthquake damage scenarios. The earthquake scenario with a maximum magnitude 8.7 and 80 km distance to the city centre (almost similar to the historical 1897 Shillong earthquake) and which can possibly be generated by the Oldham fault, will be the most devastating event compared to the other considered scenarios. This scenario is estimated as being capable to cause severe damage to many building typologies (extensive and complete damage states of up to 36% of the total building stock) which will be connected to significant economic losses and casualties in almost all districts

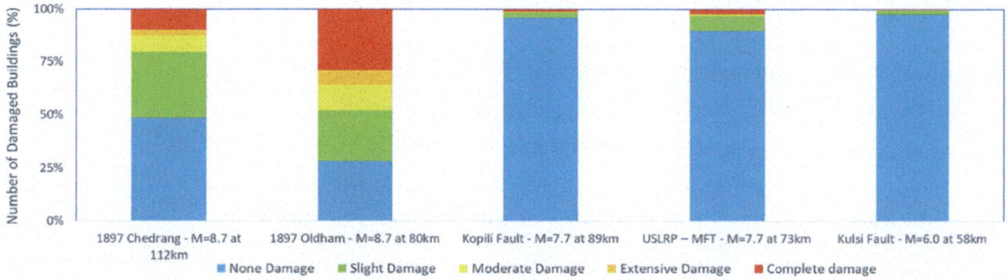

Fig. 10.10 Comparison of the estimated building damage from the different earthquake scenarios

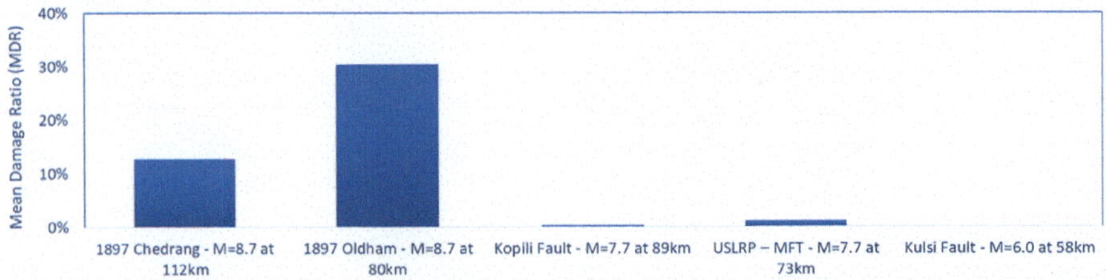

Fig. 10.11 Comparison of the estimated Mean Damage Ratio (MDR) for the different earthquake scenarios

of Guwahati. Similarly, a repeat of the 1897 earthquake, but assuming to be generated by the Chedrang fault at 112 km distance to the city centre with a magnitude of 8.7, will also cause severe damage to many building typologies (extensive and complete damage states of up to 13% of the total building stock). On the other hand, moderate to low damage and losses have been evaluated for the scenario of the MFT fault (M 7.7 in 73 km distance), as well as for the Kopili fault earthquake scenario (M 7.7 in 89 km distance), where no significant damages and losses are expected to the majority of the building stock, and only some of the most vulnerable building typologies will suffer heavy damage. For the selected hypothetical scenario earthquake, to be generated by the Kulsi fault (M 6.0 in 58 km distance), no damages and losses are expected to the building stock, maybe only slight to very slight damage can be observed in some of the most vulnerable building typologies.

The effect of magnitude as well as distance can be clearly seen in the figures. If we consider, for instance, the 1897 Chedrang earthquake and the 1897 Oldham earthquake, it can be clearly seen that the effect of magnitude as a factor could be quite significant. The resulted damage ratio and losses considering the two events are associated with a relative difference of more than 40%. On the other hand, the effect of distance can, for instance, be seen when comparing the Kopili fault earthquake (M 7.7 located at 89 km) with the USLRP-MFT earthquake (M 7.7 at 73 km). The resulted damage and losses considering the two events are associated with a relative difference of up to 25%.

10.6 Conclusion

The aim of this project was to provide local authorities with an earthquake risk model that can be used as guidance for future planning and earthquake mitigation actions for the City of Guwahati, capital of the state Assam in NE India. Guwahati is considered as one of the most rapidly growing cities in India, at the same time being the most important hub of NE India. According to the seismic zoning map of India's current seismic building code [1], Guwahati falls into the country's highest seismic zone (Zone V). Considering the history of the past earthquake disaster events that occurred in the region, five deterministic earthquake scenarios were selected and implemented for the risk assessment study. The outcomes of the risk assessment show that large portions of building typology classes which are considered to be seismically deficient and highly vulnerable would suffer severe damage. However, it is important to note that all damage and loss results presented in this study are associated with a number of epistemic uncertainties that need to be taken into account in the mitigation-planning phase.

Acknowledgements This research work has been conducted within the ELIAS project and has received funding from the Assam State Disaster Management Authority (ASDMA), India.

References

1. IS 1893 (2002). Part I: Criteria for earthquake resistant design of structures—General provisions and buildings, Bureau of Indian Standards, New Delhi, India
2. Hough SE, Martin S, Bilham R, Atkinson G (2002) The 26 january, 2001 Bhuj, India earthquake: observed and predicted ground motions. Bull Seismol Soc Am 92:2061–2079
3. Tandon AN (1954) A study of Assam earthquake of August 1950 and its aftershocks. Indian J Meteorol Geophys 5:95–137
4. AMTRON (2008) Report on seismic zonation of Guwahati Region, supported by the Seismology Division, Department of Science and Technology, Government of India
5. IS 1893 (2016) Part I: Criteria for earthquake resistant design of structures—General provisions and buildings, Bureau of Indian Standards, New Delhi, India
6. Pathak J, Meslem A, Lang D, Bora A (2019) Earthquake damage and loss model for the city of Guwahati, Assam, India, SECED 2019
7. Baruah S, Hazarika D (2008) A GIS based tectonic map of India. Curr Sci 95:176–177
8. Bilham R, England P (2001) Plateau 'pop-up' in the great 1987 Assam earthquake. Nature 410:806–809
9. Erteleva O, Aptikaev F, Saurabh B, Santanu D, Sajal K, Kayal JR (2013) Seismic treatment for a maximal credible earthquake in Guwahati city area of northeast India region. Nat Haz. https://doi.org/10.1007/s11069-013-0843-3
10. Molina S, Lang DH, Meslem A (2015) The SELENA–rise open risk package–towards the next generation of ELE software. In: SECED 2015 conference: earthquake risk and engineering towards a resilient World, 9–10 July 2015, Cambridge UK
11. Ambraseys NN, Douglas J, Sarma SK, Smit PM (2005) Equations for the estimation of strong ground motions from shallow crustal earthquakes using data from europe and the middle east: horizontal peak ground acceleration and spectral acceleration. Bull Earthq Eng 3:1–53
12. Pathak, J. Report on Seismic Vulnerability of Guwahati Region, Assam Engineering College, 2008.
13. Liel AB, Haselton CB, Deierlein GG, Baker JW (2009) Incorporating modeling uncertainties in the assessment of seismic collapse risk of buildings. Struct Saf 31:197–211
14. Federal Emergency Management Agency (FEMA P-695) (2009) Quantification of building seismic performance factors, ATC-63, Applied Technology Council, Redwood City, CA
15. Federal Emergency Management Agency (FEMA P-58) (2012) Seismic performance assessment of buildings, ATC-58, Applied Technology Council, Washington, DC
16. Deierlein GG, Liel AB, Haselton CB, Kircher CA (2007) Assessing building system collapse performance and associated requirements for seismic design. In: SEAOC convention (Tahoe, CA)
17. Wen YK, Ellingwood BR, Bracci J (2004) Vulnerability function framework for consequence-based engineering, MAE Center Project DS-4 Report

18. Federal Emergency Management Agency (2003) HAZUS-MH MR4 Technical Manual, Washington, DC
19. Lang DH, Molina-Palacios S, Lindholm CD, Balan S (2012) Deterministic earthquake damage and loss assessment for the city of Bucharest, Romania. J Seismol 16(1):67–88
20. Public Works Department Assam (2014) Schedule of rates for P.W.D. buildings 2013–2014, Government of Assam, Public Works Department (Building Wing)

Chapter 11
Probabilistic Seismic Hazard Assessment for Hydropower Project Sites in the Himalayan Region

Ambika Srivastav and Neelima Satyam

Abstract River valley projects have a lot of promise in the seismically active Himalayan orogenic region. Some hydroelectric projects are now operational, some are in the planning stages, and a few more will be built shortly. Knowing the nature of ground motion at these locations is critical. The present study uses a probabilistic seismic hazard analysis (PSHA) technique to estimate Peak Ground Acceleration (PGA) for the three hydropower projects in Uttarakhand, Himachal Pradesh, and Jammu and Kashmir (India). Given all potential earthquakes, the aim of probabilistic seismic hazard analysis (PSHA) is to quantify the rate of surpassing certain ground motion levels at the project site. Hazard curves may be used to determine the seismic design input for a location, and they can also be used to analyze the tunnel seismic reaction. The fundamental methods of PSHA are presented in this article in an attempt to offer a clear and brief introduction to the theoretical basis and implementation of PSHA in today's engineering practice.

Keywords Seismic hazard · Peak ground acceleration · Seismic input · Uniform hazard spectrum · Himalayan geology

11.1 Introduction

The Himalayas are prone to major earthquakes. The incidence of earthquakes in the Indian subcontinent has increased due to urbanization, extensive improper building, and resource extraction. Ten big earthquakes have struck the nation in the previous 15 years. According to India's current seismic zone map, 59% of the country's geographical area is under a moderate to severe seismic hazard alert, indicating that the country is vulnerable to earthquakes of magnitude 8.0 and higher. India is divided

A. Srivastav (✉)
Geotechnical Engineering Laboratory, Earthquake Engineering Research Centre, International Institute of Information Technology Hyderabad, Hyderabad, India
e-mail: ambika.srivastav@research.iiit.ac.in

N. Satyam
Discipline of Civil Engineering, Indian Institute of Technology Indore, Simrol, Indore, India

T. G. Sitharam et al. (eds.), *Theory and Practice in Earthquake Engineering and Technology*, Springer Tracts in Civil Engineering,
https://doi.org/10.1007/978-981-19-2324-1_11

into four zones according to the BIS code IS-1893 [2]. The strength of the earthquake is estimated, and these zones are subdivided into five zones: Zones 2, 3, 4, and 5. Zone 2 is the safest, while Zone 5 is the most dangerous. Moderate to major earthquakes continue to strike the country's northeastern area at regular intervals. Since 1950, the area has seen many minor earthquakes. Every year, on average, the region is struck by one earthquake with a magnitude >6.0.

With the increasing population, demand for housing, energy, water, and transportation arises. Rapid development is required in the power sector of the nation to abridge the gap between demand and supply. This tapping of the underutilized hydropower potential in a fast and sustainable way is the need of the hour. India is picking up pace in the tunneling sector over the last decade [17]. There is an increase in tunnel construction in the hydropower sector. If ongoing mitigation steps are not implemented, seismic risk will rise over time.

Engineers generally describe seismic hazards in terms of the ground motion felt at the site. In this approach, the dynamic reaction of structures may be examined to predict where they will collapse and in what conditions. As a result, site selection, design, and retrofitting techniques are aided. By integrating previous data with engineering risk estimate methods, the type and amplitude of ground motion at a location may be characterized in a probabilistic way.

Seismic hazard is defined as the probability of encountering a particular intensity of any destructive occurrence at a specific location or over an area during a certain time period. Ground tremors, liquefaction, landslides, and tsunamis are all potential dangers. Engineering demands for better designs prompted the creation of the approach for evaluating the risk of seismic hazards [3]. Long-term earthquake prediction is similar to seismic hazard assessment [6]. The Seismic Hazard Microzonation method evaluates the seismic hazard at bedrock, which entails a quantitative estimate of ground motion characteristics at a specific point in bedrock owing to a future earthquake of a specified magnitude and epicentral distance. This necessitates the identification and characterization of all potential sources of seismic activity, as well as their ability to generate future significant ground motion at the site. Geological evidence, tectonic evidence, historical seismicity, instrumental seismicity, fault activity, and other magnitude indicators are used to identify the origins.

A detailed PGA hazard map was worked out by Khattri et al. [5] and Bhatia et al. [1]. PSHA studies for the state of, northeast India (Das et al. 2006), Delhi (Iyengar and Ghosh 2004), Mumbai city [4], peninsular India [13–15], Tamil Nadu [8], the north-western region of Himalaya [7, 16], and PSHA of Vishakhapatnam city [10–12]. The National Disaster Management Agency (NDMA) [9] formed an Expert committee in 2011 to draft an All India PSHA report.

This study aims to examine the PSHA technique and discuss the estimation of seismic risk in the research regions. The suggested technique will be based on the actual case studies of three tunnel project cases in Uttarakhand, Himachal Pradesh, and Jammu & Kashmir, respectively. PSHA may appear complex since it employs several source models, recurrence laws, and attenuation laws to account for all uncertainties. PSHA is quantitative and versatile enough to accept a wide range of parameters, incorporating all available information on earthquakes and the ground motion that results.

Table 11.1 Details of Study Area

Site	State	Longitude	Lattitude	Total no. of faults considered	Total no. of epicenters considered
Vishnugarh	Uttarakhand	79.49167°E	30.51389°N	123	397
Shongtong	Himachal Pradesh	78.47°E	31.5°N	151	310
Parnai	Jammu & Kashmir	74.35222°E	33.60944°N	143	1276

11.2 Brief Description of Study Area

1. **Vishnugarh Pipalkoti Hydroelectric power plant**

The Vishnugad Pipalkoti Hydroelectric Project (4 × 111 MW) was built in Uttarakhand's Chamoli district on the Alaknanda River. Three intake tunnels, three Underground sedimentation chambers, a headrace tunnel, a surge shaft, and two pressure shafts bifurcating into four Penstocks make up the water conductor system.

2. **Shongtong karcham Hydroelectric power plant**

The Shongtong Karchham Hydro Electric Project is a run-of-river system on Satluj in Himachal Pradesh's Kinnaur district. The project entails the construction of a 102.50 m long diversion barrage near the village of Powari, as well as a water conductor system that includes an 8020 m long 10.00 m diameter circular HRT, a 39.50 m diameter surge shaft, and three 5.10 m diameter steel-lined (each 211.00 m long) pressure shafts to feed three vertical axes Francis Turbines housed in an underground powerhouse on the left bank.

3. **Parnai Hydroelectric power plant**

The Parnai Hydroelectric Project, with a capacity of 37.5 MW, is located in the district of Poonch in the Jammu and Kashmir. Details of the study regions are shown in Table 11.1.

11.3 Methodology

The simple methodology used for PSHA is as described by Kramer in 2003.

The main of PSHA is to establish magnitude-dependent features of the seismicity for each source zone by using Gutenberg–Richter relationship.

$$\log \lambda_m = E - bM$$

where λ_m = cumulative number of earthquakes of a given magnitude

'a' = logarithm of the number of earthquakes of magnitude zero or greater

'b' = slope of the curve which considers the proportion of large earthquakes to small earthquakes and M is magnitude.

The seismic details of the region which comes within a distance of 300 km around sites have been collected and re-evaluated as shown in Figs. 11.1 and 11.2. The United States Geological Survey (USGS), the Geological Survey of India (GSI), and published literature were used to compile an earthquake database for the years 1980–2015 (Rao and Rao 1984; Iyengar 1999). Windowing criteria were used to filter out foreshocks and aftershocks, and a new earthquake data catalog was created. Only earthquakes with a magnitude greater than three are considered while compiling the earthquake list. Finally, the seismicity of the research region was assessed using data from the catalog spanning the years 1980–2015. OpenJUMP software has been used for getting the epicenter and fault data for a particular location. The steps involved are shown in the flowchart.

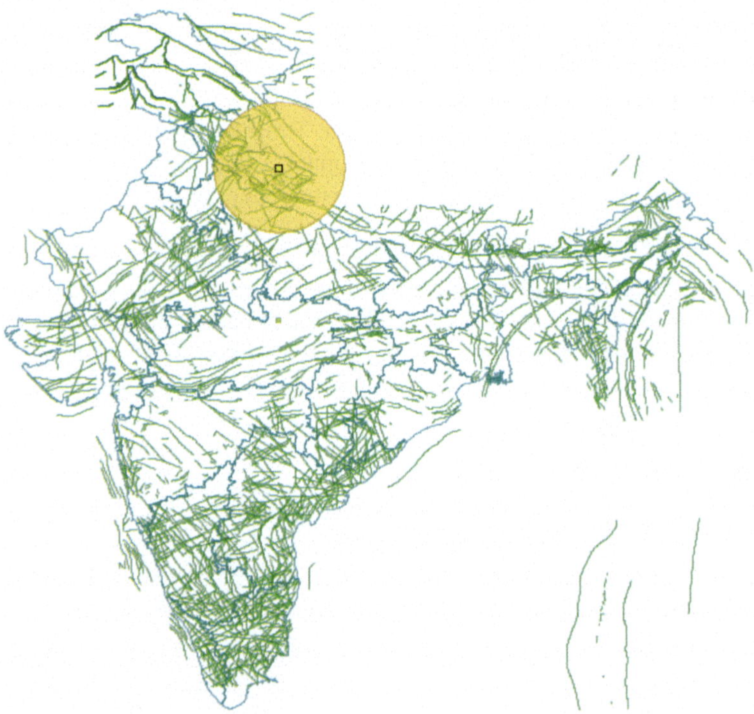

Fig. 11.1 Selecting the faults around project site within a radius (300 km)

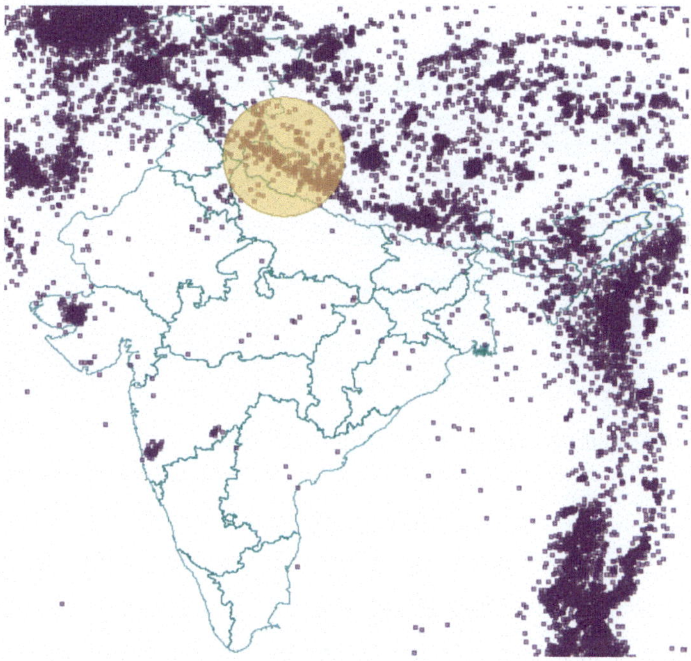

Fig. 11.2 Selecting the faults around project site within a radius (300 km)

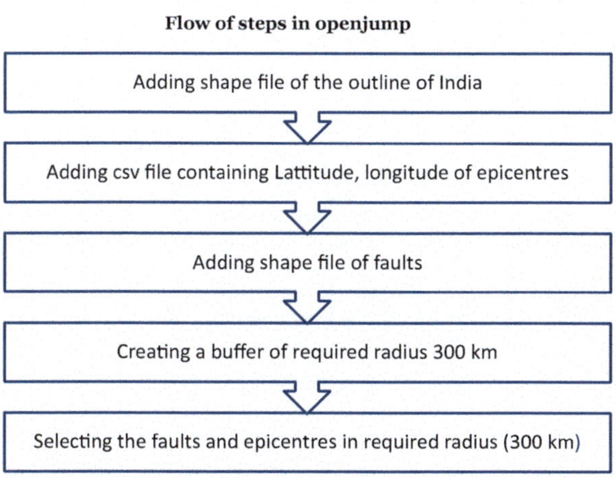

Steps involved in PSHA (Fig. 11.3):

1. The minimum distance of each fault to the site, length of the fault, and maximum moment magnitude are calculated.
2. Faults sorted and selected 30 faults according to PGA.
3. The minimum distance from each source to fault is calculated.
4. Finding the number of earthquakes occurred in the particular magnitude range.

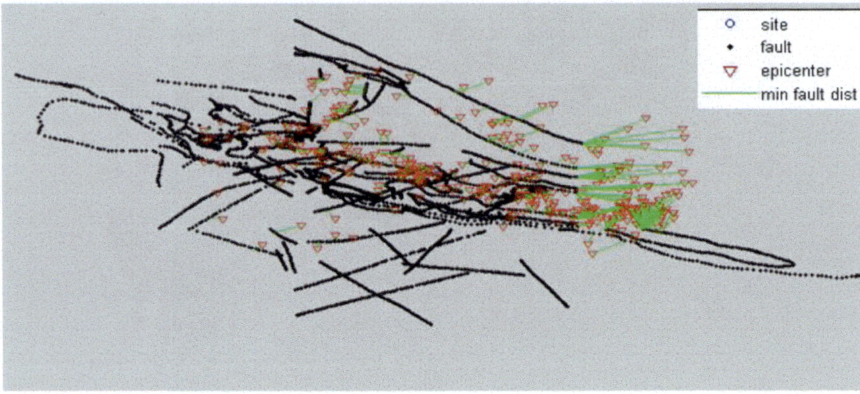

Fig. 11.3 Plotting of faults and epicenters for Vishnugarh Site

5. Finding Gutenberg–Ritcher parameters a, b.
6. Probability Calculation.

Step 1: Fault Distance and Moment Calculation

From the fault data extracted from OpenJUMP software, the minimum distance from each fault to the site is calculated. Then using the length of fault moment, the magnitude for each fault is calculated.

$$M_w = 5.08 + 1.16 \times \log_{10} L$$

where L = Length of Fault, M_w = Moment magnitude of the fault.

Step 2: PGA Calculation

PGA is calculated using the attenuation relationship proposed by Sharma for the Himalayan region of India.

$$\text{Log (A)} = -2.87 + 0.634\text{M} - 1.16\ \log\left(\text{x} + e^{(0.62*\text{M})}\right)$$

Where A = PGA (g)
M = Moment magnitude
X = Hypocentral distance (Km)

Step 3: Selection of Faults

Faults were sorted in the descending order of PGA values, and the 30 faults with max PGA values were selected. Tables 11.2, 11.3 and 11.4 represent the considered fault details and maximum PGA obtained for Vishnugarh, Parnai, and Shongtong, respectively.

Table 11.2 Faults considered with maximum PGA for Vishnugarh HEP

Fault no.	Magnitude (Mw)	Fault length (Km)	PGA (g)	Return period (years)
12	8.101575822	402.5378688	0.524831	1.093756836
120	7.979379308	315.8383927	0.368605	1.029411765
10	7.718330699	188.1148221	0.36632	1.521739263
11	7.655055804	165.9111958	0.328032	1.666666667
103	7.577972378	142.3719599	0.304073	1.75
92	8.312809674	612.2163041	0.292351	3.888888889
123	8.063282718	373.0743974	0.290109	3.181818182
15	7.529244957	129.2462843	0.266027	2.333333333
30	7.64244524	161.8096867	0.263464	2.5
90	7.80351096	222.768754	0.252212	5
34	7.966235485	307.7046352	0.245718	3.888888889
91	7.622378007	155.4909719	0.232685	17.5
100	7.912514093	276.5810965	0.229382	3.5
82	7.741124445	196.8216178	0.225271	0.603448276
113	8.166966803	458.3299933	0.223411	17.5
102	7.447243728	109.8316104	0.219155	11.66725003
13	7.753802423	201.8376069	0.214272	2.1875
72	7.933352342	288.2614069	0.207983	2.058823529
47	7.539034384	131.7823439	0.203235	1.09375
86	8.059808589	370.5104891	0.1922	5
35	7.120270931	57.39238215	0.183984	8.75
110	7.634452912	159.2628852	0.181492	11.66666667
75	7.704097041	182.8742768	0.177339	35
117	7.449572042	110.3403912	0.17184	1
121	7.960954149	304.4956995	0.168699	11.66666667
39	7.544909942	133.3283069	0.168229	0.085714286

Maximum PGA obtained for Vishnugarh (Uttarakhand) is 0.524 g

Step 4: Assigning earthquakes to faults

The minimum distance from each source to the fault is calculated according to:

a. Minimum earthquake to fault distance
b. Magnitude of each earthquake was assigned to one particular fault

Step 5: Finding the number of earthquakes occurred in the particular magnitude range

The earthquake are then categorized in different magnitude classes and number of earthquakes occurred in that magnitude class.

Table 11.3 Faults considered with maximum PGA for Parnai

Fault no.	Magnitude (Mw)	Fault length (Km)	PGA (g)	Return period (years)
104	7.916121115	278.5684935	0.445852	0.324
137	7.341705131	89.07294318	0.302791	0.555556
117	7.650818072	164.5214302	0.268832	11.66667
49	8.051119882	364.1750961	0.239119	7
141	8.214202391	503.3830786	0.227607	1.060606
128	8.074883595	381.7650813	0.227059	35
120	7.951782429	299.0022855	0.211738	0.166667
2	7.564075417	138.4982625	0.202044	0.324
99	7.655680352	166.1170066	0.200209	0.04108

Maximum PGA obtained for Parnai (Jammu & Kasmir) is 0.44 g

Step 6: Finding Gutenberg–Ritcher parameters a, b

The a, b values of the frequency-magnitude relationship are computed using the total number of occurrences. Although the value of b varies by region, it is generally between 0.6 and 1.5. Throughout the research period, the parameter 'a' indicates the total amount of seismic activity in a given location. Every year, the number of ($M_w >$ 4) earthquakes is proportional to the value of 'a.' The parameters regulate the seismic hazard at any location (a, b). Hazard curves and uniform hazard response spectra for the research region were created using MATLAB software designed specifically for PSHA. The relationship between Gutenberg and Richter has been used to describe the seismicity of the study region and estimate the hazard parameters for the study area.

Vishnugarh

The hazard parameters 'a' and 'b' have been estimated to be 3.79 and 0.72, respectively, for Vishnugarh (Fig. 11.4).

$$\log_{10}\lambda = 3.7907 - 0.7249M$$

Parnai

The hazard parameters 'a' and 'b' have been estimated to be 4.72 and 0.85, respectively, for Parnai (Fig. 11.5).

$$\log_{10}\lambda = 4.7241 - 0.8556M$$

Shongtong

The hazard parameters 'a' and 'b' have been estimated to be 3.78 and 0.76, respectively, for Shongtong (Fig. 11.6).

Table 11.4 Faults considered with maximum PGA for Shongtong

Fault no	Magnitude (Mw)	Fault length (Km)	PGA (g)	Return period (years)
103	8.101575822	402.5378688	0.38266	5.833333333
92	8.312809674	612.2163041	0.351904	4.375
58	7.921508704	281.5635741	0.32582	35
89	7.718330699	188.1148221	0.293428	3.407990004
32	7.507887239	123.8814387	0.293203	2.333333333
142	8.063282718	373.0743974	0.272028	17.5
134	7.786960331	215.569092	0.25663	1.346153846
78	7.743056145	197.5777594	0.243558	11.66666667
59	7.633600205	158.9935429	0.237306	1.590909091
20	7.539034384	131.7823439	0.231685	1.944444444
132	7.386615477	97.37817202	0.22917	0.972222222
110	7.539921786	132.0146806	0.220089	17.5
144	8.074883595	381.7650813	0.218876	5
122	7.526	129.2462843	0.207	1.060
127	7.514642875	125.5538587	0.205596	35
119	7.655055804	165.9111958	0.193151	2.916666667
2	7.933352342	288.2614069	0.192896	5.833333333
62	7.231955276	71.63641747	0.190591	11.66666667
64	7.429873615	106.1092187	0.189331	5.833333333
116	7.960954149	304.4956995	0.179892	7
133	8.087336657	391.3196001	0.177415	11.66666667
76	7.320535997	85.4076061	0.176787	2.5
25	7.340995585	88.94757768	0.175538	35
106	7.806867686	224.2580313	0.17553	2.692307692
75	7.072299012	52.17940806	0.17184	35

Maximum PGA obtained for Shongtong (Himachal Pradesh) is 0.382 g

$$\log_{10} \lambda = 3.7890 - 0.7623\text{M}$$

Once the origin of the earthquake is established and its associated source area defined, the next step is to determine the magnitude of earthquakes that may be predicted to produce the source region. Both origin areas have a total intensity of earthquakes that cannot be exceeded; for some, it can be high and for others, it can be low. A recurrence rule defines the spread of earthquake sizes over a given period. A basic assumption of PSHA is that it is possible to predict future seismicity by the recurrence rule derived from the previous seismicity.

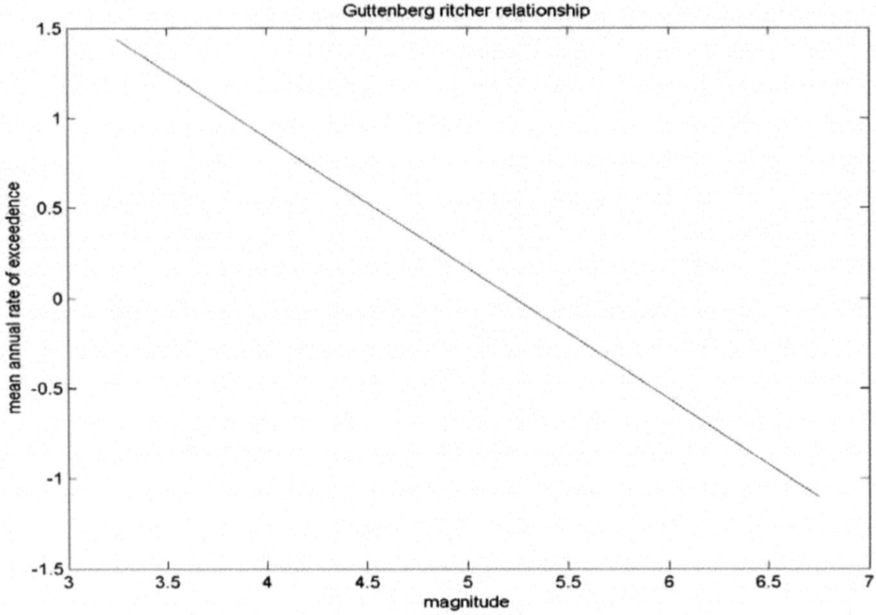

Fig. 11.4 Graph between magnitude and mean annual rate of exceedance for Vishnugarh

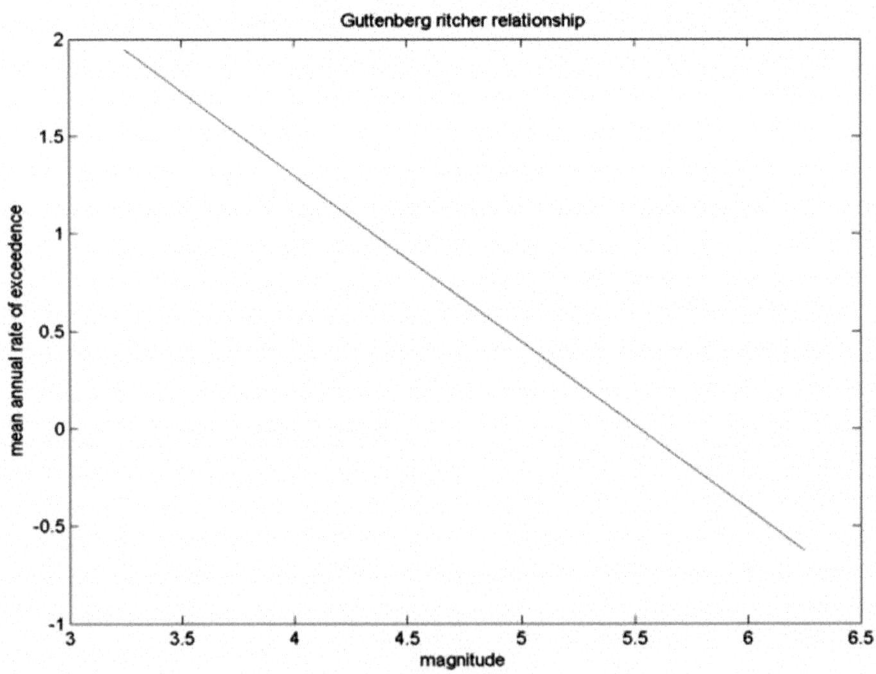

Fig. 11.5 Graph between of exceedence for magnitude and mean annual rate (Parnai)

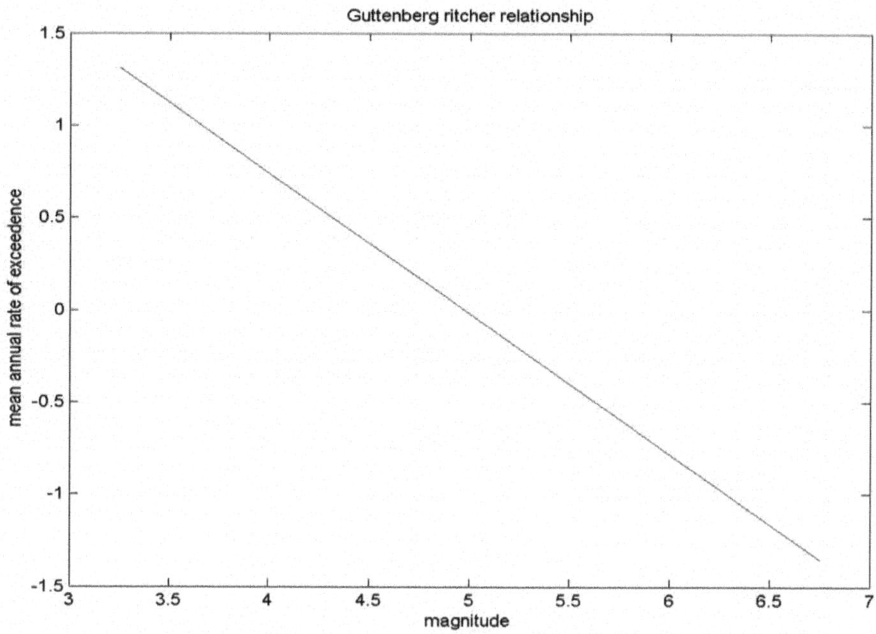

Fig. 11.6 Graph between of exceedence for magnitude and mean annual rate (Shongtong)

The upper and lower boundary CDF and PDF for the Gutenberg–Ritcher rule can be represented as

$$F_m = \frac{1 - e^{[-\beta(m-m_0)]}}{1 - e^{[-\beta(m_{\max}-m_0)]}}$$

$$F_m = \frac{\beta e^{[-\beta(m-m_0)]}}{1 - e^{[-\beta(m_{\max}-m_0)]}}$$

where $\boldsymbol{\beta} = 2.303b$
PDF = Probability Density Function
CDF = Cumulative Density Function

Figures 11.7, 11.8, 11.9, 11.10 and 11.11 illustrate the magnitude and probability density function for the three locations. To assess the chance of various risks occurring at a particular time, the distribution of seismic activity across time must be considered. Earthquakes have long been thought to occur spontaneously over time. However, an analysis of available seismicity records showed no evidence of temporal changes in the recurrence of the earthquake (when aftershocks were excluded).

The Poisson model is a basic framework for evaluating the probability of events that follow a Poisson process, which produces values for a random variable indicating the number of occurrences of a specific event over a given time interval or in a

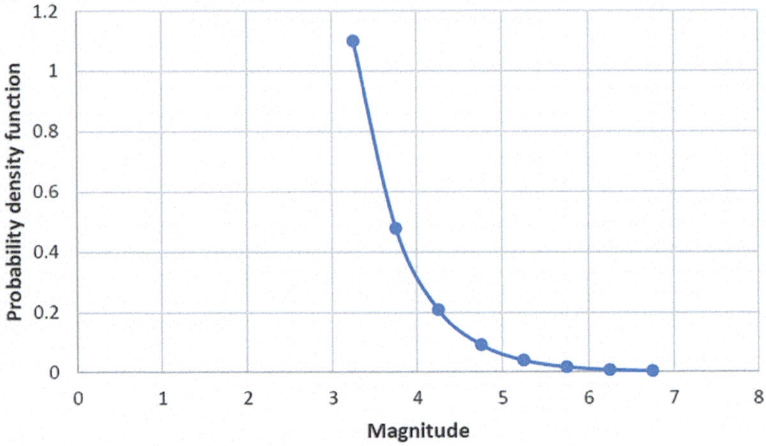

Fig. 11.7 Magnitude and probability density function for Vishnugarh

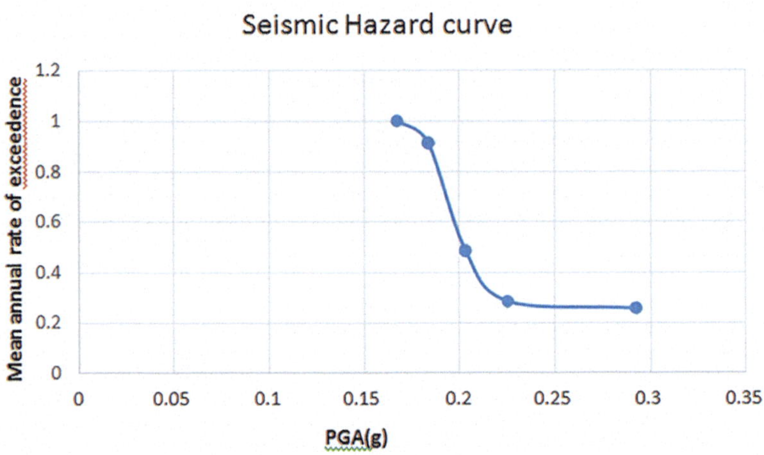

Fig. 11.8 PGA and mean annual rate of exceedance for Vishnugarh

given spatial region. The spatial implementations of the Poisson model will not be examined further since PSHAs struggle with temporal uncertainty. Such features imply that Poisson system events happen independently, with no "knowledge" of the duration, length, or position of any preceding event.

If the occurrence of significance is the excess of a single magnitude of the earthquake, the Poisson equation can be paired with an effective recurrence formula to estimate the risk of at least reaching one in a t-year cycle by the expression

$$P(N \geq 1) = 1 - e^{(-\lambda_m * t)}$$

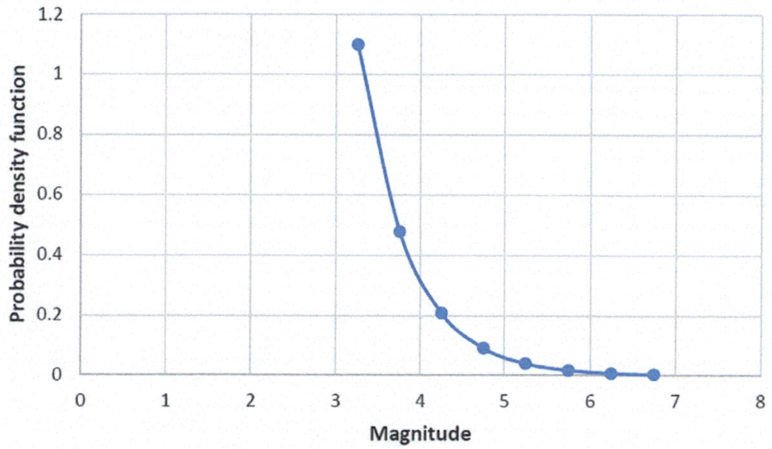

Fig. 11.9 Magnitude and probability density function for Parnai

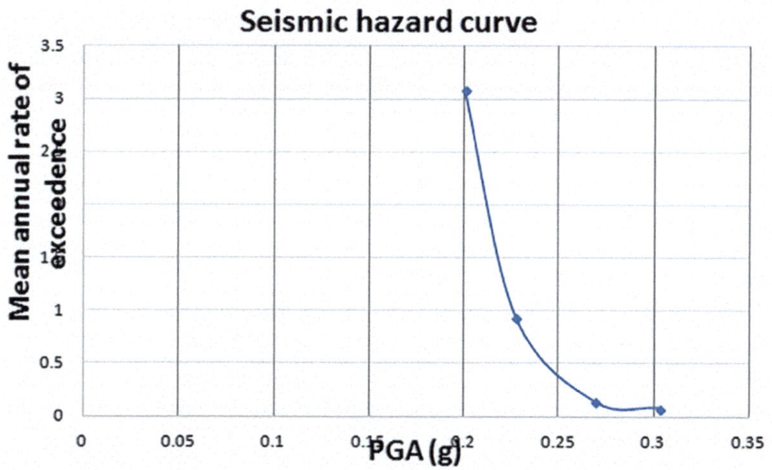

Fig. 11.10 PGA and mean annual rate of exceedance for Parnai

11.4 Results and Discussion

Seismic risk has escalated rapidly in India (Neelima and Towhata 2016). PSHA has been performed by numerous site-specific researchers and relies on a number of hypotheses (Rao et al. 2012). A seismic hazard evaluation is usually performed to determine the seismic parameters required for the seismic design or performance assessment of the tunnels. Earthquakes must be specified for analytical studies such that relevant seismic parameters (e.g., amplitude, intensity, spectral coordinates, length, etc.) may be identified. The parameters can be achieved by either a deterministic approach or a probabilistic seismic hazard assessment.

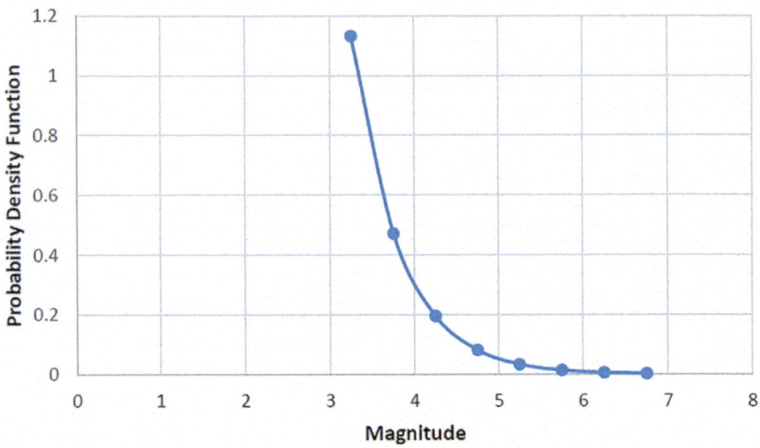

Fig. 11.11 Magnitude and probability density function for Shongtong

PSHA focuses on the uniform hazard spectrum (UHS), which is a useful tool for comparing different locations' danger representations. The spectrum amplitudes of acceleration may be estimated at all natural times with a constant probability of exceedance at a site using the PSHA approach. When a seismic wave passes through the tunnel during an earthquake, the whole tunnel experiences the same force and moves in the same direction. The ratio of peak ground acceleration to gravity acceleration can be utilized as a seismic coefficient. The PGA values can be used to calculate the horizontal seismic coefficient $K_{h,}$ which is an important factor in determining the seismic response of tunnels.

The UHS plots are designed for a 475-year return time (i.e., a 10% probability of exceedance in 50 years). The UHS plots given in Figs. 11.8, 11.10, and 11.12 represent estimates for all of the locations specified by the grid's intersecting points. The seismicity of the site, which is a function of (M_j, R_i), is assessed for this purpose by fitting the Gutenberg–Richter recurrence relation to previous earthquake data within a 300 km radius zone. The findings of hazard analysis may be utilized to create ground motion attenuation relationships for a given site, as well as simulated ground motion generation and other engineering applications. Such findings can be a great benefit in the construction of tunnels as a guide for design engineers.

11.5 Conclusion

For this study, seismic hazard parameters in terms of PGA have been estimated for three hydropower sites in the northern state of India situated in the Himalayan region. It was observed that more earthquakes occurred at smaller magnitudes compared to larger magnitudes. Faults that can induce ground motion around the site in a 300 km radius zone have been identified, and the return period of the earthquake of each

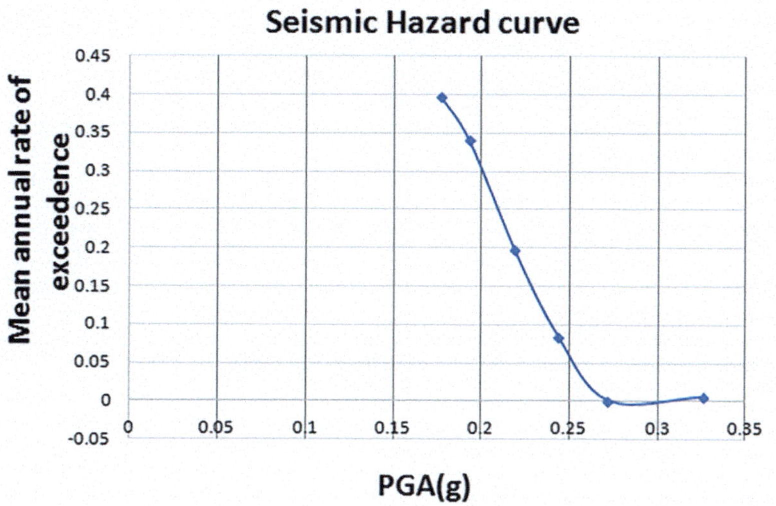

Fig. 11.12 PGA and mean annual rate of exceedance for Shongtong

magnitude from one particular fault was calculated. Guttenberg–Richter's parameter a and b was calculated. Based on the size and temporal uncertainties, probability functions were calculated. From these results, uniform hazard response spectra can be derived for the site. The results presented here may be utilized to generate a Seismic Hazard map for these specific project locations.

References

1. Bhatia SC, Ravi Kumar M, Gupta HK (1999) A probabilistic seismic hazard map of India and adjoining regions
2. BIS (2002) IS 1893 (part 1): 2002—Indian standard criteria for earthquake resistant design of structures, part 1: general provisions and buildings (Fifth Revision). Bureau of Indian Standards, New Delhi
3. Cornell CA (1968) Engineering seismic risk analysis. Bull Seismol Soc Am 58:1583–1606
4. Kanth, STGR, Iyengar RN (2006) Seismic hazard estimation for Mumbai city. Curr Sci 1486–1494
5. Khattri KN et al (1984) A seismic hazard map of India and adjacent areas. Tectonophysics 108(1–2):93–134
6. Kramer SL (1996) Geotechnical earthquake engineering. Pearson Education India
7. Mahajan AK et al (2010) Probabilistic seismic hazard map of NW Himalaya and its adjoining area, India. Nat Haz 53(3):443–457
8. Menon A et al (2010) Probabilistic seismic hazard macrozonation of Tamil Nadu in Southern India. Bull Seismol Soc Am 100(3):1320–1341
9. NDMA (2011). Geotechnical/Geophysical Investigations for seismic microzonation studies of urban centres in India. Technical Report. India
10. Putti SP, Satyam N (2020) Evaluation of site effects using HVSR microtremor measurements in Vishakhapatnam (India). Earth Syst Environ 4(2020):439–454

11. Putti SP, Satyam N (2018) Ground response analysis and liquefaction hazard assessment for Vishakhapatnam city. Innov Infrastruct Solut 3(1):1–14
12. Putti SP, Devarakonda NS, Towhata I (2019) Estimation of ground response and local site effects for Vishakhapatnam, India. Nat Haz 97(2):555–578
13. Rao KS, Neelima Satyam D (2007) Liquefaction studies for seismic microzonation of Delhi region. Curr Sci (2007): 46–654
14. Satyam DN, Rao KS (2014) Liquefaction hazard assessment using SPT and VS for two cities in India. Indian Geotech J 44:468–479. https://doi.org/10.1007/s40098-014-0098-2
15. Srikanth L, Neelima Satyam D (2014) Dynamic analysis of transmission line towers. In: International conference on civil engineering and applied mechanics (ICCEAM)
16. Sitharam TG, Kolathayar S (2013) Seismic hazard analysis of India using areal sources. J Asian Earth Sci 62(2013):647–653
17. Srivastav A, Satyam N (2020) Understanding the impact of the earthquake on circular tunnels in different rock mass: a numerical approach. Innov Infrastruct Solut 5(1):1–9

Chapter 12
On Structure-Equipment-Piping Interactions Under Earthquake Excitation

G. R. Reddy

Abstract Generally, seismic risk is discussed with respect to the structures that include lifelines. Less discussed with regard to the equipment and piping that also contribute to the total seismic risk in terms of life and economy. In view of this, in the present paper, the detailed procedure to estimate the design seismic forces in equipment and piping systems is discussed. Focus is given on structure-equipment interaction and equipment-piping interaction. Detailed explanation is made on generation inputs for the design of floor-mounted equipment and piping systems. Also, procedure for evaluating the response of piping system supported at multi locations of structure is explained.

12.1 Introduction

In this article, structure-equipment interaction is discussed considering non-industrial (residences, offices, hospitals and Institutes) and industrial facilities (manufacturing, processing and power industry). In the former case, the equipment (pumps, fans, cup boards, selves, televisions, fridges and low-pressure piping) as shown in Fig. 12.1 are light compared to the structural masses and in the later (machine tools, boilers, high capacity pumps, reactors, high-pressure piping systems) as shown in Fig. 12.2 are heavy and sometimes it has masses comparable to the mass or modal mass of the structure. Details of interaction in terms of the frequency variations of the coupled equipment and structure frequency will be discussed in detail i the forthcoming sections.

The level of risk that is a function of seismic hazard levels, capacity and status of exposure is relatively small in the case of non-industrial facilities compared to the industrial structures and equipment. To avoid failures of equipment in non-industrial facilities, some of the easy fixes those can be easily adopted are illustrated in Fig. 12.3.

G. R. Reddy (✉)
VJTI, Mumbai, India
e-mail: grrddy@yahoo.com

T. G. Sitharam et al. (eds.), *Theory and Practice in Earthquake Engineering and Technology*, Springer Tracts in Civil Engineering,
https://doi.org/10.1007/978-981-19-2324-1_12

Fig. 12.1 Equipment and piping in non-industrial facilities

Fig. 12.2 Equipment and piping in industrial facilities

Even in the hospitals, such fixes can be adopted for equipment. However, oxygen and water service lines should be anchored on concrete members than on masonry walls for better performance.

In the case of industrial facilities, there are situations where large equipment and complext piping networks will be supported on the structure as shown in Fig. 12.4. This is analogous to the situation where elephant is carrying a large mass of tree trunk as shown in Fig. 12.4. In this case, the structure analogous to elephant will surely feel reactions of the masses being carried or supported. Also, the equipment or the masses on the elephant behave based on the movement or motion of the structure or elephant. If elephant moves very fast and the rope or chain anchorage is not proper, all the material may fall and similar situation can also be seen in the equipment supported

Fig. 12.3 Easy fixes for equipment if non-industrial facilities

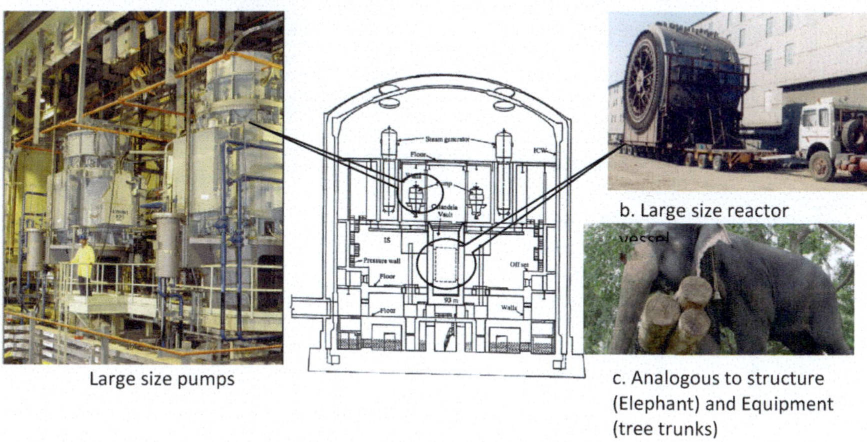

Fig. 12.4 Typical equipment in Industrial system

on the structure when the structure is excited during earthquake. If the design of equipment is made without giving due considerations to these effects, failures of equipment as shown in Fig. 12.5 are inevitable.

These effects are taken care systematically while designing the equipment or piping supported on the structure and are subjected to external excitations such as earthquake. The steps that are followed in generating the input for design of equipment are explained referring to Fig. 12.6 as follows.

Mathematical models of the structure as shown in Fig. 12.6a are developed and analysed for a given design basis earthquake time history as shown in Fig. 12.6b. The basic dynamic equilibrium equation as given below is solved using numerical integration method such as Newmark-β technique.

$$[M]\{\ddot{x}\} + [C]\{\dot{x}\} + [K]\{x\} = [M]\{\ddot{x}_g\}\{1\} \tag{12.1}$$

Fig. 12.5 Equipment failure due to earthquake

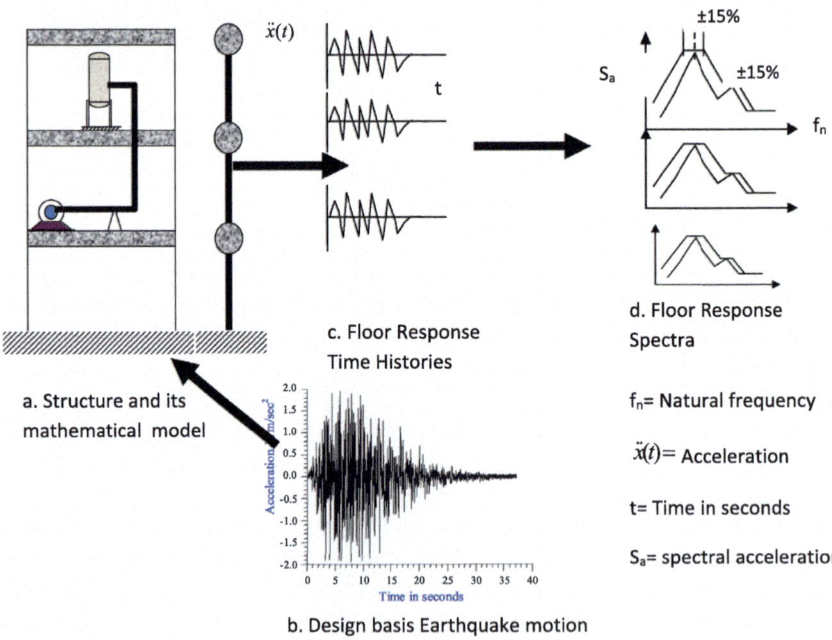

Fig. 12.6 Steps for generation inputs for designing equipment

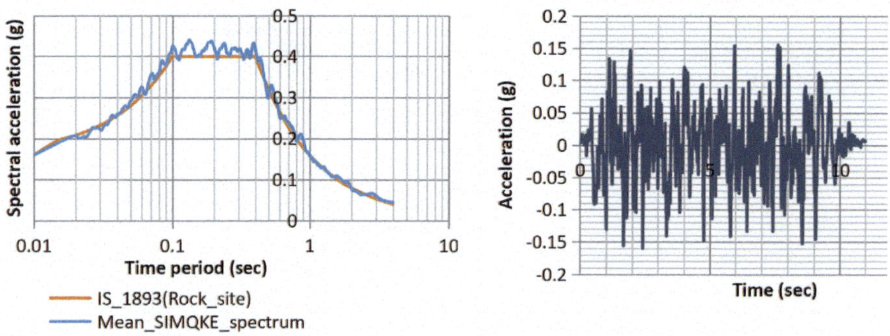

Fig. 12.7 Design Basis Ground Spectrum Compatible time History

where [M] is a mass matrix, [C] is a damping matrix and [K] is a stiffnex matric. x, x with single dot and x with double dot are response displacement, velocity and accelerations respectively as a function of time. Xg with double dot is ground acceleration. IS-1893 spectrum compatible time histories may be generated using the procedure enumerated in the references [1, 2]. Typical one of the three spectrum compatible time histories is shown in Fig. 12.7. The compatibility is checked as per the procedure given in ASCE standard [3]. {1} is the influence vector with unit values along the direction of excitation and zero values along direction of no excitation.

Response absolute accelerations are obtained at different floor levels by solving single degree of freedom (SDOF) equation of motion as given below with different frequencies for a particular damping. Floor response spectra as shown in Fig. 12.6c are obtained as the plot of maximum absolute acceleration versus a frequency for a damping value.

$$\ddot{x}_n + 2\varsigma\omega_n\dot{x}_n + \omega_n^2 x_n = -\ddot{x}_f \tag{12.2}$$

where ω_n is the natural frequency, ζ_n is the damping ratio of nth SDOF structure. x_n, x_n with single dot and x_n with double dot are response displacement, velocity and accelerations, respectively, with time. X_f with double dot is the floor acceleration.

It is very important to notice that the floor response spectrum has peaks at the natural frequencies of the structure. The frequency of the structure may change due to coupling of the equipment. This coupling effects sometimes called interaction effects and can not be over looked if the equipment and structure are tuned and have large mass (Fig. 12.4) ratio (>10%). In the case of piping systems, interaction with the structure may not affect the natural frequencies due to its light weight and generally treated as an ant sitting on elephant or birds sitting on the tress as shown in Figs. 12.8 and 12.9 respectively.

However, the requirement of coupling equipment and structure is generally decided based on ASCE [3] or ASME [4] criteria as explained in the next section.

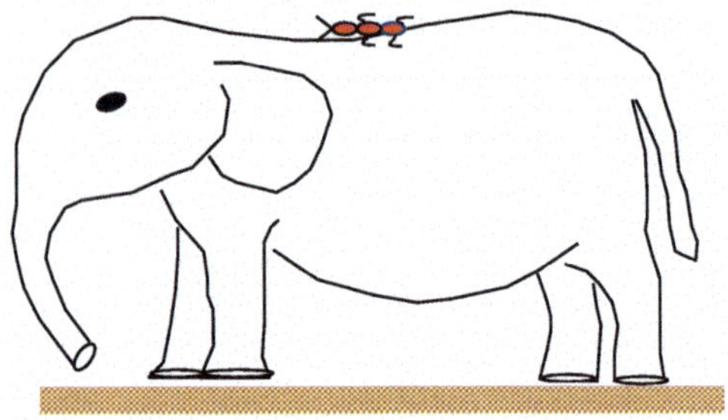

Fig. 12.8 Analogous to the structure (elephant) and piping system (ant)

Fig. 12.9 Analogous to the structure (tree) and piping system (birds)

12.2 Decoupling Criteria

Decoupling criteria are based on the frequency/modal frequency ratio R_f and mass/modal mass ratio R_m of the secondary system (SS) called equipment to the primary system (PS) called structure as given below.

$$R_f = \frac{Frequency\, of\, \text{Secondary}\, system}{Frequency\, of\, primary\, system}$$

$$R_m = \frac{Mass\ of\, \text{Secondary}\, system}{Mass\ of\, primary\, system}$$

where R_f is the ratio of frequency or modal frequency of uncoupled SS to the uncoupled PS and R_m is the ratio of mass or modal mass of the uncoupled SS to the uncoupled PS.

i. Decoupling can be done for any R_f, if $R_m < 0.01$.
ii. If $0.01 \leq R_m \leq 0.1$ decoupling can be done provided $0.8 \geq R_f \geq 1.25$.

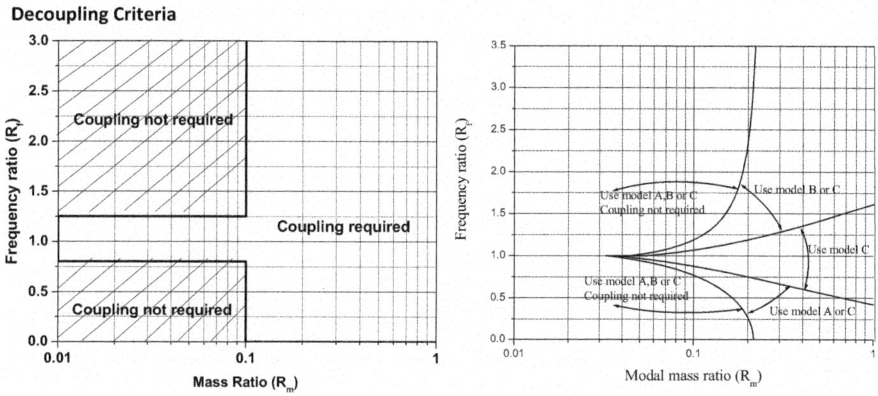

Fig. 12.10 Decoupling criteria for primary and secondary system

iii. If $R_m \geq 0.1$ and $R_f \geq 3$ (i.e. rigid secondary structure). It is sufficient to include only the mass of the system in the primary structure.
iv. If $R_m \geq 0.1$ and $R_f < 0.33$ (Flexible secondary system) decoupling can be done.
v. If $R_m \geq 0.1$ and $0.33 < R_f < 3$, coupled system analysis is required.

Note that the modes whose participation is more than 20% need to be considered in evaluating the above ratios.

Figure 12.10a and b shows the graphical representation of decoupling criteria for primary and secondary system.

One of the above criteria can be adopted for checking the requirements of coupling the equipment to the structures.

For multi-degree of freedom structure as shown in Fig. 12.12, modal frequencies and modal mass of the structure are estimated and criteria as explained above are applied considering modes of 20% mass participation. In the case of structures supported on common foundation as shown in Fig. 12.11b and supporting the equipment, degree of freedom (DOF) mass [5, 6] of the substructures are estimated and used to calculate the mass ratios and corresponding frequency ratios and applied decoupling criteria for the requirement of coupling the equipment with the structure.

Based on the coupling requirements, a coupled model is prepared considering structure and equipment and analysis as explained above is performed and inputs for design of equipment are generated. If the damping values are different for soil, structure and equipment, equivalent damping using energy principle is evaluated and used in the analysis [3, 5, 6]. There are direct method and approximate methods to generate inputs for designing floor-mounted equipment and piping systems and are described in the forthcoming sections.

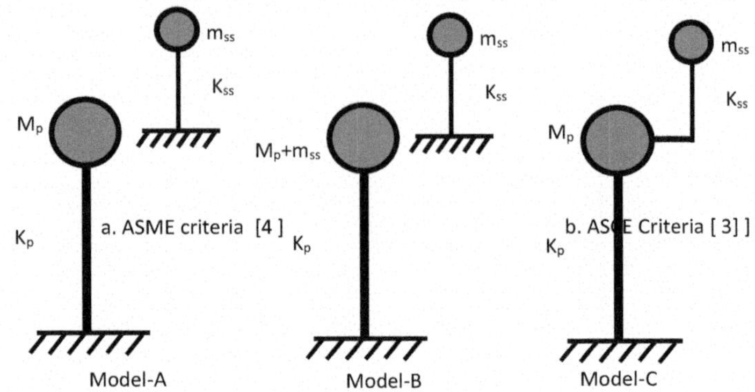

Fig. 12.11 Models adopted in ASCE criteria (Fig. 12.10b)

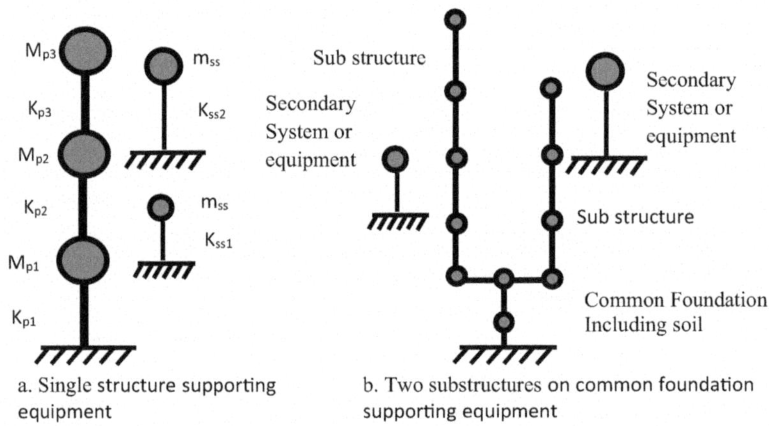

Fig. 12.12 Multi degree of freedom structure supporting equipment

12.3 Direct Method of Evaluating Floor Spectrum Using Design Ground Spectrum

Referring to IAEA-TECDOC-1347 [7], the floor response spectrum may be obtained directly using ground spectra, which correspond to the damping value of the structure including soil, and ground spectra for equipment or piping systems damping and using modal characteristics of the structure obtained in modal analysis. The various steps involved are given below:

1. Obtain the design basis ground motion called design basis response spectra corresponding to the damping value of the structure and the equipment or piping systems.
2. Generate mathematical model of the structure. The model could be beam model or 3D FE model.

3. Obtain the Eigen values and Eigen vectors of the structure by modal analysis.
4. Generate FRS by using the Eigen values and Eigen vectors of the structure.

Spectral acceleration at ith mode of the structure and at jth natural frequency of the equipment is given as follows.

$$Sa_{ij} = \frac{1}{\sqrt{\left\{1 - \left(\frac{\omega_{Ej}}{\omega_{Bi}}\right)^2\right\}^2 + 4\left(\xi_{Ej} + \xi_{Bi}\right)^2\left(\frac{\omega_{Ej}}{\omega_{Bi}}\right)^2}} \sqrt{\left\{\left(\frac{\omega_{Ej}}{\omega_{Bi}}\right)^2 Sa(\omega_{Bi}, \xi_{Bi})\right\}^2 + Sa\left(\omega_{Ej}, \xi_{Ej}\right)^2} \tag{12.3}$$

$$Sa_j = \sqrt{\sum_i^n \left(\Gamma_i \varphi_{ik} \times Sa_{ij}\right)} \tag{12.4}$$

where

Sa_j = Floor response spectrum value at jth frequency of the equipment or piping system taking into account all structural including soil modes (i = 1 to n).

Γ_i = The ith modal participation factor of structure including soil.

ϕ_{ik} = kth floor mode shape in ith mode of structure including soil.

ζ_{Ej} = Damping factor of Equipment or piping system at jth frequency.

ω_{Ej} = jth frequency of the equipment or piping system.

ζ_{Bi} = Damping factor of the structure including soil in ith mode.

ω_{Bi} = ith modal frequency of the structure including.

$Sa(\omega_{Bi}, \zeta_{Bi})$ = The standard design ground spectral value corresponding to ω_{Bi}, ζ_{Bi} of the structure including.

$Sa(\omega_{Ej}, \zeta_{Ej})$ = The standard design ground spectral value corresponding to ω_{Ej}, $\zeta_{Ej}$of the equipment or piping systems.

Notes:

(1) The mass or modal mass m_A of the equipment or piping needs to be less than 1% of the mass or modal mass m_{Bi} of the structure.
(2) The floor response spectra, obtained from the above method, need to be broadened by at least 15% to account for the uncertainty in soil-structure-interaction, equipment-structure-interaction and numerical procedures adopted in analysis.

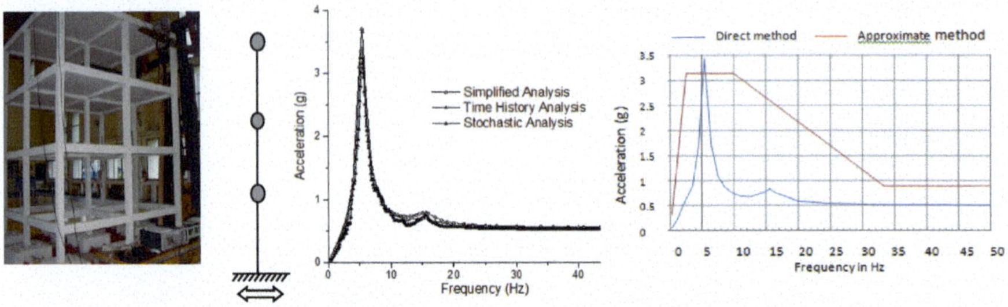

a. Structure b. Beam model c. FRS (direct method) d. FRS (approximate method)

Fig. 12.13 Comparison of floor response spectra of various methods

Figure 12.13 shows the comparison of Floor Response spectra obtained using time history and direct method. For details of stochastic method, reader is requested to see the reference [5, 6].

12.4 Approximate Method of Evaluating Floor Spectrum Using Design Ground Spectrum

12.4.1 Approximate Method

Although not a recommended procedure, the Floor Response Spectrum (FRS) as shown in Fig. 12.13d at a particular floor within a structure may be obtained by directly multiplying the design Ground Response Spectrum (GRS) by a factor depending on the height of the floor with respect to the total height of the structure. The FRS is given by:

$$S_{aj} = S_g\left(1 + c\frac{h}{H}\right) \tag{12.5}$$

where

S_{aj} = Spectral acceleration of FRS at jth equipment/piping frequency.

Sg = Spectral acceleration of Ground response spectra.

h = Height of equipment/piping support element above the ground.

H = Height of the structure.

and c = 3 for 5% damping.

12.4.1.1 Peak Broadening Floor Response Spectra

As mentioned earlier peaks of FRS occur at natural frequency predominantly and these should be evaluated as accurately as possible. However, floor response spectra need to be broadened as explained in Fig. 12.6d to account for variations of Frequencies due to uncertainty in the following.

a. Soil-structure interaction.
b. Equipment-structure interaction.
c. Numerical structural modelling and analysis procedures in estimating the natural frequencies.

12.4.1.2 Multi-supported Piping Systems

The piping system shown in Fig. 12.6a is independently shown in Fig. 12.14 and has supports at three locations such as one with top equipment and second at bottom equipment and laterally constrained at floor level. For seismic design of this piping system, the basic inputs required are floor response spectra (FRS) at support locations and support displacements. If the supports are effective in three directions, then at each support, three FRS are need to be considered and similarly three support displacements. Support displacements are also called Seismic Anchor Motions (SAM) need to be considered in design and corresponding stresses can be obtained by performing static analysis.

The piping system may be decoupled from the equipment if the following criteria are satisfied.

a. Moment of inertia of equipment is more than 100 times the moment of inertia of the piping system.
b. Equipment side ends of piping system have constraints along three directions. This requirement may be at one location or different locations.

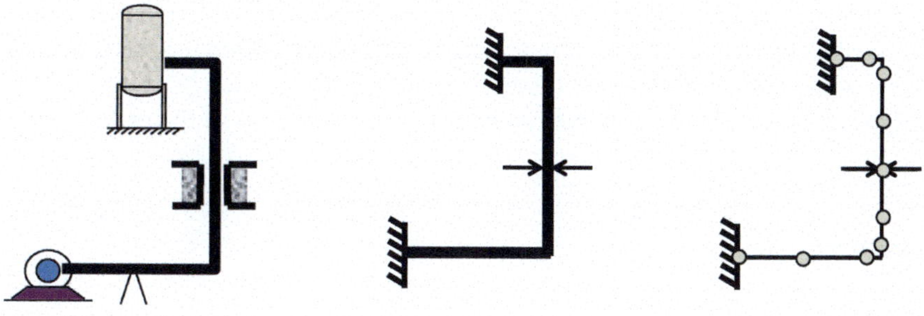

a. Typical piping system b. Uncoupled piping system c. Mathematical model of piping system

Fig. 12.14 Typical piping system connected to equipment

After finalising the boundaries as shown in Fig. 12.14b, suitable mathematical may be developed as shown in Fig. 12.14c and analysed for the forces, moment and stresses. For selecting number of elements and type of elements such as straight pipe, bend, tee etc., the reader is requested to refer [5, 6] on chapter related to multi-degree of freedom and piping design chapters. After finalising the model, the following equilibrium equation is solved considering the piping subjected to multi-support excitation.

$$[M]\{\ddot{x}\} + [C]\{\dot{x}\} + [K]\{x\} = [M]\{\ddot{x}_g\}\{I_j\} \tag{12.6}$$

It is important to note the difference of Eq. 12.6 with respect to Eq. 12.1. In the case of Eq. 12.1 for ground excitation as explained earlier, the influence vector has 1 s and 0 s and whereas the influence vector $\{I_j)$ has static displacements with unit displacement at the support j. It results in number of influence vectors equal to translational supports. Kindly note that the sum of all the influences at given node will result in unity. The above equation cam be solved either for the time wise response or frequency (mode) wise response based on the input in terms of time history and response spectrum respectively. Usually, later procedure is adopted since it is simple and Broadened FRS can be used without variations accounting for soil-structure-interaction, numerical solution variations and structure-equipment interactions. The modal response can be combined using suitable methods such as square root of sum of square, absolute sum, 10% sum or complete quadratic combination rule as explained in the reference [3, 5, 6].

12.5 Discussions and Conclusions

It is very important to perform seismic design of equipment and piping systems of non-industrial and industrial facilities to reduce the risk. Detailed procedures for obtaining the design inputs for equipment and piping system are discussed considering equipment structure interaction and equipment piping interaction. The interactions considered in the present article with respect to the variations of structural frequencies and their importance on the design basis FRS are clearly discussed.

The reader is requested to see the references for more details of the procedures.

References

1. IS 1893, Criteria for earthquake resistant design of structures Part xx: equipment and piping systems, (Draft) 2019
2. Gasparini DA, Vanmarcke EH (1976) SIMQKE, a program for artificial motion generation: user's manual and documentation, Department of Civil Engineering, MIT, USA
3. ASCE Standard, ASCE 4–98, Seismic Analysis of Safety Related Nuclear Structures, 1998
4. ASME Section III, Division 1, Appendices, 2004

5. Reddy GR (1994) Advanced approaches for the seismic analysis of nuclear power plant structure, equipment and piping systems. PhD thesis, Tokyo Metropolitan University
6. Reddy GR, Prasad H, Verma AK (2019) Textbook of seismic design (structures, piping system and components). Springer, Berlin
7. IAEA-TECDOC-1347, Consideration of external events in the design of nuclear facilities other than nuclear power plants, with emphasis on earthquakes, 2003

Chapter 13
Performance-Based Seismic Design of RC Structures

Payal Gwalani and Yogendra Singh

Abstract The concept of performance-based design (PBD) is to achieve the design of structures with a reliable understanding of the risk to life and accompanied losses that may occur due to future earthquakes. The design methodology is based on the assessment of a building's performance to determine its probability of experiencing different types of damage levels, considering a range of potential earthquakes that may affect the building structure. The performance objectives (or damage levels) such as immediate occupancy, life safety, or collapse prevention are used to define the damage state of the building. Thus, the methodology enables the owner with a means to select the desired performance goal of the building. The Indian seismic design code (IS 1893 Part 1 [6]), like most other national codes worldwide, provides simplified guidelines for the design of buildings intended to provide life safety performance level of the building against a design level earthquake. However, the prescriptive nature of these guidelines does not provide any framework to estimate the actual intended/expected seismic performance of such buildings. This article describes the technical basis of PBD and its application to building archetypes of moment frame and frame-shear wall buildings. The buildings are designed using the hazard and member design procedures of the relevant Indian codes.

13.1 Introduction

The performance of a building during an earthquake event is highly uncertain and is dependent on many factors (seismic source properties, intensity of ground motion, selection of structural system, configuration and proportion, quality of construction, reinforcement detailing, building maintenance, etc.). The effect of some of these parameters (selection of structural system, configuration and proportion, and reinforcement detailing) can be evaluated and controlled at the design stage. For this, different countries have developed seismic design guidelines based on the knowledge

P. Gwalani · Y. Singh (✉)
Department of Earthquake Engineering, IIT Roorkee, Roorkee, India
e-mail: yogendra.singh@eq.iitr.ac.in

T. G. Sitharam et al. (eds.), *Theory and Practice in Earthquake Engineering and Technology*, Springer Tracts in Civil Engineering,
https://doi.org/10.1007/978-981-19-2324-1_13

of the severity of earthquakes and structure's performance in the past events. The current guidelines in the seismic design codes, world-over, are traditionally based on the concept of force-based design. In this approach, each individual member of the structural system is proportioned for strength on the basis of internal forces developed during an elastic analysis. The inelasticity in the structure is incorporated by the use of 'response reduction factor' or 'behaviour factor', adopted based on some presumed inelastic energy dissipation capacity of the structure. The impact of building damage post-earthquake is accounted for by the use of 'importance factor'.

An effective interpretation and characterization of the damage, which occurred in the buildings during past earthquakes (in the mid and late twentieth century), revealed that the traditional seismic design methods could not efficaciously ensure the expected performance of buildings during such events [31]. The use of yielding strength (force) as the basis for design is not efficient enough, as there is no clear relation between strength and damage [29, 30]. To overcome the limitations imposed by the forced-based design methodology, an alternative design philosophy named 'Displacement-based design (DBD)' (Qi and Moehle [32]) was introduced, wherein, translational displacement, rotation, strain, etc. are included in the basic design criterion. A major development in this regard has been made by Priestley [30] and his group [31] in developing a practical methodology for DBD. In this approach, the inter-storey drifts and ductility demand are considered as governing parameters for ensuring the desired performance.

With the advancements in the field of earthquake engineering, FEMA 273 [14] introduced a new concept of seismic design, known as performance-based design (PBD), which enables the designer to design a structure for any targeted performance level. Buildings designed according to the current seismic design codes are intended to sustain some damage under a 'design level' earthquake and follow a 'No collapse' criterion for major earthquakes. However, the prescriptive nature of the code guidelines does not provide any framework for estimation of the actual intended/expected seismic performance of such buildings. On the contrary, PBD is a tool that allows the owner to set the intended performance objective (acceptable damage level) of the buildings. In PBD procedure, a realistic estimate of strength and ductility of the structural system is estimated, with the explicit consideration of non-linear deformation of the members. The performance objectives (or damage levels) are determined in terms of limits on the inelastic deformation of members (instead of force as a damage indicator in force-based design). A non-linear static analysis is imperative for such evaluation. Using this framework, the damage levels of the building can be used to estimate the potential casualties and accompanied losses using some empirical relations.

In the past two decades, the PBD procedure has undergone a significant development (FEMA 356 [15], FEMA 440 [16] ASCE 41-17 [3]) and has adopted a probabilistic framework in the last few years (FEMA P695 [17], FEMA P58 [18], FEMA P58 [19]). In its seminal form, the design methodology is based on the assessment of a building's performance to determine its probability of experiencing different types of damage levels by considering a suite of potential earthquake ground motions that may affect the building structure in future. This article provides guidance to the design professionals and individuals on the implementation of performance-based seismic design of buildings, in conjunction with the Indian codes of practice. To illustrate this, two structural systems, a moment frame building and a frame-shear wall building are considered. These buildings are designed and detailed using the current Indian seismic design codes IS 1893 Part 1 [6] and IS 13920 [5]. The seismic performance of these buildings is evaluated using non-linear static and dynamic analysis. The building models are developed and analyzed using the commercial software ETABS (CSI [11]), which is commonly used in the design offices in India.

13.2 PBD Procedure

In this study, the methodology to evaluate the performance of building models is separated into five stages: 1. Assessment and representation of ground shaking hazard, 2. Selection of performance objective(s), 3. Structural modelling of the building, 4. Non-linear analysis, and 5. Assessment of performance and iterative revision of design. A brief description of each of these stages is presented here. The process of performance-based design is briefly summarized in Fig. 13.1.

13.2.1 Assessment and Representation of Ground Shaking Hazard

Ground shaking hazard is specified in different ways depending on the type of non-linear analysis method used to quantify the building response. The PBD procedure involves defining the performance objectives in terms of performance levels of the building corresponding to different hazard levels. The hazard levels are represented in terms of the probability of occurrence of the specified intensity in the design life (usually 50 years). Two hazard levels—representing intensities corresponding to 10%

Fig. 13.1 Flowchart of the performance-based design process (FEMA P58 [18]

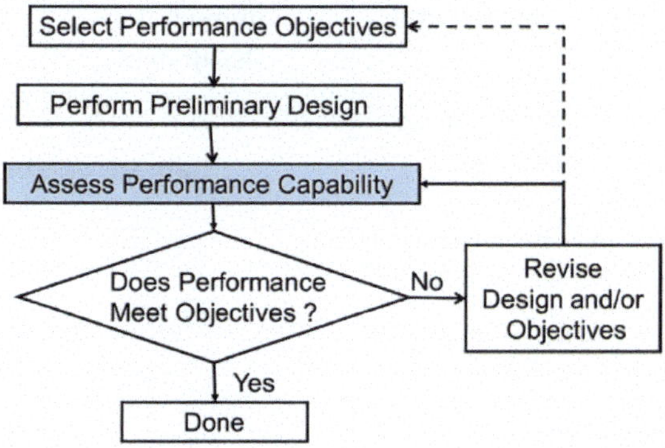

probability of exceedance in 50 years (termed as Design Basis Earthquake, DBE) and 2% probability of exceedance in 50 years (termed as Maximum Considered Earthquake, MCE)—are most commonly used across the world. Depending on the type of analysis, the ground shaking intensity corresponding to a chosen hazard level is represented by either a 5% damped elastic response spectrum (uniform hazard spectrum corresponding to a given probability of occurrence) or by a suite of ground motions, which have to be carefully selected and scaled to represent the chosen intensity.

13.2.2 Selection of Performance Objective(s)

A performance objective is the means to quantify the building's performance in terms that will be useful to the owner/designer to make the decision. ASCE 41-17 [3] defines discrete performance levels to evaluate the response of the building. These performance levels, namely, (Fig. 13.2) Operational (O), Immediate Occupancy (IO), Life safety (LS), and/or collapse prevention (CP), are defined by an appropriate acceptable range of strength and deformation demands on the structural and non-structural components of the structural system. The structural damage is represented by the inelastic rotation and deformation, while the non-structural damage is associated with the peak floor acceleration and inter-storey drift. These performance objectives are then utilized to estimate the probable losses to life, economy, and other physical losses to infrastructure/ community, using empirical relations. Table 13.1 summarizes different performance levels used in the literature.

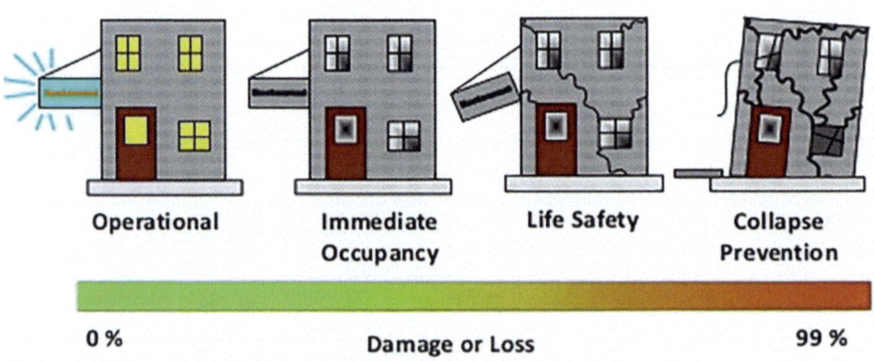

Fig. 13.2 Performance levels in performance-based design (NEHRP [25])

Table 13.1 Different performance levels considered in the literature

Document	Performance Levels	
	Structural	Non-structural
Vision 2000 (SEAOC [35])	Fully Operational Operational Life Safe Near Collapse	
SEAOC [37]	SP1, Operational SP2, Occupiable SP3, Life Safe SP4, Near Collapse SP5, Collapsed	NP1 (Damage Ratio 0%–10%) NP2 (Damage Ratio 5%–30%) NP3 (Damage Ratio 20%–50%) NP4 (Damage Ratio 40%–80%) NP5 (Damage Ratio >70%)
ATC 40 [4]	SP-1, Immediate Occupancy SP-2, Damage Control SP-3, Life Safety SP-4, Limited Safety SP-5, Structural Stability SP-6, Not Considered	NP-A, Operational NP-B, Immediate Occupancy NP-C, Life Safety NP-D, Hazard Reduced NP-E, Not Considered
BS EN 1998-3: [13]	Damage Limitation (DL) Significant Damage (SD) Near Collapse (NC)	
TSC [39]	Ready For Use Life Safety Pre-collapse Collapse	
Model Code for DBSD (Calvi and Sullivan [7])	Level 1, Serviceability Level 2, Damage Control Level 3, Collapse Prevention	
ASCE 41-17 [3]	S-1, Immediate Occupancy S-2, Damage Control (Range) S-3, Life Safety S-4, Limited Safety (Range) S-5, Collapse Prevention S-6, Not Considered	N-A, Operational N-B, Immediate Occupancy N-C, Life Safety N-D, Hazard Reduced N-E, Not Considered

13.2.3 Structural Modelling

It is to be noted that a building has some over-strength, over and above its design strength, due to the use of characteristic strength (95% confidence level) and partial safety factors in loads and materials. Hence, the actual expected strength of the building during an earthquake event is higher than the calculated design strength. Performance assessment of building structures involves evaluating the non-linear response of the structural model considering its expected strength (with the value of partial safety factors = 1). In order to evaluate the non-linear behaviour of a building

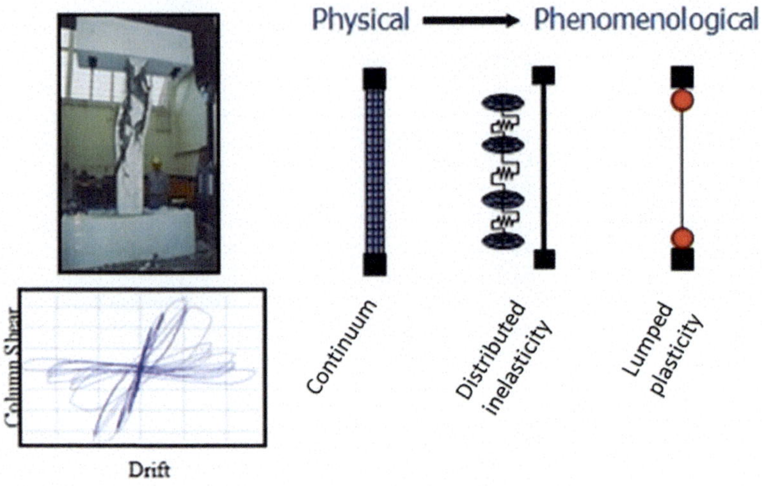

Fig. 13.3 Comparison of non-linear modelling approaches (PEER ATC-72-1) [27]

during a cyclic loading event, it becomes necessary to develop modelling strategies that can incorporate all modes that lead to the deterioration of the structural members. Three modelling approaches with different levels of accuracy and complexity are available in the literature: (i) Continuum model, (ii) Distributed inelasticity model, and (iii) Lumped plasticity model. Figure 13.3 illustrates the modelling strategy in the three different non-linear models.

In Continuum modelling, the non-linear behaviour of the structural members is evaluated by using finite element meshing throughout the section and length of the members to represent both concrete and steel reinforcement bars. The use of such explicit finite element models is mainly limited to detailed analysis of isolated critical elements (such as beam–column joints, etc.,) due to its high complexity, time consumption, and computational cost.

In distributed inelasticity model, several cross-sections with finite length are considered throughout the length of the member. At these cross sections, the non-linearity of the member is studied by discretizing the section into several fibres of confined and unconfined concrete and reinforcing steel bar. Each fibre in the cross section is assigned the corresponding material stress–strain relationship. The number of fibre cross-sections and their location is determined by the numerical integration rule, which implicitly defines the plastic hinge length over which the inelasticity of the element can occur. This model has an advantage over lumped plasticity model as it does not restrict the inelasticity at the ends of the element, but it still requires a very large computational capacity and computation time (though less than continuum models). Also, it is difficult to account for flexural–shear interaction, bond-slip, and rebar buckling and fracture using typically distributed plasticity models. However, current research is widely based on resolving these issues [8, 9, 22, 24, 28, 33], but their application is limited, in particular reference to its implementation in available commercial software packages.

Fibre elements are modelled using two different finite element approaches, namely, displacement-based approach (e.g., Zeris and Mahin [40]) and force-based approach (e.g., Neuenhofer and Filippou [26], Spacone et al. [36]). The displacement-based fibre element (DB) is based on the element's stiffness, while the force-based fibre element (FB) is based on the element's flexibility. In DB element, displacement interpolation functions are used, i.e., cubic and linear shape functions, respectively, for the transverse and axial displacements. These assumed shape functions are approximate representations of the actual displacement, and therefore, a number of elements in a structural member are required to obtain the actual response [10]. These functions are then used to determine the element displacement field by interpolating the nodal displacements. Strain at any point in the cross section is determined by nodal displacements. The stresses are then obtained from the strains using the material stress–strain relationship. This approach has the limitation that the element behaviour is characterized by linear curvature and constant axial strain fields. The non-linear curvature behaviour cannot be represented accurately. Also, the equilibrium condition in each section may not be satisfied.

In force-based fibre element, force interpolation functions (shape functions) are used to determine the section deformations. Strain in the cross section is obtained based on plane section remains plane assumption. Stress and stiffness are obtained from strains using the material stress–strain relationship. The section resisting forces are computed from the fibre stress distribution, and the stiffness matrix is obtained from the fibre stiffness. The FB elements have an advantage over DB elements, in that, the non-linear behaviour of members can be modelled using a single element. This is because the force field gives the exact solution of the governing equilibrium equations irrespective of the occurrence of plastic deformations.

In lumped plasticity model, an elastic line element is used to simulate the behaviour of the member, while yielding is assumed to take place at generalized plastic hinges of zero-length at the ends or other potential plastic hinge locations in a member. The plastic hinge modelling assumption reduces the computational demand and provides a balance between accuracy and complexity level. These models can be easily used to calibrate and capture the observed non-linear behaviour of RC members, i.e., from the start of yielding to the residual strength, including strength and stiffness degradation from concrete crushing and spalling, rebar buckling and fracture, and bond-slip. The non-linearity in lumped plasticity models can be considered by using either of the two hinge models, namely, moment-rotation hinge and fibre-hinge.

The moment-rotation hinge models are based on the definition of phenomenological relation of the overall force–deformation response or moment-rotation (curvature) of structural members (beam, column, shear wall), as observed from the experimental test results (PEER ATC-72-1 [27]. Based on various experimental results on the cyclic behaviour of beam–column elements, various documents have proposed the force–deformation cyclic backbone curves [3, 15, 21]. Figure 13.4 shows the backbone curve as defined in ASCE 41-17 [3]. The first branch of the backbone

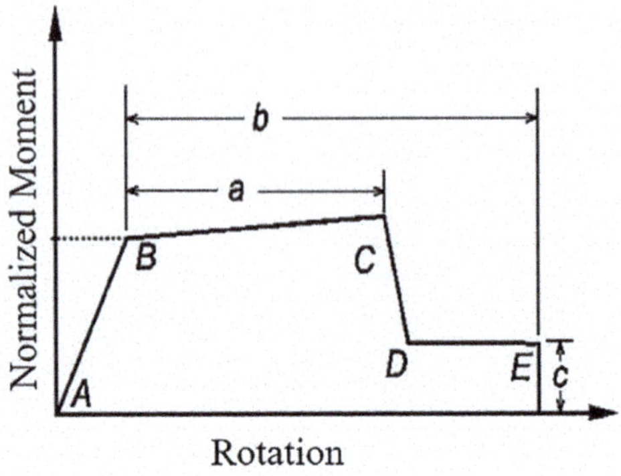

Fig. 13.4 Force–deformation curve (backbone curve) for frame elements in flexure [3]

curve (AB) represents the elastic behaviour of the member, the second branch (BC) represents the post-yield behaviour of the member, and the third branch (CDE) represents the post-peak behaviour of the member, where the strength of the member drops due to strain softening and rebar buckling.

In the fibre-hinge model, fibre-monitoring cross sections are considered at the plastic hinge locations, where the non-linearity of the member is assumed to be lumped. The discretization of monitored sections (into fibres of confined and unconfined concrete and rebars) yields an accurate response by obtaining the P-M and P-M-M interaction of beams and columns, respectively, directly by integration of the inelastic material response. The plastic hinge length for fibre-hinge is obtained from empirical relations derived from experimental test results. Figure 13.5 shows a typical discretization pattern of cross section at the fibre-hinge. The stress–strain curve of concrete should account for the effect of confinement caused by the transverse reinforcement. The cyclic response of materials can be modelled using a suitable hysteresis model duly calibrated with relevant laboratory test data.

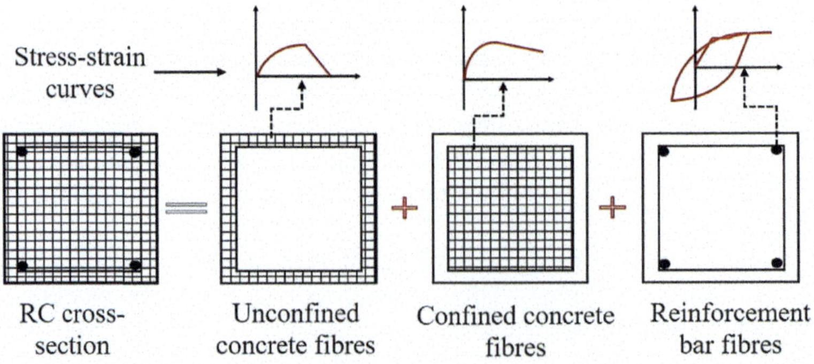

Fig. 13.5 Typical discretization pattern of a RC cross section at the fibre-hinge

13.2.4 Non-linear Analysis

In order to achieve the targeted performance objectives, PBD procedure involves estimation of the performance level of the structure. The performance of the structure can be estimated using either non-linear static analysis or non-linear dynamic analysis.

In non-linear static analysis (NSA), the capacity curve (popularly known as 'pushover curve') of the building, which represents the plot between the base shear and roof displacement, is obtained under an assumed distribution of lateral load. The lateral load distribution is considered as an approximate representation of the inertial forces developed in the structure under an actual earthquake loading event. The magnitude of the lateral load is increased monotonically to identify the critical sections (or weak links) of the building and study the failure mode. The accuracy of the pushover analysis depends upon the selection of an ideal lateral load distribution pattern (that can consider the fundamental and higher mode effects). It is a simplified method that provides useful information on the yield strength, failure mechanism, and ductility of the structure. Due to these reasons, design engineers prefer to use the pushover analysis method. However, the method is highly dependent on the loading vector and does not account for the record-to-record variability in the ground motion. Also, the method does not incorporate the biaxial effects of loading and torsional response.

The new generation performance-based design recommends the evaluation of building performance using non-linear dynamic analysis. The method provides a more accurate estimate of structural behaviour (compared to pushover analysis) since the response is obtained by direct application of earthquake ground motion. Due to the inherent variability in the nature of earthquake ground motions, NLTHA is carried out for multiple ground motions to obtain a statistically robust output. ASCE 7-16 [2] recommends a minimum no. of 11 pairs of ground motion (i.e., a 22 GM record set, having two orthogonal components of motion) for NLTHA. The selected ground motions should be representatives of the events with the same tectonic regime, consistent magnitudes, and rupture characteristics as the events that dominate the target response spectrum. Also, ground motions should match the spectral shape of the target spectrum in the range of $0.2\ T_L - 2\ T_U$, where T_L and T_U are the smaller and larger fundamental periods, respectively, in the two orthogonal directions of the building.

13.2.5 Assessment of Performance and Iterative Revision of Design

It is necessary to assess if the designed building meets the performance objectives targeted by the owner. For a building to comply with a chosen performance level, all structural members must satisfy the prescriptive acceptance criteria. For structural members, where moment-rotation hinge is used to define non-linearity, the acceptable limits are defined in terms of plastic rotation. ASCE 41-17 [3] provides these acceptable plastic rotation limits based on updated and latest experimental test studies. In case if distributed inelasticity model or fibre-hinge lumped plasticity model is used to simulate the non-linear response of RC members, where plastic rotation of members is not directly obtained, the performance can be estimated in terms of material strain in concrete and steel. These strain limits can be obtained from relevant experimental test results and analysis studies. Table 13.2 shows the limit values of strains corresponding to different performance levels taken from the Turkish seismic code [39] for beams and column members. For structural walls modelled, Akelyan et al. [1] recommends that the maximum compressive strain in concrete and the maximum axial tension strain in steel as obtained from distributed inelasticity or fibre-hinge models should not exceed 0.005 and 0.01, respectively. In addition to the member-level acceptance criteria, the global structural performance of RC buildings can also be evaluated in terms of maximum inter-storey drift. Table 13.3 provides the maximum permissible transient and permanent drift limits recommended by FEMA 356 [15].

In case the actual performance of the building does not comply with the targeted objectives, the next step would be to revise the design so as to achieve the target. Once the design is revised, the aforementioned steps are repeated again to assess the

Table 13.2 Acceptance criteria in terms of strain limit values for the concrete and reinforcing steel for distributed plasticity or fibre models for RC beams and columns [39]

Performance level	Strain limit values	
	Concrete	Reinforcement steel
Immediate occupancy (IO)	$\varepsilon_{cu} = 0.0035$	$\varepsilon_s = 0.010$
Life safety (LS)	$\varepsilon_{cg} = 0.0035 + 0.01\left(\rho_s/\rho_{sm}\right) \le 0.0135$	$\varepsilon_s = 0.040$
Collapse prevention (CP)	$\varepsilon_{cg} = 0.004 + 0.014\left(\rho_s/\rho_{sm}\right) \le 0.0180$	$\varepsilon_s = 0.050$

Note ε_{cu} = ultimate strain in unconfined concrete, ε_{cg} = ultimate strain in confined concrete, ε_s= strain in reinforcement steel, $\rho_s$and ρ_{sm} = Volumetric ratio of special seismic hoops and crossties, and transverse reinforcement, respectively

Table 13.3 Acceptance criterion in terms of maximum permissible transient and permanent drifts [15]

Structural system	IDR(%) for performance level					
	IO		LS		CP	
	Tran	Perm	Tran	Perm	Tran	Perm
RC moment frame	1.0	0.0	2.0	1.0	4.0	4.0
RC moment frame with URM infills	0.1	0.0	0.5	0.3	0.6	0.6
RC moment frame with shear wall (Dual system)	0.5	0.0	1.0	0.5	2.0	2.0

performance of the building. Hence, it is an iterative process that is followed till the targeted performance objectives are achieved.

13.3 Design Example

In the present study, two RC building models with different structural configurations, i.e., one with a moment frame and another one with frame-shear wall are designed to illustrate the above-mentioned PBD procedure. These building models are designed in confirmation with the current Indian seismic design codes—IS 1893 Part 1 [6] and IS 13920 [5]. The results from the present study illustrate the expected performance of the code-complaint buildings for the code-specified earthquake hazard level. The design and analysis of the building models are carried out in the commercial building design software ETABS 2016 [11].

13.3.1 Building Model Description

General Properties

Two 8-storey RC building models i.e., a moment frame and a frame-shear wall building, with plan dimensions as shown in Fig. 13.6, are considered. Both the models have a constant storey height of 3.3 m and a plinth level of 1.5 m. The general properties of the considered buildings are presented in Table 13.4. The period of vibration (1st and 2nd mode) and the member sizes for the considered building models are presented in Table 13.5.

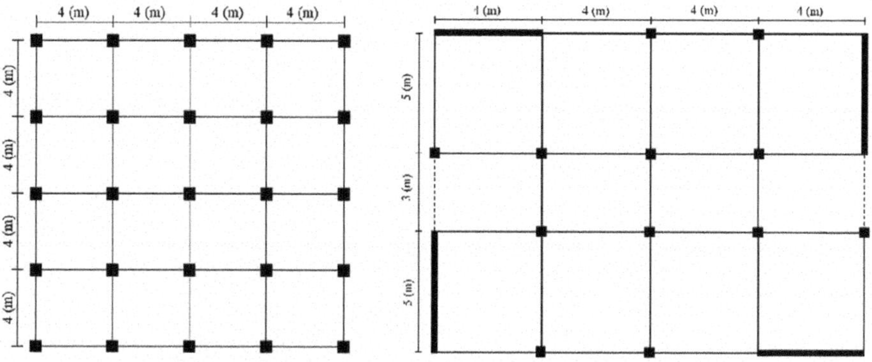

Fig. 13.6 Plan and elevation of the building models considered

Table 13.4 General properties of considered buildings

General	Design level	Bare Frame designed for Gravity and Earthquake Loads and detailed as Special Moment Resisting Frame, according to IS 1893 Part 1 [6] and IS 13920 [5] ductile detailing provisions to avoid shear failure and strong column-weak beam ratio greater than 1.4
	No. of storeys	8-storey
	cracked section properties	Considered from ASCE 41-17 [3]
Seismic Hazard	Soil type	Soil Type I (IS: 1893 [6])
	Seismic zone	Seismic Zone V (IS: 1893 [6]) ($EPGA_{MCE} = 1.5 \times 0.36$ g = 0.54 g)
Material	Concrete	Nominal (characteristic) cube strength = 40 MPa
	Steel	Nominal (characteristic) Yield Strength = 500 MPa
Loading	Dead load	Self-weight of members Weight of slabs and floor finish Weight of (230 mm exterior and 120 mm interior) infill walls Weight of 1 m high and 115 mm thick masonry parapet wall
	Live load	4 kN/m^2 on corridor 3 kN/m^2 on other floor areas 1.5 kN/m^2 on roof
	Design load combinations	According to Clause 6.3.4.1 of IS 1893 Part 1 [6]
Structural Modelling	Software used	ETABS (CSI [11])
	Structure model	Space frame model
	Element model	3D frame elements for beams and columns Wide column model for shear wall Slabs as rigid diaphragms
	P-delta effects	Considered in analysis and design

Table 13.5 Details and member sizes of the considered building models

Building model	Period of vibration (s)		Member sizes (mm)		
	X	Y	Beam	Column	Shear wall
Moment frame building	2.75	2.75	300 × 400	400 × 400 350 × 350*	–
Frame-shear wall building	1.71	1.64	300 × 400	400 × 400 350 × 350*	5000 × 150 4000 × 150

* Column size reduction after 4th floor

Non-linear Modelling

Before non-linear modelling and analysis of the buildings, the buildings are analyzed using the mode-superposition (response spectrum) method with an appropriate response reduction factor (R = 5) as per IS 1893 Part 1 [6]. In analysis, an effective cracked moment of inertia of beam–columns and shear walls is considered. It is important to consider the P-Delta effect and rigid diaphragm action of floor and roof slabs, in the design and analysis. Finite size of beam–column and beam–shear wall joints is also considered in the modelling. The beams, columns, and shear walls are designed as per relevant IS codes, and detailing is performed as per IS 13920 [5] to obtain the details of longitudinal and transverse reinforcement in each member. This information is necessary for assigning the plastic hinges for non-linear analysis.

Lumped plasticity model is used to define the non-linearity in the elements. Uniaxial moment-rotation plastic hinges (M3) are assigned at both ends of the beams (at a relative distance of 0.05 and 0.95). The backbone curve (i.e., force–deformation envelopes of beams) and the acceptable deformation limits for various performance levels (i.e., IO, LS and CP) are obtained from ASCE 41-17 [3]. The cyclic properties of the moment-rotation hinges are modelled using an energy-based degrading hysteresis model [11]. The details of the model can be found in the CSI [11] manual. Further details of the modelling and calibration can be found in Surana et al. [38]. The properties of the model (f_1, f_2 and s) for ductile beams are taken from Surana et al. [38].

The non-linearity in columns and shear walls is modelled using the fibre-hinge model. The fibre-hinge is assigned at the two ends of the element, located at the midpoint of the plastic hinge length measured from the face of the connecting element. At each plastic hinge location, the section is discretized into fibres for confined concrete, unconfined concrete, and one steel fibre per reinforcing bar. Plastic hinge length for columns is taken as half of the maximum horizontal dimension of the column. Plastic hinge length for the shear wall is taken from the empirical relation given by Priestley et al. [31]. Mander's model [23] is used for defining the stress–strain relationship between confined and unconfined concrete. The stress–strain curve of the steel reinforcing bar is assumed a bilinear elastic–plastic model with kinematic strain-hardening. The performance level (IO, LS, and CP) in case of fibre-hinge model is indicated by the strain in the extreme fibre. These acceptable values of strains corresponding to different performance levels are taken from the Turkish code [39]. The cyclic properties of the materials are modelled using an energy-based degrading hysteresis model [11]. The properties of the energy hysteresis model for concrete and steel have been calibrated for columns and shear walls using the available experimental test results of Rodrigues et al. [34] and Dazio et al. [12], respectively. Figure 13.7 shows the calibrated force–deformation hysteresis curves of column specimen PB01-N09 [34] and shear wall specimen WSH3 [12]. Further details of the calibration process can be found in Gwalani et al. [20].

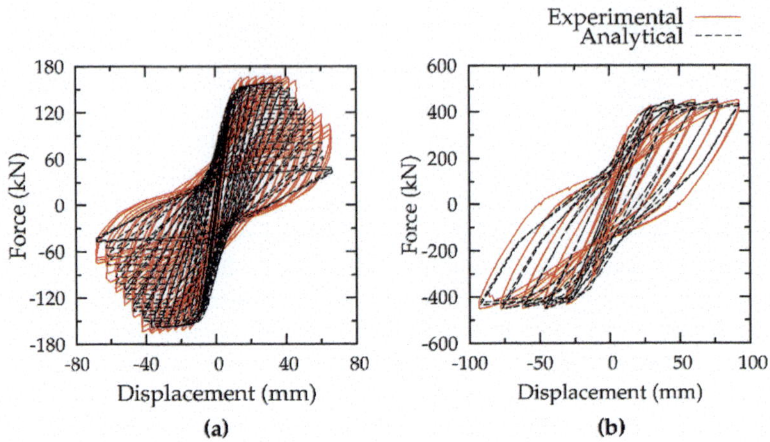

Fig. 13.7 Calibrated force–deformation response using fibre-hinge of: **a** column specimen PB01-N09 [34] and **b** shear wall specimen WSH3 [12]

Non-linear Analysis

In the present study, both Pushover analysis and NLTHA of both the building models are carried out. For the pushover analysis, the lateral load distribution proportional to the fundamental mode of the building is considered. The performance point is computed using the 'Displacement Modification Method (DMM)' specified in clause 7.4.3.3 of ASCE 41-17 [3]. For time history analysis of both the buildings, a far-field ground motion with two horizontal components, from FEMA P695 [17] and as shown in Fig. 13.8, is considered. The selected time history is scaled in the time domain (i.e., using a constant scale factor) to have the same spectral acceleration as the MCE response spectrum (corresponding to $1.5 \times Z$) at the fundamental period of the building, and is applied at the base of the building. The roof displacement and hinge pattern are evaluated to compare the results of time history with the pushover analysis. In case of NLTHA, a Rayleigh damping model is used, with a damping ratio of 5% assigned at the fundamental mode and the mode corresponding to 90% of modal mass participation. It is to be noted that as per ASCE 7-16 [2], minimum 11-time histories are to be used to estimate the average response. In the present study, only one-time history has been used for the purpose of illustration only.

13.3.2 Results

13.3.2.1 Moment Frame Building

Figure 13.9 shows the capacity curve obtained from the analysis of the moment frame building. Since it's a bisymmetric building, both directions have the same pushover

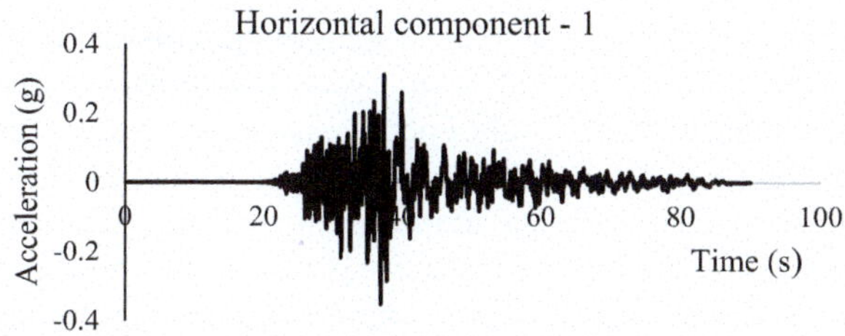

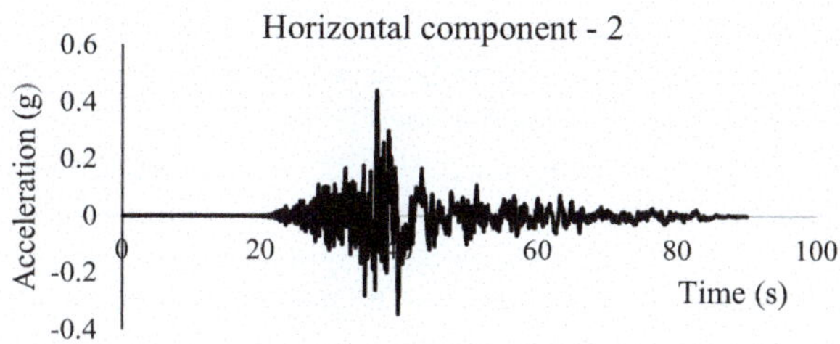

Fig. 13.8 Earthquake ground motion (Chi-Chi, Taiwan earthquake (1999)) considered in NLTHA analysis

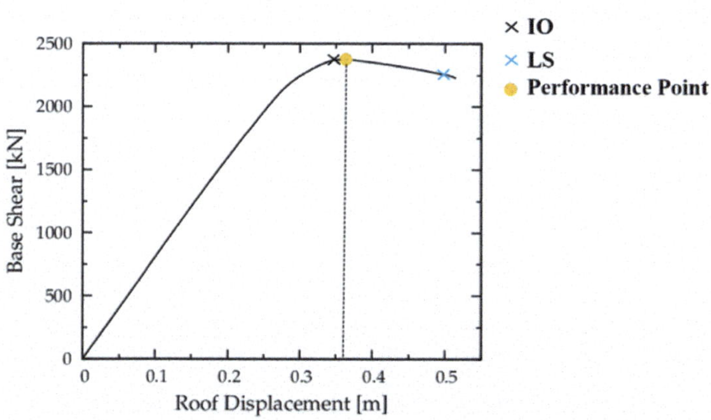

Fig. 13.9 Capacity curve of RC moment frame building obtained from pushover analysis

curve, and only one is shown here. The figure also shows the performance point at MCE and the performance levels IO and LS. The performance levels are obtained from ASCE 41-17 [3]. It is important to note that ASCE 41-17 [3] specifies the acceptable limit of rotation for each individual member, and not for the structural system as a whole. The different performance levels on the capacity curve are marked by identifying the roof displacement corresponding to the pushover step, at which the first member in the building crosses the rotation limit corresponding to the concerned performance level. It is observed that the RC bare moment frame building designed in confirmation to relevant Indian codes can sustain MCE with LS performance level. This means that the building has sufficient over-strength and ductility to survive, without collapse (in fact with LS level of performance), the MCE (1.5 × Z) level of ground shaking. Figure 13.10a and b show the hinge pattern of the moment frame building at the performance point and at the last step of the pushover curve (failure point). It is observed from the Figures that at the performance point, some beams have hinges that cross the IO performance level, whereas failure of the building occurs due to the formation of beam mechanism (formation of hinges in lower 4-storey beams and formation of hinges in ground storey columns).

Figure 13.11 shows the capacity curve of the building with the performance points obtained from NSA and NLTHA. It can be seen that there is a considerable difference in the roof displacement values obtained from pushover analysis and NLTHA. This difference is due to the fact that pushover is an approximate static analysis method that considers only the fundamental mode of vibration of the building and ignores the dynamic effect of higher modes. Figure 13.12 shows the hinge pattern at the performance point obtained from the NLTHA. A comparison of Figs. 13.10 and 13.12 highlights the difference in the performance of the building obtained using the two analyses. In case of pushover analysis, only a few beams undergo inelastic deformation, whereas a number of members develop plastic hinges in case of NLTHA. Another important observation from the hinge pattern under NLTHA is that even though the building was designed with strong columns and weak beams, with a strength ratio of 1.4, hinges are still formed in some of the columns.

Granted the results presented here are only for a single time history, but it still highlights that the response obtained from pushover analysis is not sufficient enough to adequately judge the performance of the building. A number of time history analyses should also be carried out to reliably estimate the seismic performance of the building.

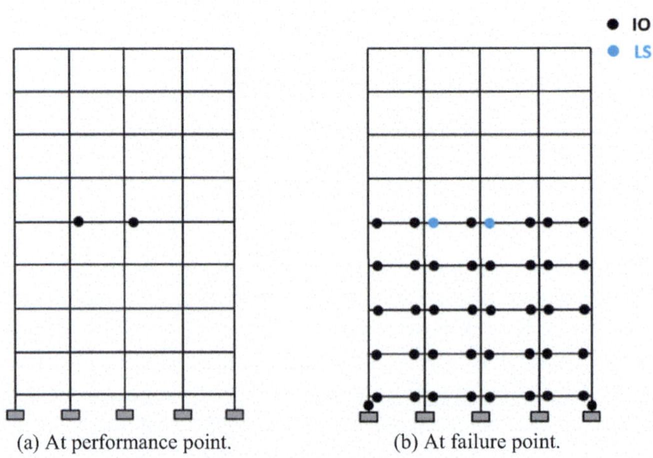

Fig. 13.10 Hinge pattern of RC moment frame building obtained from pushover analysis

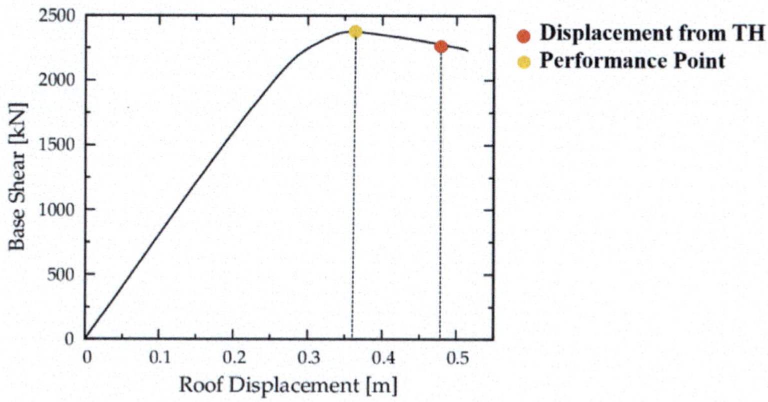

Fig. 13.11 Comparison of performance points obtained from pushover analysis and displacement from time history for RC moment frame

Fig. 13.12 Hinge pattern at Performance Point of RC moment frame building obtained from the non-linear time history analysis (NLTHA)

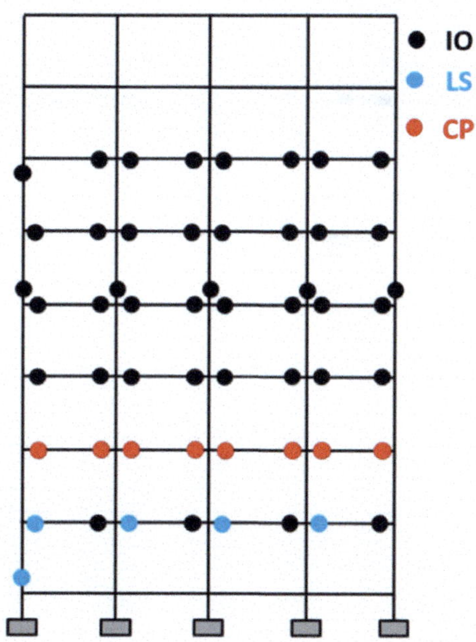

13.3.2.2 Frame-Shear Wall Building

Figure 13.13 shows the capacity curves obtained from the pushover analysis of the frame-shear wall building in the two directions. Similar to moment frame results, the figure also shows the performance point at MCE and the performance levels IO, LS, and CP. It is observed that frame-shear wall building designed conforming to Indian standard codes also has an LS level performance. The same can be observed

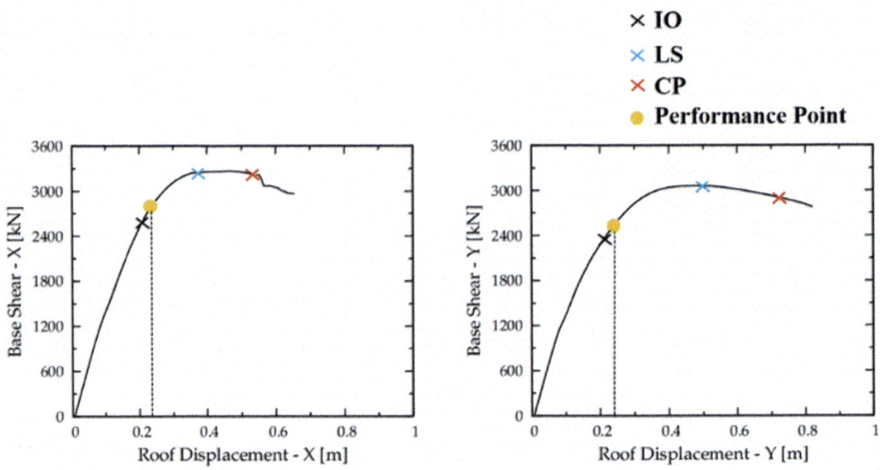

Fig. 13.13 Capacity curve of RC frame-shear wall building obtained from pushover analysis

from the hinge pattern at the performance point shown in Fig. 13.14. The yield in frame-shear wall building starts with hinge formation in the shear wall. Figure 13.15 shows the hinge pattern at the last step of the pushover curve (failure point). It can be observed that failure in case of frame-shear wall building occurs due to the failure of the shear wall (formation of CP hinge in the shear wall). Figure 13.16 shows the capacity curve of the building with the performance point obtained from NSA and

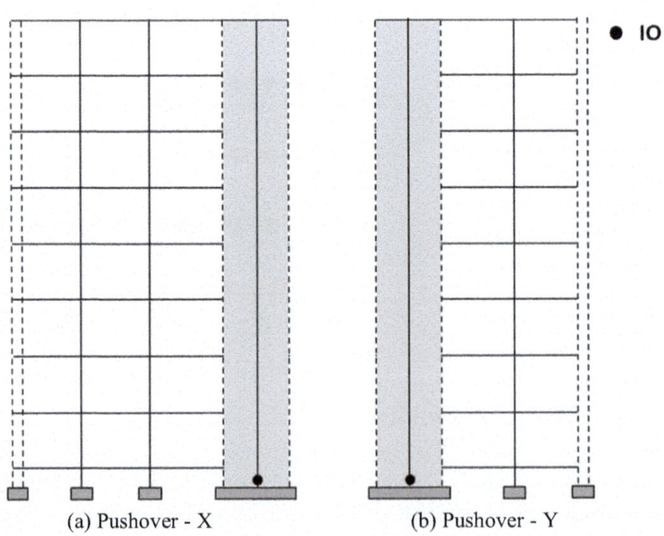

Fig. 13.14 Hinge pattern of RC frame-shear building at performance point obtained from pushover analysis

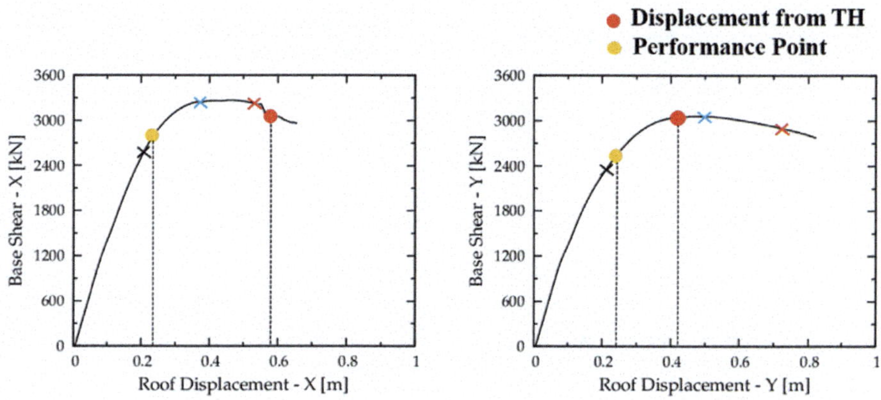

Fig. 13.15 Hinge pattern of the RC frame-shear building at failure point obtained from pushover analysis

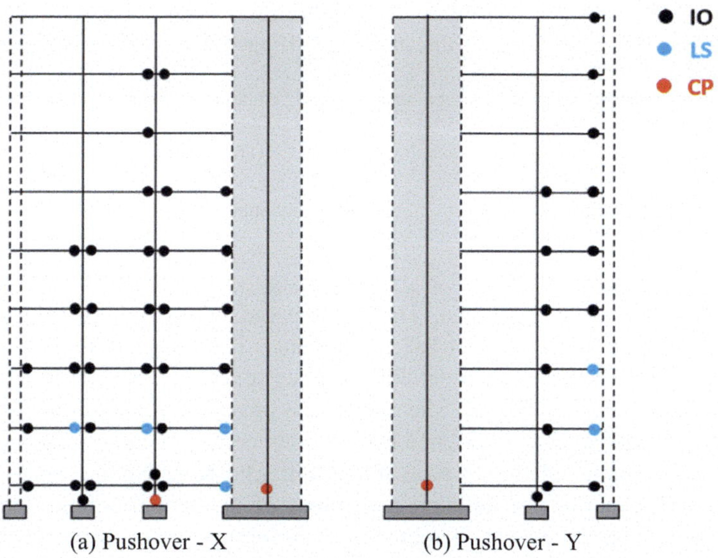

Fig. 13.16 Comparison of performance points obtained from pushover analysis and non-linear time history analysis of the RC frame-shear wall building

the NLTHA. Similar observations as in the case of frame building can be made here as well. Also, the hinge pattern in Fig. 13.17 shows that in case of NLTHA, a larger number of beams show plastic hinges, and many of those show collapse. In this case also, some of the columns develop plastic hinges, even with the strong column-weak beam ratio of 1.40, used in the design.

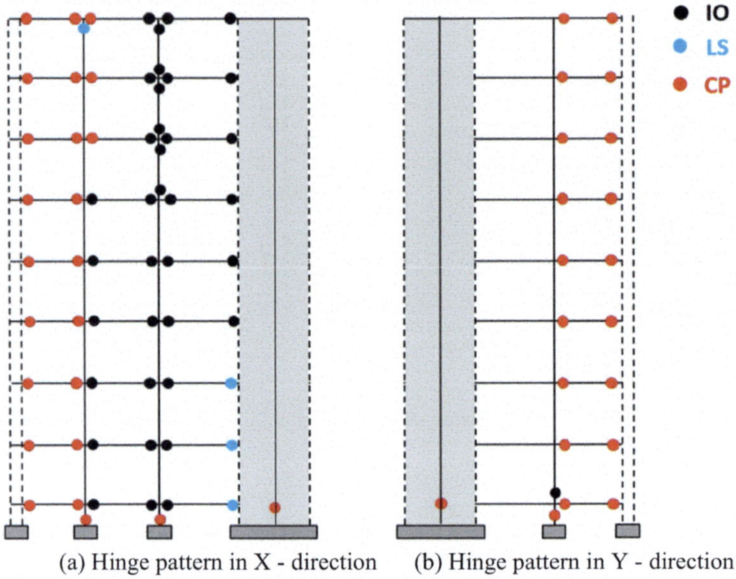

Fig. 13.17 Hinge pattern of the RC frame-shear building obtained from non-linear time history analysis

13.4 Conclusions

An iterative procedure of performance-based design of RC frame and frame-shear wall buildings has been illustrated. The relevant Indian codes have been used to estimate the hazard and design of RC components. The non-linear modelling has been performed considering the modelling and acceptance criteria of ASCE 41-17 [3] and Turkish codes (TSC [39]. The article also provides an idea of the expected performance of the buildings designed using Indian codes. The approach presented is general and is applicable to any type of building. The procedure has been demonstrated through examples of building archetypes of moment frame buildings and frame-shear wall buildings, which are the most common representatives of contemporary buildings in India. A comparison of NSA and NLTHA methods of PBD has shown that the NSA being based on the fundamental mode of structure alone is approximate and may not provide an adequate assessment of expected performance. On the other hand, NLTHA considering the effect of all the modes of vibrations provides an insight into all the possible failure modes of structure, which may not be illustrated by NSA.

References

1. Akelyan MS, Brandow G, Carpenter L, Ekwueme C, Ghodsi MT, Islam S et al. (2020) An alternative procedure for seismic analysis and design of tall buildings located in the Los Angeles region. 2020 Edition, Los Angeles Tall Buildings Structural Design Council
2. ASCE, SEI 7-16 (2016) Minimum design loads for buildings and other structures. American Society of Civil Engineers, Reston
3. ASCE, SEI 41-17 (2017) Seismic rehabilitation of existing buildings. American Society of Civil Engineers, Reston
4. ATC-40 (1996) Seismic evaluation and retrofit of reinforced concrete buildings. Applied Technology Council, Redwood City, California
5. BIS (2016) Ductile design and detailing of reinforced concrete structures subjected to seismic forces – Code of practice – IS 13920. Bureau of Indian Standards, New Delhi
6. BIS (2016b) Indian standard criteria for earthquake resistant design of structures, Part 1: General provisions and buildings (Sixth revision) IS 1893 Part 1. Bureau of Indian Standards, New Delhi
7. Calvi GM, Sullivan TJ (2009) A model code for the displacement-based seismic design of structures. IUSS Press, Pavia
8. Ceresa P, Petrini L, Pinho R (2007) Flexure-shear fiber beam–column elements for modeling frame structures under seismic loading—state of the art. J Earthq Eng 11(S1):46–88
9. Ceresa P, Petrini L, Pinho R, Sousa R (2009) A fibre flexure–shear model for seismic analysis of RC-framed structures. Earthq Eng Struct Dyn 38(5):565–586
10. Coleman J, Spacone E (2001) Localization issues in force-based frame elements. J Struct Eng 127(11):1257–1265
11. CSI (2016) ETABS 2016 Integrated building design software. Version 16.0.1, Computers and Structures Inc., Berkeley, U.S.A
12. Dazio A, Beyer K, Bachmann H (2009) Quasi-static cyclic tests and plastic hinge analysis of RC structural walls. Eng Struct 31(7):1556–1571
13. EN 1998-3 EC-8 (2004) Design of structures for earthquake resistance- Part 1: Assessment and retrofitting of buildings. European Committee for Standardization, Brusells
14. FEMA 273 (1997) Guidelines for seismic rehabilitation of buildings. Federal Emergency Management Agency, Washington, D.C., United States
15. FEMA 356 (2000) Prestandard and commentary for the seismic rehabilitation of buildings. Federal Emergency Management Agency, Washington, D.C., United States
16. FEMA 440 (2005) Improvement of inelastic analysis procedures. Federal Emergency Management Agency, Washington, D.C., United States
17. FEMA P695 (2009) Quantification of seismic performance factors. Federal Emergency Management Agency, Washington, D.C., United States
18. FEMA P58 (2012) Seismic performance assessment of buildings -Volume 1 methodology. Federal Emergency Management Agency, Washington, D.C., United States
19. FEMA P58 (2018) Seismic performance assessment of buildings -Volume 1 methodology. Federal Emergency Management Agency, Washington, D.C., United States
20. Gwalani P, Singh Y, Varum H (2021) Effect of proportioning of lateral Stiffness in orthogonal directions on seismic performance of RC buildings. J Earthq Eng. https://doi.org/10.1080/13632469.2021.1964649
21. Haselton CB, Liel AB, Taylor Lange S, Deierlein GG (2007) Beam column element model calibrated for predicting flexural response leading to global collapse of RC frame building. PEER Report 2007/03, PEER Center, University of California, Berkeley, California, United Sates
22. Kagermanov A, Ceresa P (2017) Fiber-section model with an exact shear strain profile for two-dimensional RC frame structures. J Struct Eng 143(10):04017132
23. Mander JB, Priestley MJ, Park R (1988) Theoretical stress-strain model for confined concrete. J Struct Eng 114(8):1804–1826

24. Mergos PE, Kappos AJ (2008) A distributed shear and flexural flexibility model with shear–flexure interaction for R/C members subjected to seismic loading. Earthq Eng Struct Dyn 37(12):1349–1370
25. NEHRP FEMA 451B (2006) Recommended provisions: instructional and training materials. Washington DC
26. Neuenhofer A, Filippou FC (1997) Evaluation of nonlinear frame finite-element models. J Struct Eng 123(7):958–966
27. PEER / ATC-72-1 (2010) Modelling and acceptance criteria for seismic design and analysis of tall buildings. Applied Technology Council, Redwood City, California, United Sates
28. Petrangeli M, Pinto PE, Ciampi V (1999) Fiber element for cyclic bending and shear of RC structures. I: Theory. J Eng Mech 125(9):994–1001
29. Priestley MJN (1993) Myths and fallacies in earthquake engineering. Bulletin of NZ Nat Soc for Earthq Eng 26(3):329341
30. Priestley MJN (2000) Direct displacement based design. Paper presented at 12th Conference on Earthquake Engineering, Auckland, New Zealand, 30 January-4 February 2000
31. Priestley MJN, Calvi GM, Kowalsky MJ (2007) Displacement based seismic design of structures. IUSS Press, Pavia
32. Qi X, Moehle JP (1991) Displacement design approach for reinforced concrete structures subjected to earthquakes, vol UCB/EERC-91/02 Earthquake Engineering Research Center, College of Engineering/University of California
33. Ranzo G, Petrangeli M (1998) A fibre finite beam element with section shear modelling for seismic analysis of RC structures. J Earthq Eng 2(03):443–473
34. Rodrigues H, Dias Arede A, Varum H, Costa A (2013) Experimental evaluation of rectangular reinforced concrete column behaviour under biaxial cyclic loading. Earthq Eng Struct Dyn 42(2):239–259
35. SEAOC Vision 2000 Committee (1995) Performance-based seismic engineering. Structural Engineers Association of California, Sacramento, California
36. Spacone E, Ciampi V, Filippou FC (1996) Mixed formulation of nonlinear beam finite element. Comput Struct 58(1):71–83
37. SEAOC Seismology Committee (1999) Recommended lateral force requirements and commentary. Structural Engineers Association of California, Sacramento, California
38. Surana M, Singh Y, Lang DH (2017) Seismic characterization and vulnerability of building stock in hilly regions. Nat Haz Rev 19(1):04017024
39. Turkish Seismic Code (2007) Specifications for buildings to be built in seismic zones. Ministry of Public Work and Settlement Government of Republic of Turkey, Ankara, Turkey
40. Zeris CA, Mahin SA (1988) Analysis of reinforced concrete beam-columns under uniaxial excitation. J Struct Eng 114(4):804–820

Chapter 14
Comparative Analysis of SSR and HVSR Method for Site Response Analysis

Sasanka Borah, Jayanta Pathak, and Diganta Goswami

Abstract The Northeast part of India falls under seismic Zone V (IS:1893–2002), the highest seismic activity zone. However, there is a lack of sufficient dense seismic arrays in the region to record the seismic activity. Irrespective of the availability of the seismic records this study attempts to decsribing the two site response analysis methods: Standard Spectral Ratio (SSR) and Horizontal to Vertical Spectral Ratio (HVSR). Three locations in the west Guwahati Region of the state of Assam in Northeast India, namely, Boko-Palashbari, Goalpara and Guwahati Central Region, have been considered for this study in order to analyze available strong motion data and quantify the site response. The quantification is attempted in terms of site amplification. This chapter also provides a comparison between the two methods under study. 5(five) and 15(fifteen) strong-motion recordings of earthquake events have been considered in this study for the SSR and HVSR methods respectively. The strong motion as recorded in Nongstoin has been considered as a reference site station record for the SSR study. This finding of this study compliments the findings of Field and Jacob (Field and Jacob in Bull Seismol Soc Am 85:1127–1143, 1995) that, if the SSR estimates are taken as the most reliable, the HVSR method under-predicts the site response. The results provided by HVSR are less than that of the SSR results by a factor of 2–4.

Keywords Strong motion · SSR · HVSR · Site amplification

S. Borah (✉) · J. Pathak · D. Goswami
Department of Civil Engineering, Assam Engineering College, Guwahati 781013, Assam, India
e-mail: sasankaborah.ce@aec.ac.in

J. Pathak
e-mail: jayantapathak.ce@aec.ac.in

D. Goswami
e-mail: digantagoswami2@gmail.com

T. G. Sitharam et al. (eds.), *Theory and Practice in Earthquake Engineering and Technology*, Springer Tracts in Civil Engineering,
https://doi.org/10.1007/978-981-19-2324-1_14

14.1 Introduction

The earthquakes have always surprised the engineers with their random, nature and variable damage potential for similar magnitudes and epicentral distances. It has been observed over the years that the damage potential varies from place to place within a similar radius from the source of the earthquakes. The various studies on the distribution of damages due to earthquakes have shown that the effects of the geology of the area on ground shaking represent an important factor in engineering seismology as well as in engineering design. Figure 14.1 schematically shows the general wave propagation when an earthquake occurs. It is seen in general that waves generated at the source propagate through the earth and, in general, show a trend of decreasing amplitude with increasing distance if there is no local site effect.

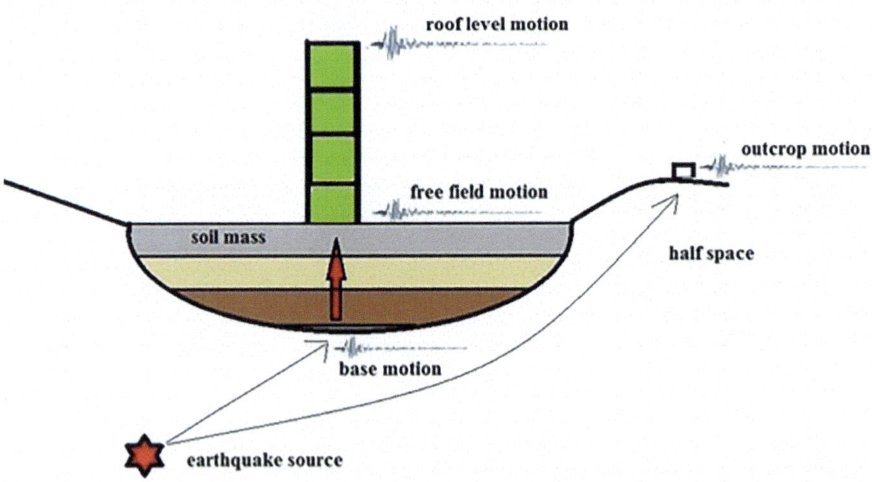

Fig. 14.1 Simplified representation of seismic wave propagation and site response

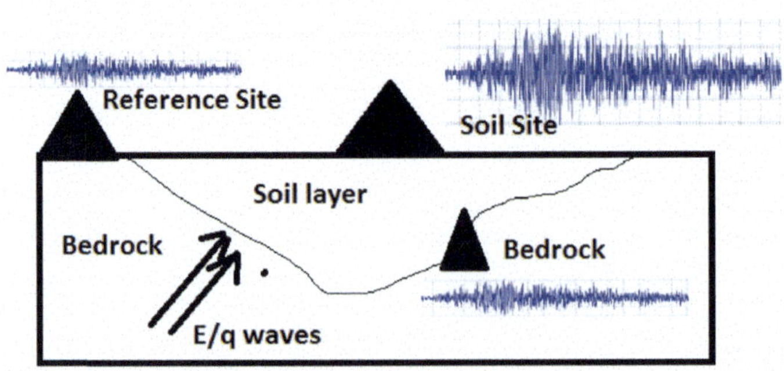

Fig. 14.2 Schematic description of the SSR method

The modification of the ground motion due to "***site response***" which is recorded at stations located in different soil conditions will be different and can be related to various factors governed by the local site conditions. In general, the main factor for such modification is the impedance contrast [43] between the soft sedimentary cover (with low density and wave propagation velocity) and the bedrock (with high density and wave propagation velocity). The impedance contrasts determine how strongly the waves at particular frequencies (the fundamental resonance frequency and the higher harmonics) are modified within the soft layer. Generally, the higher the impedance contrast, the greater the trapped energy, therefore, **the** higher the modification of the seismic waves within the layer. This modification results in various intensities of the earthquake ground shaking at various locations depending on the soil conditions and in turn may result in various categories of damage.

Estimation of site response can be done by various site response analysis methods. The suitability of most of the methods is based on the availability of strong-motion data. The present work adopted the two most widely accepted site response analysis methods namely—Standard Spectral Ratio (SSR) and Horizontal to Vertical Spectral Ratio (HVSR) of strong-motion data. The SSR and HVSR method require the availability of recorded strong ground motions. The SSR and the HVSR method involve waveform analysis. These methods were chosen to assess the site-specific response of a soil deposit in the Western Guwahati region because the recorded waveform data-set in the form of three-component time-history was available (for SSR and HVSR analysis).

It is to be noted that while the SSR method is a reference site method involving waveform analysis, the HVSR method is a non-reference site method (single station method) involving waveform analysis. The SSR and HVSR method involve the determination of site response in the form of ratios in the frequency domain. The site response data obtained is defined as the "*site transfer function*" or "*amplification ratio*" plot for the site concerned. The data of these plots can be used to obtain the time history of the site, modified due to site response.

This chapter attempts to compare the results obtained by the two methods and give a discussion on the results obtained. The area of the study includes three locations in the west Guwahati Region of the state of Assam in India, namely, Boko-Palashbari, Goalpara and Guwahati Central Region.

14.2 Standard Spectral Ratio (SSR Analysis)

One of the most popular, widely used and accepted techniques, which involves standard reference sites to characterize site amplification, is Standard Spectral Ratio (SSR) technique. It has been used by various authors to determine the site response [8, 18, 30, 47, 49]. The SSR is a technique where the site response is defined as the ratio of the Fourier amplitude spectrum of ground motions recorded at a soil-site to that of ground motions recorded at a rock-site record located nearby, from the same earthquake and component of motion [6].

In this method it is assumed that earthquake records obtained from the reference site (an earthquake recording station located on a hard rock outcrop) are free from site effects and contain the same source properties, and when the two sites are closely located, the path or propagation effects are also same for the pair of records. Hence, the ratio of the Fourier amplitude spectra of the sedimentary site to the reference site expresses the local site effects or in other terms amplification at the sedimentary or soil site.

Alternatively, it can be said that the technique depends on two basic assumptions [27]:

(a) The source and paths effects at both sites are similar.
(b) The reference site has a negligible site response, that is, its spectrum is flat.

The first assumption requires that the distance between both sites is limited in order to have a similar wave field (with similar incidence angles and azimuths) arriving at the two sites. This effect can also be reached with increasing source-receiver distance compared to the distance between the two recording sites [17].

The second assumption implies that the record on the rock site, which usually is a free field station on the ground surface, is equivalent to the input motion at some depth, not taking into account the free-surface effects [31].

Obtaining pure outcropping rock sites consisting of geological bedrock which is close to basins of sedimentary deposits or soft alluvial sites, as well as unweathered outcropping sites at the ground surface, are very hard to find [27, 47]. Difficulties in the application of SSR are discussed by Safak [49] and Steidl et al. [53]. Safak [49] pointed out problems in the application of the spectral ratio method to real records because the spectral ratios could be significantly influenced by their processing procedure. Apart from the theoretical background, one of the major problems when applying the standard spectral ratio method is identifying a pure bedrock site that meets the mentioned pre-requisites even when bedrock sites are available. However, the assumed condition that an outcropping rock reference site is free of any site effects is often not valid because a seemingly good rock site may have an amplification of its own. Therefore, a careful examination of the reference site is necessary to accurately estimate amplification/site response in sedimentary sites [53]. Steidl et al. [53]concluded that surface rock sites are inevitably affected by amplification at a frequency as low as 4–5 Hz because of the thin weathered layer that is almost always present on rock sites. They advocated the use of seismic stations at the bottom of deep boreholes. Steidl et al. [53] showed that deep bore hole records, when corrected for the effects of downward propagating waves, are more reliable records to be used as reference site motion. This conclusion is well-substantiated but impractical in most places, especially in developing countries because of the large cost involved.

14.2.1 Standard Spectral Ratio (SSR) Method

The Standard Spectral Ratio (SSR) method is defined as the ratio of ground motions records of that of a soil site to that of a rock site, where the rock site is used as a reference site (Fig. 14.2). It is assumed that earthquakes recorded on the reference site (i.e. recorded on a station placed on outcropping hard bedrock) contain the same source and propagation effects as that of the records from the other sites that is

$$Standard\ Spectral\ Ratio\ (SSR) = \frac{U_{ij}(x, y)}{U_{ik}(x, y)} = \frac{S_i(x)Z(x)_j A_{ij}(x, y)}{S_i(x)Z(x)_k A_{ik}(x, y)} \quad (14.1)$$

where

$U_{ij}(x,y)$ = Fourier Amplitude of the ground motion observed at a soil site j for an event i,

$U_{ik}(x,y)$ = Fourier Amplitude of the ground motion observed at reference site k for an event i,

$S_i(x)$ = source function,

$Z_i(x)$ = response of the site,

$A_{ij}(x,y)$ = Attenuation function of soil site,

$A_{ik}(x,y)$ = Attenuation function of reference site,

x = the frequency,

y = distance from epicenter and.

k = reference station.

It is to be noted that similar to what is schematically indicated in Fig. 14.1, the ground motion recorded at one station is considered as the convolution of three terms, specifically the source $S_i(f)$, the path $A_{ij}(f,r)$ and site response $Z_j(f)$. If the stations are nearby, and the reference station is not affected by site effects, then $A_{ij}(f,r) \approx A_{ik}(f,r)$. Thus, the spectral ratio directly provides the site response at the non-reference stations. The major drawback of this method is that rock stations may also have their own response. Furthermore, a good reference site might be located too far away from the target site, therefore, not allowing the assumption of similarity in the wave paths towards the two stations.

14.2.2 Selection of Reference Site

Of the two datasets considered for the study and of all the recording stations, Nongstoin station recordings were considered as reference site motion. Table 14.1

Table 14.1 Time histories recorded at Nongstoin

Earthquake event	Epicenter	Region	Magnitude (Mw)	Focal depth (km)	Epicentral distance (km)
11 Aug. 2009	24.4 N 94.8E	Myanmar	5.6	22	378
03 Sep. 2009	24.3 N 94.6E	Myanmar	5.9	100	363
21 Sep. 2009	27.3 N 91.5E	Bhutan	6.2	8	197
29 Oct. 2009	27.3 N 91.4E	Bhutan	5.2	5	197
29 Oct. 2009	26.6 N 90.0E	Kokrajhar	4.2	10	174

gives the details of the recordings at Nongstoin station (Table 14.2).

The basic conditions for the application of SSR are [45]:

(a) the existence of simultaneous recordings at a soil site and at the reference site,
(b) the reference site has to be free of any site effects, and
(c) the distance between the soil site and the reference one ought to be small (i.e. smaller than the epicentral distance), to consider that the effect of the propagating path of the seismic energy is the same for the two sites.

It can be clearly seen from Table 14.3 that the first condition of application of SSR technique, i.e., the existence of simultaneous recordings at a soil site and the reference site for the same earthquake event, is satisfied by the recording of Boko-Palashbari, Guwahati-Central, and Goalpara. Moreover, Nongstoin is characterized as Site Class A with site geology consisting of granite [34]. Therefore, recordings of this station can safely be assumed to be that of a reference site that satisfies the second condition. Further, as indicated in Table 14.4, the inter-station distance between Nongstoin and Boko-Palashbari, Guwahati-Central, Goalpara stations is less as compared to the epicentral distance as is required by the third condition, for each earthquake considered for SSR analyses. Thus it can be conservatively assumed that the effect of the propagating path of the seismic energy is the same for the two sites considered respectively.

14.2.3 Dataset and Data Processing of Earthquake Records

From the 5(five) earthquake events, a total of 16 earthquake recordings having three components of motion, i.e. E-W(East–West), N-S(North–South) and VT(Vertical) wereselected for the analyses. Of these 11(eleven) were related to soil sites (Boko, Goalpara, and Guwahati) and 5(five) were related to reference site (Nongstoin).

The complete duration of all the records was Fourier transformed to obtain the Fourier Amplitude Spectra (FAS) of the respective components by using the software code DEEPSOIL [21].

14.2.4 Analyses and Discussion

The analyses were carried out for 3(three) soil sites with reference to a single rock/reference site. A total of 5(five) earthquake events qualified for the SSR evaluation technique as given in Table 14.2 within the area of study. SSR analysis of all three components of the records i.e. EW, NS, and VT was carried out. SSR method employs a comparison of round records in the frequency domain. Additionally, in this study, the properties of ground motions of the soil sites were compared with that of the reference site in the time domain as well. The comparison in the time domain was carried out in the form of ratios of soil versus reference site of the PGA, PGV and PGD respectively and presented in Table 14.13.

Table 14.2 Properties of recorded motions (Nongstoin station)

Earthquake event	Maximum acceleration (g)			Maximum velocity (m/s)			Maximum displacement (m)			Duration (second)
	EW	NS	VT	EW	NS	VT	EW	NS	VT	
11Aug. 2009	0.014	0.016	0.015	0.004	0.004	0.002	0.003	0.003	0.006	73.040
03 Sep. 2009	0.009	0.010	0.007	0.003	0.003	0.002	0.002	0.008	0.008	73.475
21 Sep. 2009	0.028	0.019	0.016	0.012	0.009	0.013	0.017	0.009	0.014	106.335
29 Oct. 2009	0.010	0.008	0.05	0.003	0.003	0.002	0.007	0.009	0.009	67.080
29 Oct. 2009	0.006	0.010	0.003	0.002	0.003	0.001	0.009	0.001	0.004	65.985

Table 14.3 Selected reference site and soil site motions and epicentral distances

Earthquake event	Epicenter	Focal depth (km)	Boko-Palashbari (soil site)	Guwahati-central (soil site)	Goalpara (soil site)	Nongstoin (Ref. site)
			25.976 N 91.230E	26.190 N 91.746E	26.152 N 90.627E	25.522 N 91.264E
11 Aug. 2009	24.4 N 94.8E	22	400	366	463	378
03 Sep. 2009	24.3 N 94.6E	100	388	356	450	363
21 Sep. 2009	27.3 N 91.5E	8	148	125	153	197
29 Oct. 2009	27.3 N 91.4E	5	+	+	147	197
29 Oct. 2009	26.6 N 90.0E	10	+	+	80	174

Table 14.4 Inter-station distances (in km)

Stations	Nongstoin	Boko-Palashbari	Goalpara	Guwahati-Central
Nongstoin	0	50	90	80
Boko	50	0	60	50
Goalpara	90	60	0	120
Guwahati	80	50	120	0

Table 14.5 Properties of recorded motions (Guwahati-central station)

Earthquake events	Maximum acceleration (g)			Maximum velocity (m/s)			Maximum displacement (m)			Duration (second)
	EW	NS	VT	EW	NS	VT	EW	NS	VT	
11 Aug. 2009	0.013	0.011	0.013	0.006	0.004	0.004	0.003	0.004	0.001	73.040
03 Sep. 2009	0.012	0.007	0.007	0.007	0.003	0.003	0.004	0.006	0.004	71.065
21 Sep. 2009	0.028	0.024	0.013	0.011	0.009	0.006	0.016	0.038	0.021	107.280

14.2.4.1 Guwahati-Central Site

The Guwahati-Central station has been classified as Site Class ‘C’ [34]. The station was triggered and had recorded 3(three) events out of the total of the 5(five) events that had been considered for the study. Table 14.5 shows the duration and peak values of acceleration, velocity and displacement of the recorded ground motions of Guwahati-Central site. Figures 14.3, 14.4 and 14.5 represent the SSR plots of EW, NS and VT components respectively, for records of the 3(three) earthquake events of Guwahati-Central site with respect to Nongstoin station. From Figs. 14.3, 14.4 and 14.5 it can be seen that the SSR plot of the components was flatter in respect of the 21-Sep-2009 earthquake event as compared to the 11-Aug-2009 and 03-Sep-2009 earthquake events. The 11-Aug-2009 and 03-Sep-2009 earthquake events showed comparable plots for all the components. However, the 03-Sep-2009 earthquake event showed an SSR plot slightly having a higher amplitude in the frequency range of 2.0–3.0 Hz.

14.2.4.2 Boko-Palashbari Site

Boko station has been classified as Site Class ‘C’ [34]. The station was triggered and had recorded 3(three) events out of the total of the 5(five) events that had been considered for the study. Table 14.6 shows the duration and peak values of acceleration, velocity and displacement of the recorded ground motions of Boko-Palashbari site. Figures 14.6, 14.7 and 14.8 represent the SSR plots of EW, NS and VT components

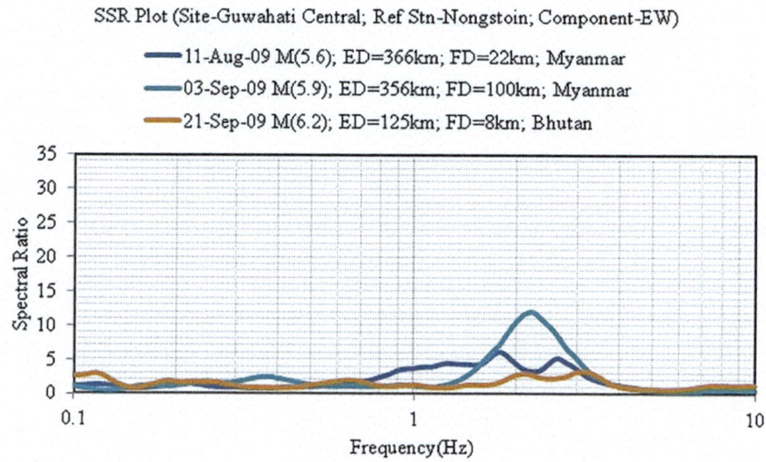

Fig. 14.3 SSR Plot of Guwahati-Central station, EW-component

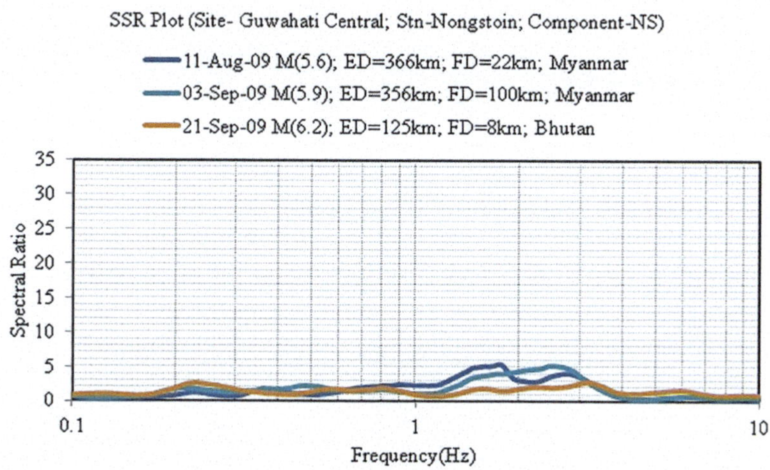

Fig. 14.4 SSR Plot of Guwahati-Central station, NS-component

respectively, for records of the 3(three) earthquake events of the Boko-Palashbari site with respect to Nongstoin station. The SSR plot as shown in Figs. 14.6, 14.7 and 14.8 shows that the maximum amplitude for the horizontal components i.e. the EW and NS components falls in the frequency range of 0.5–1.0 Hz. The vertical component, however, shows higher SSR amplitude in the frequency range of 1.0–3.0 Hz.

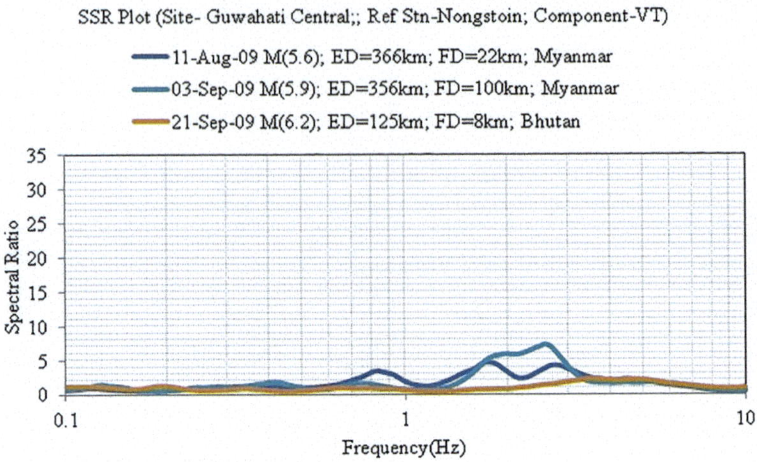

Fig. 14.5 SSR Plot of Guwahati-Central station, VT-component

Table 14.6 Properties of recorded motions considered (Boko-Palashbari station)

Earthquake event	Maximum acceleration (g)			Maximum velocity (m/s)			Maximum displacement (m)			Duration (second)
	EW	NS	VT	EW	NS	VT	EW	NS	VT	
11 Aug 2009	0.016	0.017	0.008	0.006	0.004	0.002	0.004	0.008	0.011	77.550
03 Sep 2009	0.014	0.009	0.005	0.003	0.003	0.002	0.003	0.004	0.007	72.675
21 Sep 2009	0.021	0.021	0.013	0.012	0.010	0.008	0.019	0.025	0.005	113.875

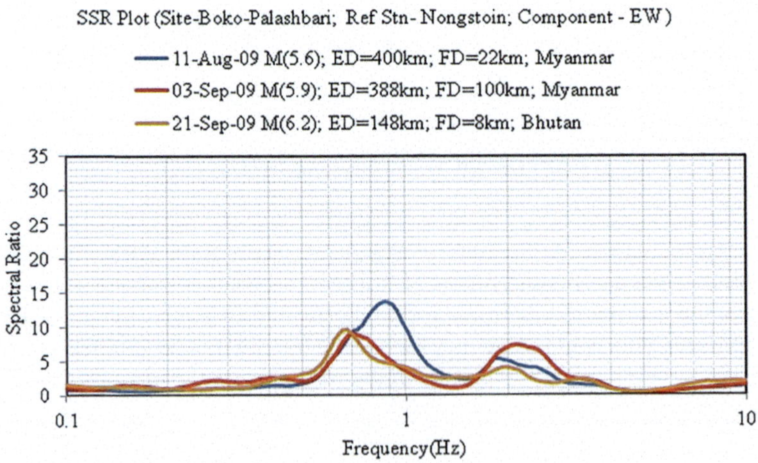

Fig. 14.6 SSR plot of Boko-Palashbari station, EW-component

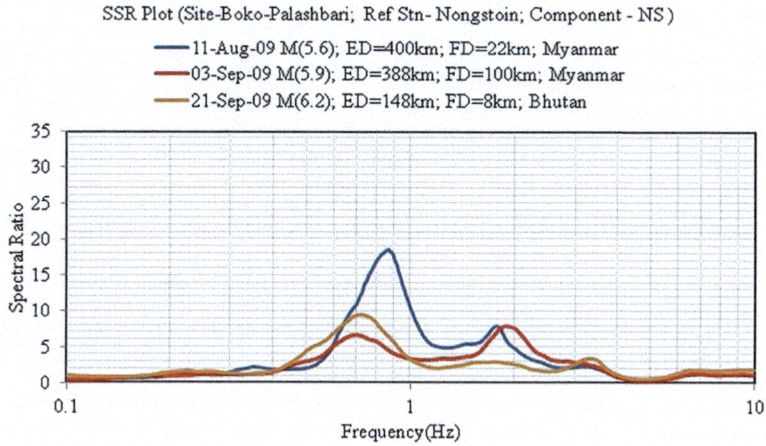

Fig. 14.7 SSR plot of Boko-Palashbari station, NS-component

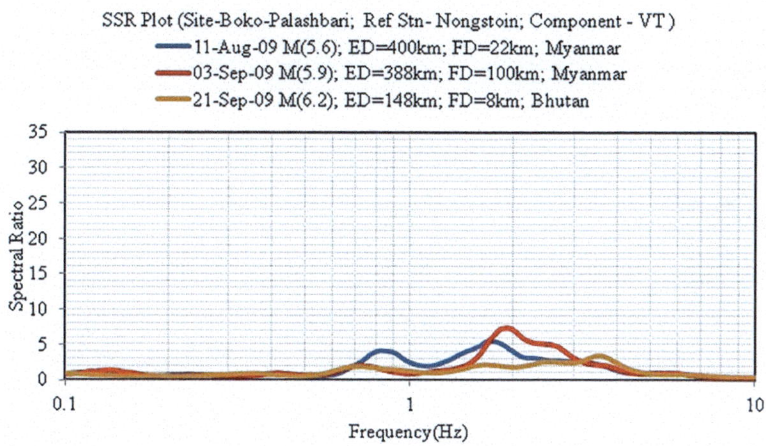

Fig. 14.8 SSR plot of Boko-Palashbari station, VT-component

14.2.4.3 Goalpara Site

Goalpara station has been classified as Site Class 'C' [34]. The station was triggered and recorded 5(five) events out of the total of the 5 events that had been considered for the study. Table 14.7 shows the duration and peak values of acceleration, velocity and displacement of the recorded ground motions of Goalpara site. Figures 14.9, 14.10 and 14.11 represent the SSR plots of EW, NS and VT components respectively, for records of the 5(five) earthquake events of Goalpara site with respect to Nongstoin station. From Figs. 14.9, 14.10 and 14.11 it can be seen that the maximum amplitude

Table 14.7 Properties of recorded motions considered (Goalpara station)

Earthquake event	Maximum acceleration (g)			Maximum velocity (m/s)			Maximum displacement (m)			Duration (second)
	EW	NS	VT	EW	NS	VT	EW	NS	VT	
11 Aug 2009	0.016	0.013	0.008	0.005	0.005	0.004	0.005	0.003	0.004	73.965
03 Sep 2009	0.006	0.006	0.004	0.002	0.002	0.002	0.005	0.003	0.007	66.620
21 Sep 2009	0.042	0.041	0.024	0.033	0.030	0.013	0.012	0.022	0.099	113.83
29 Oct 2009	0.015	0.011	0.007	0.006	0.006	0.004	0.003	0.019	0.036	75.405
29 Oct 2009	0.015	0.022	0.011	0.004	0.007	0.003	0.002	0.007	0.036	68.415

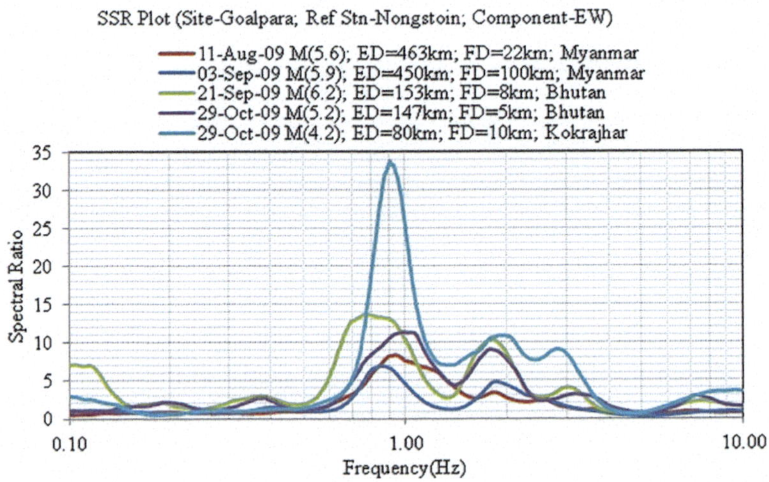

Fig. 14.9 SSR plot of Goalpara station, EW-component

of the SSR plot is in the frequency range of 0.60–2.0 Hz for the horizontal components, whereas for the vertical component the maximum values lie in a frequency range of 1.0–3.0 Hz.

14.2.5 Site Amplification

In this study, the definition of site amplification from SSR study is adopted as the maximum SSR amplitude. The corresponding frequency at which the maximum amplification is obtained is termed the fundamental frequency of the site. The site

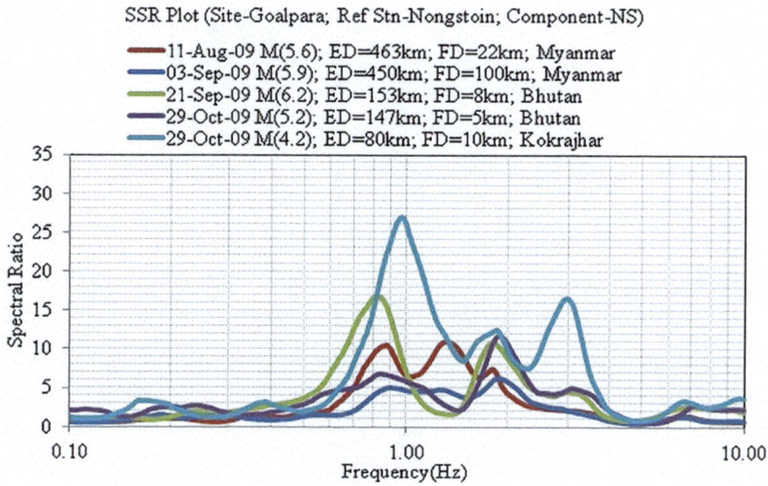

Fig. 14.10 SSR plot of Goalpara station, NS-component

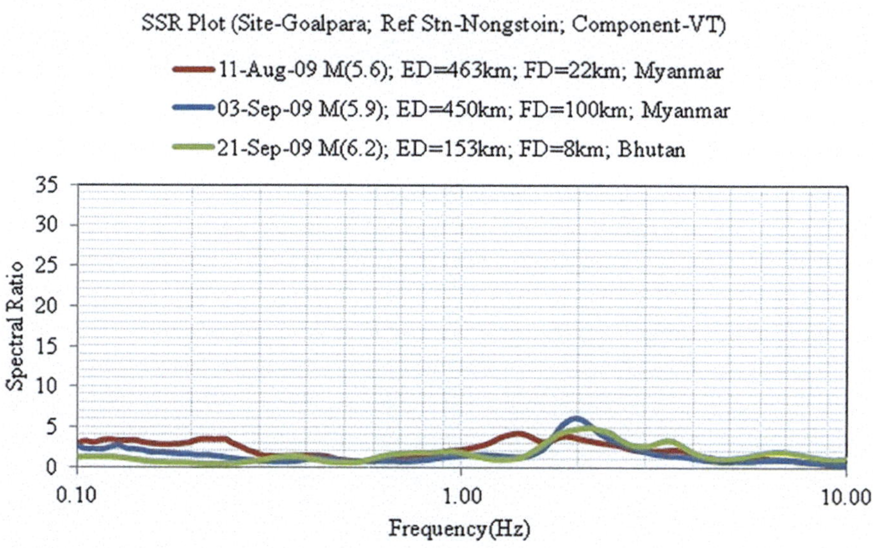

Fig. 14.11 SSR plot of Goalpara station, VT-component

amplification observed at Boko, Guwahati and Goalpara from the SSR analysis is provided in Table 14.8.

Table 14.8 Maximum SSR amplitude and corresponding frequency

Earthquake event	Magnitude (Mw)	Focal depth (km)	(SSR) site amplification/frequency								
			Boko-Palashbari (soil site)			Guwahati-Central (soil site)			Goalpara (soil site)		
			EW	NS	VT	EW	NS	VT	EW	NS	VT
11 Aug. 2009	5.6	22	13.60/0.87	18.48/0.85	5.54/1.69	5.99/1.81	5.34/1.77	4.61/1.75	8.32/0.91	10.77/1.28	4.27/1.38
03 Sep. 2009	5.9	100	9.04/0.69	7.99/1.89	7.42/1.87	12.11/2.20	5.19/2.48	7.19/2.58	6.78/0.82	6.29/1.88	6.19/1.93
21 Sep. 2009	6.2	8	9.66/0.66	9.54/0.72	3.55/3.49	3.30/3.05	2.79/3.13	2.19/3.44	12.90/0.76	16.85/0.82	5.00/2.08
29 Oct. 2009	5.2	5	–	–	–	–	–	–	11.29/1.04	11.78/1.88	4.30/1.75
29 Oct. 2009	4.2	10	–	–	–	–	–	–	33.74/0.70	27.14/0.96	9.69/1.99

14.2.6 Observations

The results of the SSR analysis have been provided in Table 14.8. At the Boko-Palashbari site, the amplification varied from 7.99 to 18.59 while for Guwahati-Central and Goalpara sites the same varied from 2.19–12.16 and 4.27–33.74. It is observed that, for all the sites, the maximum amplification was for the horizontal component as compared to the vertical.

Site amplification for the Boko-Palashbari site showed maximum amplitude for the horizontal components of the 11th August 2009 earthquake which had the lowest magnitude of 5.6 (Mw) among the three and had the maximum epicentral distance of 400 km. The fundamental frequency for the Boko site for the horizontal components ranged from 0.66 to 1.89 Hz, which reflects the characteristics of a soil site with a soft consistency.

Site amplification for Guwahati-Central showed maximum amplitude for the horizontal components of the 3rd September 2009 earthquake which had a magnitude of 5.9 (Mw) and the highest focal depth of 100 km and an epicentral distance of 356 km. The fundamental frequency of Goalpara was found to be in the range of 1.75–3.44 Hz.

Site amplification for Goalpara showed maximum amplitude for the horizontal components of the lowest magnitude earthquake of 4.2 (Mw) at an epicentral distance of 80 km and focal depth of 10 km. The fundamental frequency of Goalpara was found to be in the range of 0.70–1.88 Hz.

Mittal et al. [33] observed that the vertical component also undergoes significant amplification in order 20. Similar observations of amplification of vertical component in the present study were also observed at Boko-Palashbari showing amplification of the vertical component of the order 7.42 which is almost similar to that observed for the NS-component of the same earthquake of 3rd September 2009. Guwahati-Central showed amplification of vertical component which exceeded that of the NS-component of the 3rd September 2009 earthquake by a value of 2, which is significant. In general, Guwahati-Central showed amplification of the vertical component of the maximum order 7.19. The Goalpara site also showed a significant amplification of the vertical component to the order 9.69. However, the amplification of the vertical component of the 3rd September 2009 earthquake showed similar values for the horizontal as well as the vertical components. Of all the earthquakes considered for the study, it is observed that the 3rd September 2009 earthquake which had a magnitude of 5.9 (Mw) and focal depth of 100 km produced higher amplification of the vertical component with values exceeding that of the horizontal component for Guwahati site.

Further, peak site amplification is given in Tables 14.9, 14.10 and 14.11 for the frequency ranges 0.1–1.0, 1.0–3.0 and 3.0–10.0 Hz for EW, NS and VT components respectively. These frequency ranges are considered based on the fact that the natural frequencies of multistoried buildings (3–10 floors) range between 1 and 3 Hz, whereas the natural frequencies for 1–3 storey buildings vary from 3 to 10 Hz. The frequency range 0.1–1.0 Hz is the natural frequency of more than 10 storey buildings

Table 14.9 Site amplification in frequency range for earthquake events (EW-component)

Earthquake event	Peak amplification in frequency range (EW component)								
	Boko-Palashbari (soil site)			Guwahati-Central (soil site)			Goalpara (soil site)		
	0.1–1.0 Hz	1.0–3.0 Hz	3.0–10.0 Hz	0.1–1.0 Hz	1.0–3.0 Hz	3.0–10.0 Hz	0.1–1.0 Hz	1.0–3.0 Hz	3.0–10.0 Hz
11 Aug. 2009	13.60	10.15	1.71	3.67	6.07	3.54	8.45	7.67	1.44
03 Sep. 2009	8.94	7.35	2.64	2.54	12.16	5.14	6.90	4.81	1.55
21 Sep. 2009	9.67	4.23	2.48	2.98	3.22	3.32	13.77	10.47	4.10
29 Oct. 2009	**	**	**	**	**	**	11.32	11.30	3.17
29 Oct. 2009	**	**	**	**	**	**	33.74	25.38	8.44

Table 14.10 Site amplification in frequency range for earthquake events (NS-component)

Earthquake	Peak Amplification in frequency range (NS component)								
	Boko-Palashbari (soil site)			Guwahati-Central (soil site)			Goalpara (soil site)		
	0.1–1.0 Hz	1.0–3.0 Hz	3.0–10.0 Hz	0.1–1.0 Hz	1.0–3.0 Hz	3.0–10.0 Hz	0.1–1.0 Hz	1.0–3.0 Hz	3.0–10.0 Hz
11 Aug 2009	18.59	11.33	2.40	2.48	5.34	3.57	10.46	10.91	2.18
03 Sep 2009	6.71	7.99	2.92	2.29	5.19	3.68	5.04	6.30	2.12
21 Sep 2009	9.56	3.50	3.44	2.61	2.67	2.80	17.00	10.95	4.51
29 Oct 2009	**	**	**	**	**	**	6.86	11.79	5.02
29 Oct 2009	**	**	**	**	**	**	27.14	25.79	16.60

Table 14.11 Site amplification in frequency range for earthquake events (VT-component)

Earthquake	Peak amplification in frequency range (VT component)								
	Boko-Palashbari (soil site)			Guwahati-Central (soil site)			Goalpara (soil site)		
	0.1–1.0 Hz	1.0–3.0 Hz	3.0–10.0 Hz	0.1–1.0 Hz	1.0–3.0 Hz	3.0–10.0 Hz	0.1–1.0 Hz	1.0–3.0 Hz	3.0–10.0 Hz
11 Aug 2009	4.15	5.44	2.73	3.35	4.61	3.59	3.54	4.28	2.17
03 Sep 2009	1.99	7.47	3.04	1.82	7.21	3.91	2.71	6.24	1.98
21 Sep 2009	1.95	2.66	3.56	1.23	1.77	2.21	2.06	5.04	3.48
29 Oct 2009	**	**	**	**	**	**	8.02	4.32	3.98
29 Oct 2009	**	**	**	**	**	**	7.78	9.72	6.77

Table 14.12 Peak site amplification in frequency range

Site	Peak site amplification in frequency range			Peak spectral ratio (amplification)
	0.1–1.0 Hz	1.0–3.0 Hz	3.0–10.0 Hz	
Boko –Palashbari	18.59	11.33	3.56	18.59
Guwahati-Central	3.67	12.16	5.14	12.16
Goalpara	33.74	25.79	16.60	33.74

Table 14.13 SSR results

Site	Amplification parameters			
	Peak SSR (amplification)	PGA (PGA_{soil}/PGA_{rock})	PGV (PGA_{soil}/PGA_{rock})	PGD (PGA_{soil}/PGA_{rock})
Boko –Palashbari	18.59	1.56	1.50	2.78
Guwahati-Central	12.16	1.33	2.33	4.22
Goalpara	33.74	3.67	3.33	9.00

[52]. It can be seen in Table 14.12 that site amplification for Guwahati site is highest at the frequency range 1–3 Hz while for Boko-Palashbari and Goalpara regions the frequency range 0.1–1.0 Hz shows the highest site amplification thus governing the design considerations of their respective buildings categories respectively.

Additionally, Table 14.13 provides the various amplification parameters in terms of SSR, PGA, PGV, and PGD amplification respectively.

14.2.7 Summary

In this chapter an introduction to the SSR method is provided. The SSR method is further applied to the study region by employing recorded strong motions of Guwahati-Central, Boko-Palashbari and Goalpara sites. Site response in terms of various amplification parameters is provided in Table 14.13.

From the observations and results obtained a general conclusion may be assumed that earthquakes originating at a greater focal depth tend to amplify the vertical component. This contradicts the study by Lang [26] where it was stated that SSR does not depend on the focal depth, however, agreement with the statement that SSR does not depend on the magnitude or epicentral distance [20] is achieved in this study.

14.3 Horizontal to Vertical Spectral Ratio (HVSR) Analysis

14.3.1 Introduction

Horizontal to Vertical Spectral Ratio (HVSR) method which is a non-reference site technique was originally proposed by Nakamura [38] to interpret microtremors measurements and provides an alternative for the SSR method. Since this method requires no measurement at a reference site, this method overcomes the basic difficulty faced by the SSR technique which is the availability of a reference station. This method rests on the hypothesis that the vertical component of ground motion contains more information about the source of ground motion than does the horizontal components [29]. This technique, originally applied to microtremors [16, 26, 41], has also been applied to weak ground motion studies [12, 15, 29] and, in some cases, to strong ground motion studies [11, 29, 57].

Non-reference site technique or single station Horizontal to Vertical Spectral Ratio [29] follows the same idea of Nakamura technique, i.e., in the case of a soft layer that overlaps a generic stiff bedrock the incident vertical wave field does not undergo significant modification along with the whole source to site path with respect to the horizontal one. In this wave supposing a 1D configuration of the considered site, the simple ratio between the Fourier Amplitude Spectrum (FAS) of the horizontal component and FAS of the vertical component (both selected in the S-phase) allows us to detect the real response of the site (due to the body wave only).

The H/V method has been used for microzonation studies to predict site response to earthquake seismicity [25, 38, 48] and also as a method to estimate unconsolidated sediment thickness, map the bedrock surface, and infer fault locations (e.g., [9, 22, 42]).

A large number of experiments [14, 19, 29, 54] have shown that the HVSR procedure can be successfully applied for the identification of the fundamental frequency of sedimentary deposits. These observations were supported by several theoretical 1-D investigations [16, 26, 30, 60, 57]. Although the HVSR method has been able to identify the fundamental frequency of soil sites having relatively simple geology, the amplification factor of HVSR is highly debated [1, 2, 5, 32, 37, 50], and a comprehensive conclusion as to how the amplification can be explained using this method is yet to be defined. However, owing to its simplicity this method has been used for various microzonation studies [11] to have a preliminary assessment of the seismic vulnerability of a site due to seismic amplification.

14.3.2 HVSR Technique

The technique was originally proposed by Nakamura [38] to interpret microtremor measurements. The method states that microtremor energy consists mainly of Rayleigh waves and that site amplification is due to the presence of the surface of a soft layer overlying a half-space. Under these conditions, there are four components of ground motion involved: horizontal and vertical components in the half-space, and horizontal and vertical components at the surface. According to Nakamura, it is possible to estimate the amplitude effect of the source, A_S, by the ratio,

$$A_S = \frac{V_S}{V_B} \tag{14.2}$$

where V_S = FAS of the vertical component of motion at the surface,

V_B = FAS of the vertical component of motion at the half-space.

Therefore, the definition of an estimate of site effect of interest in earthquake engineering, S_E, can be given as the ratio.

$$S_E = \frac{H_S}{H_B} \tag{14.3}$$

where H_S = FAS of the horizontal component of motion at the surface, and,

H_B = FAS of the horizontal component of motion at the base of the soil layer.

Now, to compensate S_E by the effect of the source, a modified site effect function, S_M, is calculated, where.

$$S_M = \frac{S_E}{A_S} \tag{14.4}$$

Which is equivalent to.

$$S_M = \frac{\left(\frac{H_S}{V_S}\right)}{\left(\frac{H_B}{V_B}\right)} \tag{14.5}$$

Now, if it is accepted that the ratio H_B/V_B is equal to unity, the site effect function, corrected by the source term, may be written as.

$$S_M = \frac{H_S}{V_s} \tag{14.6}$$

The assumption that H_B/V_B is equal to unity (within a factor of 2, the usual uncertainty when using spectral ratios, e.g. Tucker and King [58], Tucker et al. [59] was verified by Nakamura experimentally, using microtremor measurements at various depths in the borehole.

Mucciarelli and Gallipolli [36] reported that the HVSR technique as is generally known is not a single entity but several variations of the HVSR method are present. In this study, the HVSR methodology of Microzonation of Guwahati City [11] has been adopted for the analysis in an attempt to supplement the site response study carried out earlier in the greater Guwahati region.

14.3.3 HVSR Methodology Adopted

In the present study, the $HVSR_{ij}(f_k)$ is computed at each j site for the ith event at the central frequency f_k from the root mean square average of the amplitude spectra [11] as.

$$HVSR_{ij}(f_k) = \frac{\frac{1}{\sqrt{2}}\sqrt{abs\left|H_{ij}(f_k)\right|_{NS}^2 + abs\left|H_{ij}(f_k)\right|_{EW}^2}}{abs\left|V_{ij}(f_k)\right|}. \tag{14.7}$$

where

$\left|H_{ij}(f_k)\right|_{NS}$ = FAS of horizontal NS component.

$\left|H_{ij}(f_k)\right|_{EW}$ = FAS of horizontal EW component.

$V_{ij}(f_k)$ = FAS of vertical component.

Finally, the event average $HVSR_{ij}^{ave}$ (f_k) [18] is computed at each site for the kth frequency to consider the contribution of all the seismic events recorded at that station.

Additionally, amplification of PGA, PGV and PGD is also determined by obtaining the ratio of the horizontal to the vertical component.

14.3.4 Selection of Records

Earthquake records (accelerogram) considered for HVSR analysis were selected only for those stations which were of importance with respect to the present study. The selected sites were Guwahati, Boko-Palashbari, and Goalpara. Out of the 30(thirty) earthquake events available, it was found that 15(fifteen) events (as given in Table 14.14) were responsible for triggering the selected 3 nos. of stations.

All the recordings had three components of motion, two orthogonal and one vertical. Of the selected 15(fifteen) events, the Boko-Palashbari station was triggered by 6(six) events, Guwahati by 13(thirteen) events and Goalpara by 9(nine) events. Thus, a total of 28(twenty-eight) three-component motions were recorded.

All the 74 (28 × 3) records were Fourier transformed and smoothed. The individual properties of each component of the recordings are indicated in Tables 14.15, 14.16 and 14.17 for the respective stations.

Table 14.14 Selected earthquake motions and epicentral distances

Earthquake event	Epicenter	Magnitude (Mw)	Focal depth (km)	Boko-Palashbari (soil site)	Guwahati Central (soil site)	Goalpara (soil site)
				25.976 N 91.230E	26.190 N 91.746E	26.152 N 90.627E
25 Apr. 2009	26.4 N 91.7E	4	10	66	24	**
11 Aug. 2009	24.4 N 94.8E	5.6	22	400	366	463
19 Aug. 2009	26.6 N 92.5E	4.9	20	152	88	**
30 Aug. 2009	25.4 N 94.8E	5.3	85	364	318	**
03 Sep. 2009	24.3 N 94.6E	5.9	100	388	356	450
21 Sep. 2009	27.3 N 91.5E	6.2	8	148	125	153
29 Oct. 2009	27.3 N 91.4E	5.2	5	**	**	147
29 Oct. 2009	26.6 N 90.0E	4.2	10	**	**	80
29 Dec. 2009	24.5 N 94.8E	5.5	80	**	318	426
31 Dec. 2009	27.3 N 91.4E	5.5	7	**	127	147
26 Feb. 2010	28.5 N 86.7E	5.4	28	**	561	468
12 Mar. 2010	23.0 N 94.5E	5.6	96	**	450	526
26 Jul. 2010	26.5 N 91.3E	4.1	31	**	56	**
11 Sep. 2010	25.9 N 90.2E	5	20	**	158	**
21 Nov. 2010	25.1 N 95.3E	5.8	80	**	377	**

14.3.5 Selection of Time Window

While determining HVSR the entire horizontal and vertical accelerogram are divided into a fixed number of time-windows and then the average HVSR of each time-window is considered for interpretation. Sairam et al. [52] used a time window of 10.50 s, starting from 0.50 s before S-wave phase arrival. This time-window was chosen to best contain most of the high-amplitude of the S-wave energy of earthquake

Table 14.15 Properties of recorded motions (Boko-Palashbari station)

Earthquake event	Maximum acceleration (g)			Maximum velocity (m/s)			Maximum displacement (m)			Duration (second)
	EW	NS	VT	EW	NS	VT	EW	NS	VT	
25 Apr. 2009	0.011	0.009	0.010	0.004	0.002	0.001	0.005	0.005	0.003	70.320
11 Aug. 2009	0.016	0.017	0.008	0.006	0.004	0.002	0.004	0.008	0.011	77.550
19 Aug. 2009	0.008	0.019	0.005	0.002	0.004	0.001	0.003	0.008	0.003	66.725
30 Aug. 2009	0.007	0.012	0.005	0.002	0.003	0.002	0.002	0.007	0.004	66.725
03 Sep. 2009	0.014	0.009	0.005	0.003	0.003	0.002	0.003	0.004	0.007	72.675
21 Sep. 2009	0.021	0.021	0.013	0.012	0.010	0.008	0.019	0.025	0.005	113.875

Table 14.16 Properties of recorded motions (Goalpara station)

Earthquake event	Maximum acceleration (g)			Maximum velocity (m/s)			Maximum displacement (m)			Duration (second)
	EW	NS	VT	EW	NS	VT	EW	NS	VT	
11 Aug. 2009	0.016	0.013	0.008	0.005	0.005	0.004	0.005	0.003	0.004	73.965
03 Sep. 2009	0.006	0.006	0.004	0.002	0.002	0.002	0.005	0.003	0.007	66.620
21 Sep. 2009	0.042	0.041	0.024	0.033	0.030	0.013	0.012	0.022	0.099	113.830
29 Oct. 2009	0.015	0.011	0.007	0.006	0.006	0.004	0.003	0.019	0.036	75.405
29 Oct. 2009	0.015	0.022	0.011	0.004	0.007	0.003	0.002	0.007	0.036	68.415
29 Dec. 2009	0.008	0.010	0.006	0.003	0.004	0.002	0.002	0.006	0.020	70.055
31 Dec. 2009	0.010	0.009	0.009	0.006	0.005	0.004	0.013	0.013	0.032	82.350
26 Feb. 2010	0.005	0.005	0.003	0.002	0.002	0.002	0.004	0.017	0.016	89.900
12 Mar. 2010	0.004	0.005	0.003	0.02	0.002	0.003	0.008	0.019	0.056	95.660

Table 14.17 Properties of recorded motions (Guwahati-Central station)

Earthquake event	Maximum acceleration (g)			Maximum velocity (m/s)			Maximum displacement (m)			Duration (second)
	EW	NS	VT	EW	NS	VT	EW	NS	VT	
25 Apr. 2009	0.013	0.014	0.015	0.002	0.002	0.002	0.003	0.007	0.004	66.185
11 Aug. 2009	0.013	0.011	0.013	0.006	0.004	0.04	0.003	0.004	0.001	73.040
19 Aug. 2009	0.020	0.030	0.021	0.006	0.010	0.005	0.014	0.002	0.003	71.660
30 Aug. 2009	0.015	0.009	0.010	0.005	0.003	0.003	0.004	0.004	0.004	68.450
03 Sep. 2009	0.012	0.007	0.007	0.007	0.003	0.003	0.004	0.006	0.004	71.065
21 Sep. 2009	0.028	0.024	0.013	0.011	0.009	0.006	0.016	0.038	0.021	107.280
29 Dec. 2009	0.011	0.005	0.009	0.004	0.002	0.003	0.003	0.014	0.008	80.765
31 Dec. 2009	0.010	0.007	0.006	0.004	0.002	0.002	0.015	0.006	0.005	80.470
26 Feb. 2010	0.003	0.002	0.001	0.001	0.001	0.001	0.004	0.008	0.010	82.785
12 Mar. 2010	0.006	0.005	0.005	0.002	0.003	0.001	0.015	0.009	0.004	91.615
26 Jul. 2010	0.022	0.016	0.017	0.005	0.004	0.004	0.002	0.006	0.010	94.720
11 Sep. 2010	0.004	0.005	0.003	0.002	0.002	0.001	0.008	0.004	0.003	90.895
21 Nov. 2010	0.009	0.009	0.007	0.005	0.005	0.002	0.005	0.027	0.012	112.435

records. In Microzonation of Guwahati city [11] a time window of 5.0-s duration starting from the onset of S-wave arrival and containing the maximum of S-wave arrival was considered for analysis. Bonilla et al. [4] stated that using a longer time results in better spectral resolution. Borcherdt et al. [6] also mentioned that the most stable estimate of the site response is generally provided by spectra computed from the entire seismogram as opposed to some portion.

However, it was also stated by Bonilla et al. [4] that, using longer time windows will result in the inclusion of scattered and reflected energy and also the effect of surface waves in the spectra. Bonilla et al. [4] and Field and Jacob [18] found no statistical variation in site response computed with spectra of different time windows. However, Castro et al. [7] suggested that S-waves could be contaminated at a larger epicentral distance, which demands variable time windows for the estimation of the HVSR using S-waves.

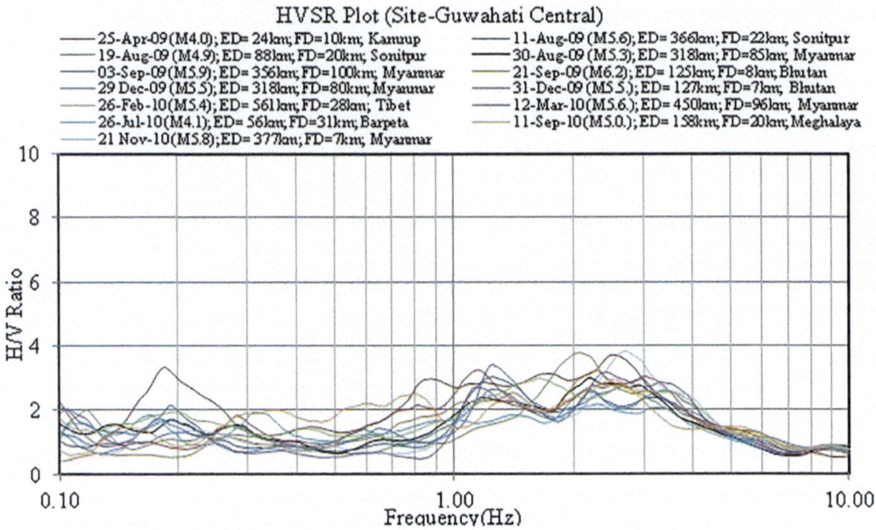

Fig. 14.12 HVSR plot of earthquake recordings of Guwahati-Central station

In this study, the entire accelerogram is used for analysis considering it as a single-window in order to obtain a better spectral resolution and stable estimate of site response.

14.3.6 Analysis Results

All the records considered were analyzed using spreadsheets developed for the purpose. The records were analyzed for the usual frequency range of 0.1–10.0 Hz, which is the frequency range of practical interest in engineering [53]. The HVSR versus Frequency data were plotted on a semi-log plot. Figures 14.12, 14.13 and 14.14 show the HVSR plots for Guwahati-Central, Boko-Palashbari and Goalpara respectively.

14.3.7 Site Amplification from HVSR of Strong Motion

Site Amplification from HVSR has often been a debated topic with various quarters putting forth various reasoning regarding its adequacy in determining site amplification. Since the application of HVSR to earthquake motions, various scholars [26, 28, 39] have used the term amplification to define the site effects of the HVSR method. Lang [26] studied the qualitative evaluation of different empirical methods on instrumented seismic data as per scientific literature and commented that the HVSR method

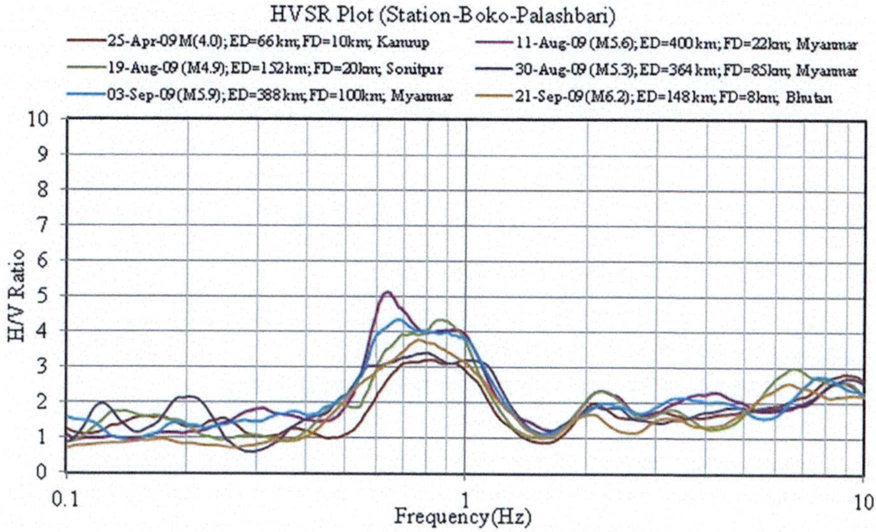

Fig. 14.13 HVSR plot of earthquake recordings of Boko-Palashbari station

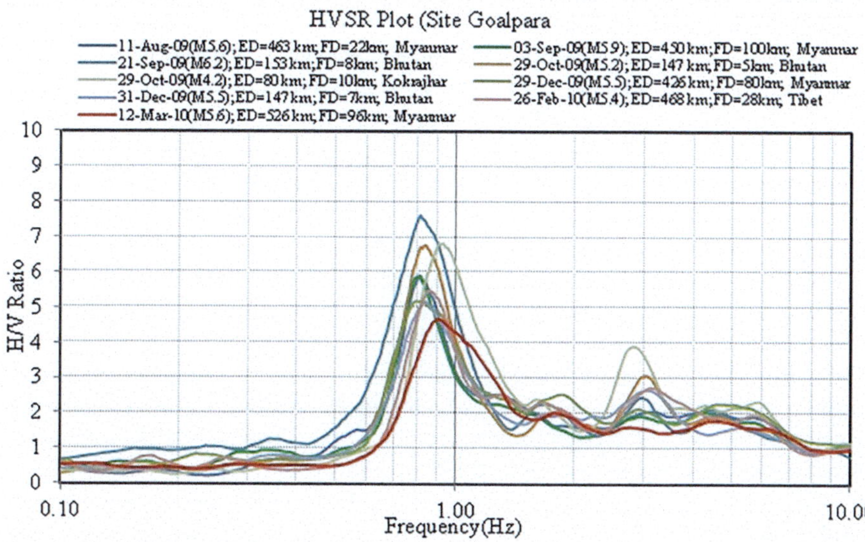

Fig. 14.14 HVSR plot of earthquake recordings of Goalpara station

when applied to earthquake data provided the predominant site frequency (Fundamental Frequency) and level of site amplification with high accuracy. It is however not quantified, as to how and to what degree the accuracy was applicable. Following Lang [26] an attempt has been made in this study to define the site amplification.

Generally, site amplification from HVSR is defined as the maximum amplitude of the HVSR curve obtained. The results obtained from this study, where three sites have been evaluated are tabulated in Table 14.18.

The HVSR curve from 13(thirteen) earthquake events used in the case of Guwahati site produced curves that are consistent in shape and have shown site amplification with a minimum of 2.16 and a maximum of 3.81. Goalpara site showed site amplification within the range of 4.68–7.60 for the 9 earthquake events considered for the site, while the Boko-Palashbari site indicated site amplification in the range of 3.21–5.15 for the 6 earthquake events considered.

14.3.8 Determination of Fundamental Frequency

The Fundamental Frequency (FF) is defined as the frequency at which the HVSR curve has a peak value, or, in other terms, it is the frequency (for a particular earthquake) or frequency range (for a set of earthquakes) at which the amplitude ratio value of the curve is maximum and is a characteristic of a site. Nakamura [37] suggested that this peak value ratio is at the S-waves resonance frequency of the layered structure and the location of the peak does not depend on the source characteristics. This suggestion is further strengthened by Lachet and Bard [25], who reported that the H/V spectral ratios obtained for different source characteristics clearly exhibit a peak whose position was constant regardless of the source type and the source function. Lachet and Bard [25] further based on their study on the peak frequency of HVSR for various geological structures suggested that H/V spectra were a reliable indication of the fundamental resonance frequency of a horizontally layered structure which has been asserted by many authors [27, 47]. The fundamental frequencies as obtained in this study for the three sites are indicated in Table 14.18.

The Fundamental Frequency range of Guwahati site as obtained from the HVSR curves is in the frequency band of 1.14–2.69 Hz, for the Goalpara site it is 0.80 Hz to 0.93 Hz while for Boko-Palashbari it is 0.63–0.82 Hz.

From Fig. 14.12, multiple HVSR peaks in Guwahati as compared to Boko-Palashbari and Goalpara are observed which may be attributed to Rayleigh wave ellipticity for the site with high-impedance contrasts at two very different scales. This has resulted in the fundamental frequency band being wider as compared to Boko-Palashbari and Goalpara respectively. From Figs. 14.13 and 14.14, for Boko-Palashbari and Goalpara respectively, where the bedrock depth was almost at depth $> = 100$ m clear peaks were identifiable.

14.3.9 Determination of Bedrock Depth

The depth of bedrock is related to the fundamental frequency by the relation [53].

Table 14.18 Fundamental frequency (FF) and corresponding amplification (Am)

Earthquake event	Epicenter	Magnitude (Mw)	Focal depth (km)	Boko-Palashbari		Guwahati Central		Goalpara	
				FF (Hz)	Am	FF (Hz)	Am	FF (Hz)	Am
25 Apr. 2009	26.4 N 91.7E	4	10	0.82	3.21	2.50	3.69	**	**
11 Aug. 2009	24.4 N 94.8E	5.6	22	0.63	5.15	1.25	3.02	0.81	5.89
19 Aug. 2009	26.6 N 92.5E	4.9	20	0.85	4.32	2.06	3.80	**	**
30 Aug. 2009	25.4 N 94.8E	5.3	85	0.79	3.42	2.20	3.01	**	**
03 Sep. 2009	24.3 N 94.6E	5.9	100	0.68	4.33	1.25	2.65	0.81	5.90
21 Sep. 2009	27.3 N 91.5E	6.2	8	0.76	3.78	2.26	3.22	0.81	7.60
29 Oct. 2009	27.3 N 91.4E	5.2	5	**	**	**	**	0.84	6.78
29 Oct. 2009	26.6 N 90.0E	4.2	10	**	**	**	**	0.93	6.80
29 Dec. 2009	24.5 N 94.8E	5.5	80	**	**	1.26	3.42	0.80	5.18
31 Dec. 2009	27.3 N 91.4E	5.5	7	**	**	1.14	3.23	0.84	5.16
26 Feb. 2010	28.5 N 86.7E	5.4	28	**	**	1.63	3.00	0.85	5.47
12 Mar. 2010	23.0 N 94.5E	5.6	96	**	**	2.39	3.16	0.90	4.68
26 Jul. 2010	26.5 N 91.3E	4.1	31	**	**	2.30	2.16	**	**
11 Sep. 2010	25.9 N 90.2E	5	20	**	**	2.50	2.95	**	**
21 Nov. 2010	25.1 N 95.3E	5.8	80	**	**	2.69	3.81	**	**

** not analyzed

Table 14.19 Depth of bedrock

Site	Fundamental Frequency (FF) band (Hz)	Shear wave velocity range, versus (m/s)	Depth to bedrock (m)
Boko–Palashbari	0.80–0.93	200–375	53.76–117.18
Guwahati-Central	1.14–2.69	200–375	18.59–2.24
Goalpara	0.63–0.82	200–375	60.98–148.81

$$FF = \frac{V_s}{4H} \tag{14.8}$$

where FF = Fundamental Frequency

V_s = Shear wave velocity

H = Depth to bedrock

Using the relation the bedrock to for all the three sites is determined. The Shear wave velocities are adopted from Mittal et al. [33], where it is classified that Guwahati, Boko-Palashbari, and Goalpara sites fall under the site class C (alluvium) and had a shear wave velocity range of 200–375 m/s. However it must be mentioned that the shear wave velocity as mentioned by Mittal et al. [33] is for the top 30 m of the soil strata, and its applicability for the calculation of the depth to bedrock is questionable. Despite this fact, it could be assumed that in the case where the shear wave velocity profile is not available, a good preliminary estimate of the bedrock depth could be made from the above assumption. The calculations of depth to bedrock are summarized in Table 14.19.

14.3.10 Observations and Discussions

In this chapter, an attempt to determine site transfer function, fundamental frequency and depth of bedrock was carried out. From the results obtained for Guwahati station which is situated in greater Guwahati and the other two stations, which lie in the study region, it can be seen that as we travel towards the western part of Guwahati the fundamental frequency range shifts to the lower range indicating trapping of seismic energy by the soft alluvial layer. Also, it is seen that the amplification ratio increases significantly (from 2.16 to 7.60) as we move towards the western part of the Guwahati region. This perhaps explains the reason why site response analysis should be taken up for the western part of Guwahati. It is observed that the peak amplification in Guwahati-Central is observed to be 3.81. Nath et al. [39] reported a peak amplification value of 6.5 for the same area. The deviation of results can be attributed to various reasons. For example, first of all, the derivation of the present peak amplification factor is based on the evaluation of 13(thirteen) earthquake strong-motion recordings, while that of Nath et al. [39] is based on 2(two) recordings. Furthermore, Nath

Table 14.20 Site amplification in frequency range for earthquake events

Earthquake event	Peak amplification in frequency range (Hz)								
	Boko-Palashbari			Guwahati-Central			Goalpara		
	0.1–1.0	1.0–3.0	3.0–10.0	0.1–1.0	1.0–3.0	3.0–10.0	0.1–1.0	1.0–3.0	3.0–10.0
25 Apr. 2009	3.21	2.90	2.82	3.32	3.70	2.87	**	**	**
11 Aug. 2009	5.15	3.92	2.61	2.13	3.02	2.37	5.89	3.14	2.12
19 Aug. 2009	4.32	3.70	2.98	2.03	3.80	2.44	**	**	**
30 Aug. 2009	3.42	3.21	2.71	1.88	3.01	2.71	**	**	**
03 Sep. 2009	4.33	3.81	2.76	1.72	2.65	2.62	5.90	3.04	2.05
21 Sep. 2009	3.78	3.13	2.57	2.18	3.22	2.60	7.60	4.84	2.42
29 Oct. 2009	**	**	**	**	**	**	6.78	4.04	3.08
29 Oct. 2009	**	**	**	**	**	**	6.80	6.11	3.61
29 Dec. 2009	**	**	**	1.51	3.42	2.93	5.18	3.18	2.26
31 Dec. 2009	**	**	**	2.49	3.23	3.02	5.16	3.73	2.66
26 Feb. 2010	**	**	**	1.95	3.00	2.62	5.47	3.66	2.72
12 Mar. 2010	**	**	**	2.04	3.16	2.85	4.68	4.28	1.69
26 Jul. 2010	**	**	**	1.94	2.16	2.06	**	**	**
11 Sep. 2010	**	**	**	2.52	2.95	2.26	**	**	**
21 Nov. 2010	**	**	**	1.64	3.81	3.39	**	**	**

et al. [39] adopted a cubic spline interpolation smoothing technique, while in this study a log-triangle smoothing procedure as developed by David Boore and adapted by DEEPSOIL [21] is applied. Additionally, Numerical instability as reported by Kausel and Assimiki [23] could be another reason for the varying results. However, considering all these uncertainties, it can be said that the method has predicted fairly reasonable results of peak amplification in the case of Nath et al. [39] and this study (Tables 14.20 and 14.21).

Further site amplification ranges are given in 20, 21 and 22 for the frequency ranges 0.1–1.0 Hz, 1.0–3.0 Hz and 3.0–10.0 Hz. These frequency ranges are considered based on the fact that the natural frequencies of multistoried buildings (3–10 floors)

Table 14.21 Site amplification range in frequency range

Site	Amplification range in frequency range			Peak amplification
	0.1–1.0 Hz	1.0–3.0 Hz	3.0–10.0 Hz	
Boko–Palashbari	3.21–5.15	2.90–3.92	2.57–2.98	5.15
Guwahati-Central	1.51–3.32	2.16–3.81	3.06–3.39	3.81
Goalpara	4.68–7.60	3.04–6.11	1.69–3.69	7.60

range between 1 and 3 Hz, whereas the natural frequencies for 1–3 storey buildings vary from 3 to 10 Hz. The frequency range 0.1–1.0 Hz is the natural frequency of more than 10 storey buildings. It can be easily seen that the frequency range which will govern the design of buildings in Boko-Palashbari and Goalpara is 0.1–1.0 Hz, while for Guwahati it is critical that buildings falling in the frequency range of 1.0–3.0 Hz are likely to be affected more.

Additionally, Table 14.22 provides the various amplification parameters in terms of HVSR, PGA, PGV, and PGD amplification respectively.

A comparison of the results of SSR (Table 14.13) and that of HVSR (Table 14.18) shows that the FAS amplification is of greater magnitude in the case of SSR as compared to HVSR. However, PGA, PGV and PGD amplification can be considered broadly to be within the same range.

14.3.11 Summary

In this chapter, an introduction to the HVSR analysis of strong-motion is provided. The HVSR analysis of three soil sites is carried out, and the results are presented. Also, the determination of FF and bedrock depth is attempted from the results of the HVSR analysis. Further, the results are provided in the form of peak HVSR ratio, PGA, PGV and PGD amplification respectively. Lermo and Chavez-Garcia [29], Mucciarelli and Gallipolli [35] etc. are a few of the studies which have reported that the HVSR method is known to be able to identify the fundamental frequency of a site under observation. It has been found in this study that the fundamental frequency range of the Boko-Palashbari site is 0.80–0.93 Hz,Guwahati-Central is 1.14–2.69 Hz and that of Goalpara is 0.63–0.82 Hz. DST [11] has reported a fundamental frequency of 1.5–2.5 Hz which compares well with the results of the present study. Further, basement depth obtained from the HVSR method implies that Boko-Palashbari, Guwahati Central and Goalpara may have bedrock at a depth ranging from 53.76–117.18 m, 18.59–82.24 m and 60.98–148.81 m respectively. This range of values of bedrock depth compares well with that of the bedrock depth of <100 m for Guwahati-Central and >100 m for Boko Palashbari and Goalpara respectively. Further, the site amplification of the western Guwahati region as per the HVSR method can be ranged between 5.1. and 7.60.

Table 14.22 HVSR results

Site	Amplification parameters			
	Peak HVSR (amplification)	PGA ($PGA_{EW, NS}/PGA_{VT}$)	PGV ($PGV_{EW, NS}/PGV_{VT}$)	PGD ($PGD_{EW, NS}/PGD_{VT}$)
Boko–Palashbari	5.15	2.70	3.00	4.40
Guwahati-Central	3.81	2.50	2.50	2.50
Goalpara	7.60	1.86	3.67	6.56

14.4 Comparision of Results

Several studies have attempted to compare the results of site response obtained by various methods. Field and Jacob [18], Theodulidis et al. [56], Lachet et al. [25], Riepl et al. [47], Nath et al. [39], Mucciarelli et al. [36], Ergin et al. [13], Molnar et al. [34], Pilz et al. [43] are a few studies, where comparison of site response analysis methods has been carried out in the past.

The results obtained by employing the SSR and HVSR methods in this study are provided in terms of FAS amplification only and can be seen in Table 14.23. A simple observation of the results shows that the FAS amplification is highest in the case of SSR as compared to HVSR.

The amplification as observed consisted of a higher value for SSR as compared to HVSR for all the site, which is in agreement with the observations reported by several authors [18, 25, 54, 56] that HVSR provides a lower value of amplification as compared to SSR.

From the results of HVSR and SSR as given in Table 14.23, it can be observed that the ratio of SSR to that of HVSR for Boko-Palashbari is 3.61, for Guwahati Central is 3.19 and for Goalpara is 4.44, which is within the range as observed and reported by Theodulidis et al. [56]. Theodulidis et al. [56] reported that the most prominent difference between the HVSR and the SSR was in their absolute levels, with the latter being 2 (two) to about 6 (six) times higher. The reason for systematically lower amplitude values of HVSR as compared to SSR could be attributed to the relative enrichment of the vertical component as reported by Raptakis et al. [45].

In the study, it was also attempted to determine the fundamental frequency of the soil sites from the waveform analysis. The term "*fundamental frequency*" denotes the frequency at which the Fourier amplitude ratio is maximum for the concerned earthquake event and is used in many studies (e.g., [10, 25, 54]. It also has to be mentioned that various authors have used various terms while defining the frequency with respect to the maximum amplitude. For example Taber [54] has termed it as "*predominant frequency*", Field and Jacob [16] reported it to be the "*fundamental resonant frequency*", Lachet and Bard [25] termed it as "resonance frequency".

However, soil is a highly complex material and as seen from the results provided in Table 14.18, has provided various fundamental frequencies for various earthquake events but within a certain frequency range. As such, the fundamental frequency is best defined in the form of a frequency range termed as "*fundamental frequency band*". Therefore, in this study, a range of frequency is provided which is the frequency band between which the soil site understudy has shown maximum amplification.It can be observed from Table 14.24 that Boko-Palashbari has a fundamental

Table 14.23 Peak amplification from SSR and HVSR

Site	SSR	HVSR
Boko-Palashbari	18.59	5.15
Guwahati-Central	12.16	3.81
Goalpara	33.74	7.6

Table 14.24 Fundamental frequency band

Site	Fundamental Frequency (FF) band (Hz) HVSR	Fundamental Frequency (FF) band (Hz) SSR
Boko–Palashbari	0.80–0.93	0.66–1.89
Guwahati-Central	1.14–2.69	1.69–3.49
Goalpara	0.63–0.82	0.70–1.88

frequency band of 0.80–0.93 Hz obtained by the HVSR method while for the SSR method it is 0.66–1.89 Hz. For Guwahati-Central, HVSR provided a range of 1.14–2.69 Hz while SSR provided a range of 1.69–3.49 Hz. For Goalpara the HVSR reported a frequency range of 0.63–0.82 Hz and SSR reported a range of 0.70–1.88 Hz. It is clearly identified that the fundamental frequency ranges as provided by both the methods compared well, which are in agreement with Pilz et al. [43], Lachet et al. [25], Taber [54], Ergin et al. [13]. Additionally, it can also be seen that the fundamental frequency band, as provided by the SSR method is comparatively wide, when compared to the HVSR method. This observation is in agreement with the observation made by Ibs-Vohn Seht and Wolhenderg that the SSR method is less reliable as compared to HVSR in determining the fundamental frequency of the subsoil. In this study, the fundamental frequency as provided by the HVSR method is considered to be representative of the sites investigated.

14.5 Summary

In this chapter a discussion on the results of the various site response analysis methods employed in this study viz. SSR and HVSR is presented. During the discussion, it is seen that the SSR results were consistently higher than that of the HVSR analysis. Pursuant to the statements made by Beresnev and Wen [3] and Rodriguez and Midorikawa [48] it can be asserted that the SSR method provides a more accurate estimate of site amplification as compared to HVSR. Based on the observations of Beresnev and Wen [3], Rodriguez and Midorikawa [48] and Mucciarelli et al. [36] in their study, it can be summarised that, for the Western Guwahati region, SSR provides a more reliable estimate of the site response.

This finding of this study is also complimentary to the findings of Field and Jacob [18] that, if the SSR estimates are taken as the most reliable, then the HVSR method under predicts the site response. The results provided by HVSR are less than that of the SSR results by a factor of 2–4.

SSR and HVSR analysis is carried out in the frequency domain and does not provide a seismic time history, which is required for dynamic analysis of structures. Further, SSR and HVSR analysis is dependent on the availability of recorded ground motions.

The Northeast part of India falls under seismic Zone V (IS:1893–2002), the highest seismic activity zone. However, there is a lack of sufficient dense seismic arrays in the region to record the seismic activity. Irrespective of the availability of the seismic records, studies related to possible seismic hazards have to be carried out in the region.

References

1. Al Yuncha Z, Luzon F (2000) On the horizontal-to-vertical spectral ratio in sedimentary basins. Bull Seismol Soc Am 90:1101–1106
2. Bard P-Y (1998) Microtremor measurements: a tool for site effect estimation? Proceeding of the second international symposium on the effects of surface geology on seismic motion, Yokohama, Japan 3:1251–1279
3. Beresnev IA, Wen KL (1996) The accuracy of soil estimates using soil-to-rock spectral ratios. Bull Seismol Soc Am 86(2):519–523
4. Bonilla LF, Steidl JH, Lindley GT, Tumarkin AG, Archuleta RJ (1997) Site amplification in the San Fernando valley, California: variability of site effect estimation using S-wave, coda, and H/V methods. Bull Seismol Soc Am 87(3):710–730
5. Bour M, Fouissac D, Dominique P, Martin C (1998) On the use of microtremor recordings in seismic microzonation. Soil Dyn Earthq Eng 17:465–474
6. Borcherdt RD (1970) Effects of local geology on ground motion near San Francisco Bay. Bull Seismol Soc Am 60:29–61
7. Castro RR, Mucciarelli M, Pacor F, Petrungaro C (1997) S-Wave site response estimates using horizontal-to-vertical spectral ratios. Bull Seismol Soc Am 87:256–260
8. Chavez-Garcia FJ, Cuenca J (1998) Site effects and microzonation in acapulco. Earthq Spectra 14(1):75–94
9. Delgado J, Lopez CC, Estevez A, Giner J, Cuenca A, Molina S (2000) Mapping soft soils in the Segura river valley (SE Spain): a case of study of microtremors as an exploration tool. J Appl Geophys 45:19–32
10. Dimitriu P, Kalogeras I, Theodulidis N (1999) Evidence of nonlinear site response in horizontal-to-vertical spectral ratio from near-field earthquakes. Soil Dyn Earthq Eng 18:423–435
11. DST (2008) Report on seismic microzonation of Guwahati region. http://www.am.tron.in/microzonation
12. Duval AM (1994) Determination de la reponse d'un site anx seismes al'aide du bruit de fond: evaluation experimentale, PhD thesis, Universite Paris 6, 265
13. Ergin M, Ozalaybey S, Aktar M, Yalcin MN (2004) Site amplification at Avcilar, Istanbul. Tectonophysics 391:335–346
14. Fah D (1997) Microzonation of the city of Basel. J Seismolog 1:87–102
15. Field EH (1994) Earthquake site response estimation, PhD thesis, Columbia University, New York, p 303
16. Field EH, Jacob KH (1993) The theoretical response of sedimentary layers to ambient seismic noise. Geophys Res Lett 20:2925–2928
17. Field EH, Jacob KH, Hough SE (1992) Earthquake site response estimation: a weak-motion case study. Bull Seismol Soc Am 82:2283–2307
18. Field EH, Jacob KH (1995) A comparison and test of various site response estimation techniques, including three that are not reference-site dependent. Bull Seismol Soc Am 85(4):1127–1143
19. Gitterman Y, Zaslavsky Y, Shapira A, Shtivelman V (1996) Empirical site response evaluations: case studies in Israel. Soil Dyn Earthquake Eng 15:447–463

20. Gutierrez C, Singh K (1992) A site effect study in Acapulco, Guerrero, Mexico: comparison of results from strong-motion and microtremor data. Bull Seismol Soc Am 82(2):642–659. https://doi.org/10.1785/BSSA0820020642
21. Hashash YMA, Musgrove MI, Harmon JA, Groholski DR, Phillips CA, Park D (2015) DEEPSOIL 6.1, User manual, p 135
22. Ibs- Von Seth M, Wohlenberg J (1999) Microtremor measurements used to map thickness of soft sediments. Bull Seismol Soc Am 89:250–259
23. Kausel E, Assimaki D (2002) Seismic simulation of inelastic soils via frequency-dependent moduli and damping. J Eng Mech 128(1):34–47
24. Konno K, Ohmachi T (1998) Ground-motion characteristics estimated from spectral ratio between horizontal and vertical components. Bull Seismol Soc Am 88(1):228–241
25. Lachet C, Hatzfeld D, Bard PY, Theodilidis N, Papaioannou C, Savvaidia A (1996) Site effects and microzonation in the city of Thessaloniki (Greece), comparison of different approaches. Bull Seismol Soc Am 86(6):1692–1703
26. Lang DH (2004) Damage potential of seismic ground motion considering local site effects, PhD thesis, Bauhaus-University Weimar
27. Lebrun (1997) Les effects de site: etude experimentale et simulation de trios configurations. In: These de doctorat del Úniversite J. Fourier de Grenoble, France, p 208
28. Lermo J, Chavez-Garcia EJ (1993) Site effect evaluation using spectral ratios with only one station. Bull Seismol Soc Am 83:1574–1594
29. Lermo J, Chavez Garcia FJ (1994) Site effect evaluation at Mexico City: dominant period and relative amplification from strong motion and microtremor records. Soil Dyn Earthq Eng 13:413–423
30. Lermo J, Chávez-García FJ (1994) Are microtremors useful in site response evaluation? Bull Seismol Soc Am 84(5):1350–1364
31. Maresca R, Castellano M, De Matteis R, Saccorotti G, Vaccariello P (2003) Local site effects in the town of Benevento (Italy) from noise measurements. Pure Appl Geophys 160:1745–1764
32. Mittal H, Kamal K, Kumar A, Singh SK (2013) Estimation of site effects in Delhi using standard spectral ratio. Soil Dyn Earthq Eng 50:53–61
33. Mittal H, Kumar A, Ramhmachuani R (2012) Indian national strong motion instrumentation network and site characterization of its stations. Int J Geosci 3:1151–1167
34. Molnar S, Cassify JF, Dosso SE (2004) Site response studies in Victoria, B.C., Analysis of Mw 6.8 nisqually earthquake recordings and shake modelling. In: Proceedings of the 13th world conference on earthquake engineering, Vancouver, B.C., August 1–6, 2004, paper No 2121
35. Mucciarelli M, Gallipoli MR (2001) A critical review of 10 years of microtremor HVSR technique. Bollettino Di Geofisica Teorica Ed Applicata. 42(3–4):255–266
36. Mucciarelli M, Gallipoli MR, Arcieri M (2003) The stability of the horizontal to vertical spectral ratio of triggered noise and earthquake recordings. Bull Seismol Soc Am 93(3):1407–1412
37. Nakamura Y (1989) A method for dynamic characteristics estimation of subsurface using microtremors on the ground surface. Quarterly Reports of the Railway Technical Research Institute Tokyo 30:25–33
38. Nath SK, Raj A, Sharma J, Thingbaijan KKS, Kumar A, Nandy DR, Yadav MK, Dasgupta S, Majumdar K, Kayal JR, Shukla AK, Deb SK, Pathak J, Hazarika PJ, Paul DK, Bansal BK (2008) Site amplification, Qs, and source parameterization in Guwahati region from seismic and geotechnical analysis. Seismol Res Lett 79(4):526–539
39. Nath SK, Sengupta P, Srivastav SK, Bhattacharya SN, Dattatrayam RS, Prakash R, Gupta HV (2003) Estimation of S-wave site response in and around Delhi region from weak motion data. Proc Indian Acad Sci (Earth Planet Sci) 112(3):441–462
40. Ohmachi T, Nakamura Y, Toshinawa T (1991) Ground motion characteristics of the San Francisco bay area detected by microtremor measurements. Second international conference on recent advances in geotechnical earthquake engineering and soil dynamics, St Louis 2:1643–1648
41. Parolai S, Bormann P, Mikereit C (2002) "New relationships between Vs, thickness of sediments, and resonance frequency calculated by the H/V ratio of seismic noise for the Cologne area (Germany). Bull Seismol Soc Am 92:2521–2527

42. Parolai S (2012) Investigation of site response in urban area using earthquake data and seismic noise. https://doi.org/10.2312/GFZ.NMSOP-2_ch14. www.gfzpublic.gfz-potsdam.de
43. Pilz M, Parolai S, Leyton F, Campos J, Zschau J (2009) A comparison of site response techniques using earthquake data and ambient seismic noise analysis in the large urban areas of Santiago de Chile. Geophys J Int 178(2):713–728. https://doi.org/10.1111/j.1365-246X.2009.04195.x,2009
44. Pitilakis K (2004) "Site Effects", Recent advances in earthquake geotechnical engineering and microzonation, Ansal A (ed). Kluwer Academic Publishers, Springer Science, Netherlands, pp 139–197
45. Raptakis D, Theodulidis N, Pitilakis K (1998) Data analysis of the Euroseistest strong motion array in Volvi (Greece): Standard and horizontal-to-vertical spectral ratio techniques. Earthq Spectra 14(1):203–223
46. Rial JA, Saltzman NG, Ling H (1992) Earthquake-induced resonance in sedimentary basins. Am Sci 80:566–578
47. Riepl J, Bard P-Y, Hatzfeld D, Papaioannou C, Nechstein N (1998) Detailed evaluation of site-response estimation methods across and along the sedimentary valley of Volvi (Euroseistest). Bull Seismol Soc Am 88(2):488–502
48. Rodriguez HS, Midorikawa S (2003) Comparison of spectral ratio techniques for estimation of site effects using microtremor data and earthquake motions recorded at the surface and in boreholes. Earthquake Eng Ans Struct Dyn 32:1691–1714
49. Safak E (1991) Problems with using spectral ratios to estimate site amplification. In: Proceedings of the fourth international conference on seismic zonation, vol. II. EERI, Oakland, pp 277–284
50. Sairam B, Rastogi BK, Aggarwal S, Chauhan M, Bhonde U (2011) Seismic site characterization using Vs30 and site amplification in Gandhinagar region, Gujarat, India. Curr Sci 100(5):754–761
51. Salazar W, Brown L, Mannette G (2013) Surface soil effects studies based on H/V ratios of microtremors at Kingston metropolitan Area, Jamaica. J Civil Eng Architect 7(10) (Serial No. 71):1301–1322
52. Seekins LC, Wennerberg L, Marghereti L, Liu H-P (1996) Site amplification at five locations in San Fransico, California: a comparison of S waves, codas, and microtremors. Bull Seismol Soc Am 86:627–635
53. Steidl JH, Tumarkin AG, Archuleta RJ (1996) What is a reference site? Bull Seismol Soc Am 86:1733–1748
54. Taber JJ (2000) Comparison of site response determination techniques in the Wellington region, New Zealand. In: Proceedings of the 12th world conference on earthquake engineering, New Zealand
55. Theodulidis N, Bard P-Y (1995) Horizontal to vertical spectral ratio and geological conditions: an analysis of strong motion data from Greece and Taiwan (SMART-I). Soil Dyn Earthquake Eng 14:177–197
56. Theodulidis N, Bard PY, Archuleta R, Bouchon M (1996) Horizontal-to-vertical spectral ratio and geologic conditions: the case of garner valley downhole array in Southern California. Bull Seismol Soc Am 86(2):306–319
57. Tokeshi JC, Sugimura Y (1998) On the estimation of the natural period of the ground using simulated microtremors. In: Proceedings of the second international symposium on the effects of surface geology on seismic motion, Yokohama, Japan, vol 2, pp 651–664
58. Tucker BE, King JL (1984) Dependence of sediment-filled valley response on input amplitude and valley properties. Bull Seismol Soc Am 74:153–165
59. Tucker BE, King JL, Hatzfeld D, Nersesov IL (1984) Observations of hard-rock site effects. Bull Seismol Soc Am 74:121–136
60. Wakamatsu K, Yasui Y (1996) Possibility of estimation for amplification characteristics of soil deposits based on ratio of horizontal to vertical spectra of microtremors. In: Proceedings of the 11th world conference on earthquake engineering. Acapulco, Mexico

The manufacturer's authorised representative in the EU is Springer Nature Customer Service Centre GmbH, Europaplatz 3, 69115 Heidelberg, Germany. If you have any concerns regarding our products, please contact ProductSafety@springernature.com

Printed and bound by CPI Group (UK) Ltd, Croydon, CR0 4YY
30/09/2026
02225315-0008